INTRODUCTORY MATHEMATICS

INTRODUCTORY MATHEMATICS

Fourth Edition

Nigel P. Cook

PEARSON

Prentice
Hall

Upper Saddle River, New Jersey
Columbus, Ohio

Editor in Chief: Stephen Helba
Senior Acquisitions Editor: Gary Bauer
Editorial Assistant: Natasha Holden
Production Editor: Louise N. Sette
Production Coordinator: Holly Henjum, Carlisle Publishers Services
Design Coordinator: Diane Ernsberger
Cover Designer: Jason Moore
Cover Art: Photonica
Production Manager: Brian Fox
Marketing Manager: Leigh Ann Sims

This book was set in Times Roman by Preparé. It was printed and bound by Banta Book Group. The cover was printed by Phoenix Color Corp.

Pearson Education Ltd.
Pearson Education Singapore Pte. Ltd.
Pearson Education Canada, Ltd.
Pearson Education—Japan

Pearson Education Australia Pty. Limited.
Pearson Education North Asia Ltd.
Pearson Educación de Mexico, S. A. de C.V.
Pearson Education Malaysia Pte. Ltd.

10 9 8 7 6 5 4 3 2 1
ISBN: 0-13-045284-X

To Dawn, Candy, and Jon

Books by Nigel P. Cook

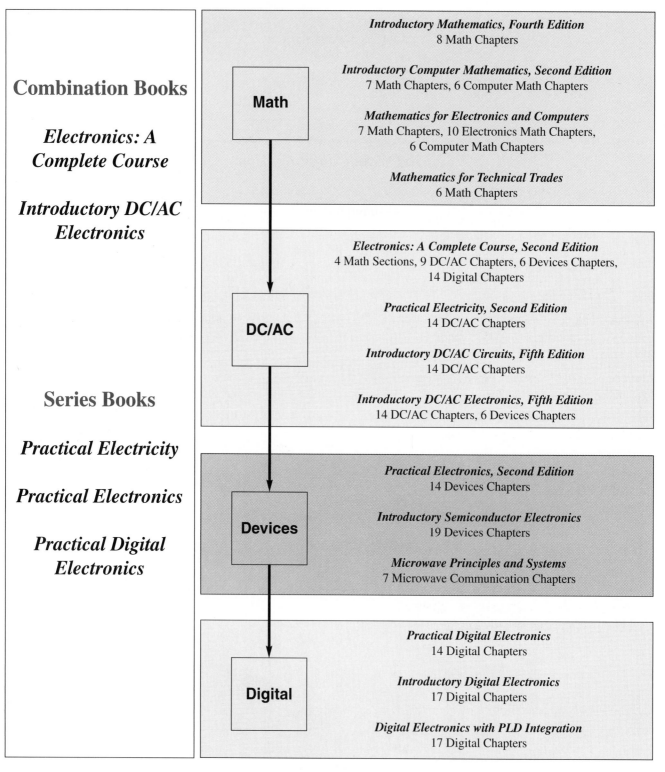

Combination Books

Electronics: A Complete Course

Introductory DC/AC Electronics

Math

Introductory Mathematics, Fourth Edition
8 Math Chapters

Introductory Computer Mathematics, Second Edition
7 Math Chapters, 6 Computer Math Chapters

Mathematics for Electronics and Computers
7 Math Chapters, 10 Electronics Math Chapters,
6 Computer Math Chapters

Mathematics for Technical Trades
6 Math Chapters

DC/AC

Electronics: A Complete Course, Second Edition
4 Math Sections, 9 DC/AC Chapters, 6 Devices Chapters,
14 Digital Chapters

Practical Electricity, Second Edition
14 DC/AC Chapters

Introductory DC/AC Circuits, Fifth Edition
14 DC/AC Chapters

Introductory DC/AC Electronics, Fifth Edition
14 DC/AC Chapters, 6 Devices Chapters

Series Books

Practical Electricity

Practical Electronics

Practical Digital Electronics

Devices

Practical Electronics, Second Edition
14 Devices Chapters

Introductory Semiconductor Electronics
19 Devices Chapters

Microwave Principles and Systems
7 Microwave Communication Chapters

Digital

Practical Digital Electronics
14 Digital Chapters

Introductory Digital Electronics
17 Digital Chapters

Digital Electronics with PLD Integration
17 Digital Chapters

For more information and a desk copy of any of these textbooks by Nigel P. Cook, call 1-800-228-7854, visit the Prentice Hall website at www.prenhall.com, or ask your local Prentice Hall representative.

Preface

INTRODUCTION

Mathematics is interwoven into the very core of science, and therefore an understanding of mathematics is imperative for anyone pursuing a career in technology. As with any topic to be learned, the method of presentation can make a big difference between clear comprehension and complete confusion. Employing an "integrated math applications" approach, this text reinforces all math topics with extensive applications to show the student the value of math as a tool; therefore, if the need is instantly demonstrated, the tool is retained.

OUTLINE

After 14 textbooks, 21 editions, and 19 years of front-line education experience, best-selling author Nigel Cook has written *Introductory Mathematics,* Fourth Edition, as a complete math course for technology students.

Chapter 1   Whole Numbers and Fractions
Chapter 2   Decimal Calculation
Chapter 3   Positive and Negative Numbers
Chapter 4   Exponents and the Metric System
Chapter 5   Algebra, Equations, and Formulas
Chapter 6   Geometry and Trigonometry
Chapter 7   Logarithms and Graphs
Chapter 8   Computer Math

SUPPLEMENTS

- Instructor's Manual (ISBN 0-13-045285-8)
- PH TestGen (ISBN 0-13-114156-2)
- Companion Website: http://www.prenhall.com/cook

*New Geometry Chapter,
New Computer Math Chapter,
and Twice the Math
Applications*

6–2–1 *Quadrilaterals*

A **polygon** is a plane geometric figure that is made up of three or more line segments. Quadrilaterals, such as the square, rectangle, rhombus, parallelogram, and trapezoid, are all examples of four-sided polygons.

The information given in Figure 6–2 summarizes the facts and formulas for the square, which has four right angles and four equal length sides, and is a **parallelogram** (which means it has two sets of parallel sides). The box in Figure 6–2 lists the formulas for calculating a square's perimeter, area, and side length.

Polygon
A plane geometry figure bounded by three or more line segments.

Parallelogram
A quadrilateral in which both pairs of opposite sides are parallel.

■ **INTEGRATED MATH APPLICATION: ELECTRONICS**

The capacitor shown has two same-size metal foil plates that are separated by a nonconductive insulator called a *dielectric*. Unwrapped, the foil strips measure 6.2 centimeters by 1.3 centimeters. What is the capacitor's plate area?

■ *Solution:*

$$\text{Area}(A) = \text{length }(l) \times \text{width }(w)$$
$$= 6.2 \text{ square centimeters(cm}^2)$$

SQUARE

Right Angle (90°)

Has four sides of equal length ($l = w = x$).
Is a quadrilateral (4-sided figure).
Is a parallelogram (both sets of opposite sides are parallel).
Has vertices (points where two sides meet) of 90°.

Width (w)

$$P \text{ (Perimeter)} = 4 \times x$$
$$A \text{ (Area)} = x \times x \text{ or } x^2$$
$$x \text{ (side)} = \sqrt{A}$$

Perimeter   Area   Side

Length (l)

Length (l) = Width (w) = Side (x)

FIGURE 6–2 The Square.

TABLE

Input	Output
−4	−3
−3	−1
−2	−0.5
−1	−0.25
0	0
+1	0
+2	0
+3	+0.5
+4	+3.5

Positive y-axis (Positive output values)

As an example, an input of −3 will yield an output of −1.

(+4, +3.5)
(+3, +0.5)

Negative x-axis (Negative input values)

Positive x-axis (Positive input values)

(−3, −1)
(−4, −3)

Negative y-axis (Negative output values)

As an example, an input of +4 will yield an output of +3.5.

A rectangular coordinate such as (4, 3.5), also known as a Cartesian coordinate, is a point on a graph that has two perpendicular (at 90°) axes that will form a rectangle with the axes (shown shaded in graph).

5–4–3 *Graphing Linear Equations*

To combine our understanding of linear equations and graphs, let us step through the process of converting an equation to a graph.

Consider the linear equation:

$$y = 2x + 2$$

This equation is, in fact, the formula you would use to determine the output (y) from a device for any given input (x). For example, if you applied 3 to the input ($x = 3$), the output would be:

x → [DEVICE $y = 2x + 2$] → y

$$y(\text{output}) = 2(3) + 2 \qquad x(\text{input}) = 3$$
$$= 6 + 2$$
$$= 8$$

So how can we graph the response of this device? The answer is to pick some example input values, calculate the outputs for these inputs, and plot the points on a graph.

To find the *x-intercept* (the point where the line passes through the x-axis), let $y = 0$.
$$y = 2x + 2$$
$$0 = 2x + 2 \quad (-2)$$
$$-2 = 2x \quad (+2)$$
$$-1 = x$$
The x-intercept, therefore, is at the coordinate (−1, 0).

To verify a line's accuracy, the line should pass through at least three points. To identify another line point, let x equal some value and then solve for y.
$$y = 2x + 2 = 2(x) + 2 = 6$$

To find the *y-intercept* (the point where the line passes through the y-axis), let $x = 0$.
$$y = 2x + 2$$
$$y = 2(0) + 2$$
$$y = 0 + 2$$
$$y = 2$$
The y-intercept, therefore, is at the coordinate (0, 2).

(2, 6)

5–4–4 *Slope of a Line*

The steepness of the line in a graph is an indication of how much or little the output is changing for a given input. For example, in the distance versus time graph that follows, we can see how fast we were going by seeing how steep the line is.

*Conversational Writing Style,
Visually Descriptive Layout,
and Humanistic Vignettes*

Now that we have a formula for voltage ($V = I \times R$), we can complete this problem:

$$V = I \times R \qquad \text{(Voltage = current} \times \text{resistance)}$$
$$= 2\,A \times 3\,\Omega$$
$$= 6\,V$$

This equation means that a voltage or electrical pressure of 6 volts would cause 2 amperes of electric current in a circuit having 3 ohms of resistance.

To test if these three formulas are correct, we can now plug into each formula the values of the preceding example, as follows:

$$I = \frac{V}{R} \qquad\qquad V = I \times R \qquad\qquad R = \frac{V}{I}$$

$$2\,A = \frac{6\,V}{3\,\Omega} \qquad 6\,V = 2\,A \times 3\,\Omega \qquad 3\,\Omega = \frac{6\,V}{2\,A}$$

As you can see from these formulas, the transposition has been successful.

■ INTEGRATED MATH APPLICATION: ELECTRICAL

Figure 5–3 shows a basic circuit for a home electrical system. If each household light and appliance has 120 volts connected across it, how much current will pass through lamp 1, lamp 2, the hair dryer, and the space heater?

■ *Solution:*

$$\text{Current through lamp 1 } (I_{L1}) = \frac{V_{L1}}{R_{L1}} = \frac{120\,V}{125\,\Omega} = 960\,mA$$

$$\text{Current through lamp 2 } (I_{L2}) = \frac{V_{L2}}{R_{L2}} = \frac{120\,V}{250\,\Omega} = 480\,mA$$

$$\text{Current through hair dryer } (I_{HD}) = \frac{V_{HD}}{R_{HD}} = \frac{120\,V}{40\,\Omega} = 3\,A$$

$$\text{Current through space heater } (I_{SH}) = \frac{V_{SH}}{R_{SH}} = \frac{120\,V}{12\,\Omega} = 10\,A$$

FIGURE 5–3 Home Electrical System.

Integrated Math Applications, Worked-Out Example Section, and Self-Test Evaluation Points

■ [INTEGRAT]ED MATH APPLICATION: METALWORKING

[...conv]erses of the following pieces of sheet metal to their fractional equivalents

b. 0.859375 inch

b. 55.64 inch

■ [INTEGRAT]ED MATH APPLICATION: AUTOMOTIVE

[...me]asures $2\frac{9}{16}$ inches in diameter before it is rebored to increase its size by 0.05 inch. What is the new size of the cylinder as a decimal value?

■ *Solution:*

$$2\frac{9}{16} = \quad 2.5625 \text{ inches}$$
$$+\ 0.05 \text{ inch}$$
$$\overline{\quad 2.6125 \text{ inches}}$$

Calculator Sequence Examples, Extensive End-of-Chapter Problems, and Margin Term Definitions

CALCULATOR KEYS

Name: Conversion Functions
Display the answer as a fraction (▶ Frac).
Display the answer as a decimal (▶ Dec).

Function:
Value ▶ Frac (display as a fraction) displays an answer as its written fraction equivalent.
Value ▶ Dec (display as a decimal) displays an answer as its decimal equivalent.

Example: $\frac{1}{2} + \frac{1}{3}$ ▶ Frac
$$\tfrac{5}{6}$$

Ans ▶ Dec
$$.8333333333$$

SELF-TEST EVALUATION POINT FOR SECTION 2–1

Now that you have completed this section, you should be able to:

■ *Objective 3.* *Define the function of the decimal point.*

■ *Objective 4.* *Describe the positional weight of decimal fractions.*

■ *Objective 5.* *Demonstrate how to convert written fractions to decimal fractions, calculate reciprocals, and convert decimal fractions to written fractions.*

Use the following questions to test your understanding of Section 2–1.

1. What are the individual digit weights of the number 178.649?

2. Convert the following written fractions to decimal fractions.
   a. $1\frac{5}{2}$ **b.** $192\frac{3}{4}$ **c.** $67\frac{6}{9}$

3. Calculate the reciprocal of the following numbers.
   a. 2500   **b.** 0.25   **c.** 0.000025

4. Convert the following decimal fractions to written fractions.
   a. $7.25 = \frac{?}{16}$ **b.** $156.90625 = \frac{?}{32}$

Contents

1

Whole Numbers and Fractions 2

Vignette: There's No Sleeping When He's Around 2
Outline and Objectives 3
Introduction 4
1-1 Whole Numbers and Counting 4
1-2 Representing Fractions 6
1-3 Adding Fractions 9
1-4 Subtracting Fractions 18
1-5 Multiplying Fractions 20
1-6 Dividing Fractions 23
1-7 Canceling Fractions 26
Review Questions 28

2

Decimal Calculation 30

Vignette: An Apple a Day 30
Outline and Objectives 31
Introduction 32
2-1 The Decimal Point 32
2-2 Calculating in Decimal 42
Review Questions 60

3

Positive and Negative Numbers 62

Vignette: Blaise of Genius 62
Outline and Objectives 63
Introduction 64
3-1 Expressing Positive and Negative Numbers 65
3-2 Adding Positive and Negative Numbers 67
3-3 Subtracting Positive and Negative Numbers 70
3-4 Multiplying Positive and Negative Numbers 74
3-5 Dividing Positive and Negative Numbers 76
3-6 Order of Operations 79
Review Questions 82

4

Exponents and the Metric System 84

Vignette: Not a Morning Person 84
Outline and Objectives 85
Introduction 86
4-1 Raising a Base Number to a Higher Power 86
4-2 Powers of 10 94
4-3 Weights and Measures 100
Review Questions 116

5

Algebra, Equations, and Formulas 118

Vignette: Back to the Future 118
Outline and Objectives 119
Introduction 120
5-1 The Basics of Algebra 120
5-2 Transposition 126
5-3 Substitution 146
5-4 Equations and Graphing 153
5-5 Rules of Algebra—A Summary 163
Review Questions 167

6

Geometry and Trigonometry 170

Vignette: Finding the Question to the Answer! 170
Outline and Objectives 171
Introduction 172
6-1 Basic Geometric Terms 172
6-2 Plane Figures 174
6-3 Trigonometry 187
6-4 Solid Figures 198
Review Questions 203

7

Logarithms and Graphs   206

Vignette: The First Pocket Calculator 206
Outline and Objectives 207
Introduction 208
7-1 Common Logarithms 208
7-2 Graphs 213
Review Questions 221

8

Computer Math   224

Vignette: Moon Walk 224
Outline and Objectives 225
Introduction 226

8-1 Analog and Digital Data and Devices 226
8-2 Analog and Digital Signal Conversion 230
8-3 The Binary Number System 232
8-4 The Hexadecimal Number System 238
8-5 Binary Codes 243
8-6 Logic Gates and Boolean Expressions 247
8-7 Binary Arithmetic 256
Review Questions 272

Appendices   275

A Answers to Self-Test Evaluation Points 275
B Answers to Odd-Numbered Problems 282

Index   287

INTRODUCTORY MATHEMATICS

Whole Numbers and Fractions

There's No Sleeping When He's Around!

Carl Friedrich Gauss was born April 30, 1777, to poor, uneducated parents in Brunswick, Germany. He was a child of precocious abilities, particularly in mental computation. In elementary school he soon impressed his teachers, who said that mathematical ability came easier to Gauss than speech.

In secondary school he rapidly distinguished himself in ancient languages and mathematics. At 14, Gauss was presented to the court of the duke of Brunswick, where he displayed his computing skill. Until his death in 1806, the duke generously supported Gauss and his family, encouraging the boy with textbooks and a laboratory.

In the early years of the nineteenth century, Gauss's interest was in astronomy, and his accumulated work on celestial mechanics was published in 1809. In 1828, at a conference in Berlin, Gauss met physicist Wilhelm Weber, who would eventually become famous for his work on electricity. They worked together for many years and became close friends, investigating electromagnetism and the use of a magnetic needle for the measurement of current. In 1833 they constructed an electric telegraph system that could communicate across Göttingen from Gauss's observatory to Weber's physics laboratory. (This telegraph system of communication was later developed independently by U.S. inventor Samuel Morse.)

Gauss conceived almost all of his fundamental mathematical discoveries between the ages of 14 and 17. There are many stories of his genius in his early years, one of which involved a sarcastic teacher who liked giving his students long-winded problems and then resting, or on some occasions sleeping, in class. On his first day with Gauss, who was 8 years old, the teacher began, as usual, by telling the students to find the sum of all the numbers from 1 to 100. The teacher barely had a chance to sit down before Gauss raised his hand and said "5050." The dumbfounded teacher, who believed Gauss must have heard the problem before and memorized the answer, asked Gauss to explain how he had solved the problem. He replied: "The numbers 1, 2, 3, 4, 5, and so on to 100 can be paired as 1 and 100, 2 and 99, 3 and 98, and so on. Since each pair has a sum of 101, and there are 50 pairs, the total is 5050."

Outline and Objectives

VIGNETTE: THERE'S NO SLEEPING WHEN HE'S AROUND!

INTRODUCTION

1–1 WHOLE NUMBERS AND COUNTING

Objective 1: Describe the origination of the decimal number system.

1–1–1 Base or Radix of a Number System

Objective 2: Explain the basic process of counting in decimal, and the following terms.
 a. Base or radix
 b. Positional weight
 c. Reset and carry

1–1–2 Positional Weight

1–1–3 Reset and Carry

1–2 REPRESENTING FRACTIONS

Objective 3: Describe what a fraction is, and name its elements.

Objective 4: Demonstrate how fractions are represented.

1–3 ADDING FRACTIONS

Objective 5: Describe how fractions are added.

1–3–1 Method 1 for Finding a Common Denominator

Objective 6: Show two methods used for calculating the lowest common denominator.

1–3–2 Method 2 for Finding a Common Denominator

1–3–3 Improper Fractions and Mixed Fractions

Objective 7: Define a(n):
 a. Proper fraction
 b. Improper fraction
 c. Mixed fraction

1–3–4 Adding Mixed Numbers

Objective 8: Show how mixed fractions are added.

1–3–5 Reducing Fractions

Objective 9: Describe how to reduce fractions.

1–4 SUBTRACTING FRACTIONS

Objective 10: Show how to subtract proper and mixed fractions.

1–4–1 Subtracting Mixed Fractions

1–4–2 Borrowing

1–5 MULTIPLYING FRACTIONS

Objective 11: Show how to multiply proper and mixed fractions.

1–5–1 Multiplying Mixed Fractions

1–6 DIVIDING FRACTIONS

Objective 12: Show how to divide proper and mixed fractions.

1–6–1 Dividing Mixed Fractions

1–7 CANCELING FRACTIONS

Objective 13: Demonstrate how to use cancellation to simplify fraction multiplication or division.

Introduction

Since this is your first chapter, let me try to put you in the right frame of mind before you begin. Most people have trouble with mathematics because they never mastered the basics. As you proceed through this chapter and the succeeding chapters, it is imperative that you study every section, example, self-test evaluation point, and end-of-chapter question. If you cannot understand a particular section or example, go back and review the material that led up to the problem and make sure that you fully understand all the basics before you continue. Since each chapter builds on previous chapters, you may find that you need to return to an earlier chapter to refresh your understanding before moving on with the current chapter. This process of moving forward and then backtracking to refresh your understanding is very necessary and helps to engrave the basics of mathematics in your mind. Try never to skip a section or chapter because you feel that you already have a good understanding of that material. If it is a basic topic that you have no problem with, read it anyway to refresh your understanding about the steps involved and the terminology because these may be used in a more complex operation in a later chapter.

1–1   WHOLE NUMBERS AND COUNTING

Thousands of years ago, people used their fingers and thumbs to keep track of objects, and so it is not surprising that our number system has 10 numerals or digits. In fact, the word "digit" means "finger," and the decimal numbers system (from the Latin word *decimus*, which means ten) uses the following digits:

$$0, 1, 2, 3, 4, 5, 6, 7, 8, \text{and } 9$$

Counting was probably the first application for a number system such as decimal because people needed some way of keeping track of things. *Decimal is based on the number 10.* Let us first examine what is meant by the *base* of a number system.

1–1–1   *Base or Radix of a Number System*

Base or Radix

A number equal to the number of units in a given number system. The decimal system uses a base of 10.

Subscript

A distinguishing letter or numeral written immediately below and to the right or left of another character.

The key element that distinguishes one number system from another is the **base** *or* **radix** *of the number system.* The base describes the number of digits that are used. The decimal number system has a base of 10, which means that it makes use of 10 digits (0 through 9) to represent the value of a quantity. A **subscript,** which is a smaller number written to the right or left of and below the main number, is sometimes included to indicate the number's base. For example, all the numbers following have the subscript "$_{10}$" to indicate that they are base 10, or decimal, numbers.

Large numbers indicate value of quantity. 4956_{10} $235{,}163_{10}$ 8_{10} 11_{10} Subscript indicates base of number.

If a subscript is not present, the number is assumed to be a base 10, or decimal, number. For example, the following are all decimal numbers, even though they do not include the subscript 10.

$$23 \quad 101 \quad 3{,}634{,}987 \quad 47$$

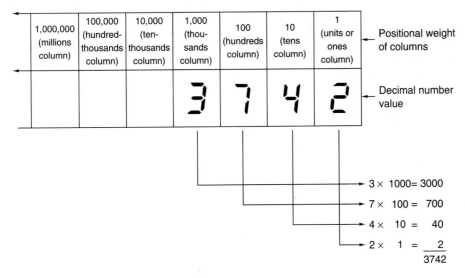

FIGURE 1–1 **Positional Weight of Each Decimal Column.**

1–1–2 *Positional Weight*

The position of each digit of a decimal number determines the weight of that digit. For example, a 2 by itself indicates only a value of two, whereas a 2 to the left of three zeros (2000) indicates a value of two thousand. The position of a decimal digit therefore determines its weight, or value, with digits to the left carrying the greater weight. For example, the decimal number 3742 contains three thousands (3×1000), seven hundreds (7×100), four tens (4×10), and two ones (2×1). Figure 1–1 shows this decimal number with the positional weight of each decimal column.

You may have noticed that the weight of each decimal (base 10) column increases by a factor of 10 as you move to the left. For example, the units column has a weight of 1. The next column to the left of the units column (tens column) has a weight that is ten times ($\times 10$) larger than the units column—$10 \times \text{one} = 10$. The next column to the left of the tens column (hundreds column) has a weight ten times ($\times 10$) larger than the tens column—$10 \times \text{ten} = 100$. Therefore, each column in the decimal number system increases by ten times ($\times 10$) as you move to the left.

■ **INTEGRATED MATH APPLICATION: MACHINING**

What is the distance between holes in the pictured machined part?

a. *x* and *y*?   **b.** *y* and *z*?

■ *Solution:*

a. *x* to *y* = 24 centimeters   **b.** *y* to *z* = 36 centimeters

1–1–3 *Reset and Carry*

Figure 1–2 shows the reset and carry action that occurs when we count in decimal. *This* **reset and carry action** *occurs whenever a column reaches the highest decimal digit of 9.* For example, when we begin to count from zero (0), the units column continues to advance by one or **increment** until it reaches 9. An increment beyond this point causes the units column to be reset to zero and a one to be carried into the tens column, producing 10 (ten). As

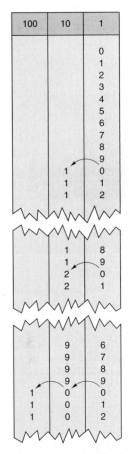

FIGURE 1–2 **The Reset and Carry Action That Occurs When Counting in Decimal.**

Reset and Carry Action

An operation that occurs when counting, in which a digit resets after advancing beyond its maximum count and carries a one into the next higher order column.

Increment

The action or process of increasing a quantity or value by one.

this count continues, the units column will again repeat the cycle going from 0 to 9, which with the one in the tens column will be a count of 10 (ten) to 19 (nineteen). An increase by one beyond 19 will again cause the units column to reset to zero and carry a one into the tens column, resulting in a count of 20 (twenty). The units column will continue to cycle from 0 to 9 and reset and carry into the tens column until the tens column reaches the highest decimal digit, 9. As you can see in Figure 1–2, an advance by one beyond a count of 99 will cause the units column to reset and carry one into the tens column, and because the tens column is at its maximum, 9, it will also reset to zero and carry one into the hundreds column. This "reset to zero and carry one into the next left column action" will continue to occur in a right-to-left motion whenever a column reaches the maximum digit 9. In other words, the units column will reset and carry into the tens column, the tens column will reset and carry into the hundreds column, the hundreds column will reset and carry into the thousands column, and so on.

All of the counting numbers discussed so far are in fact whole numbers, since a whole number, by definition, is a single digit or multiple digit complete number such as 7,236, or 15,250.

Some whole numbers are followed by a fraction. For example, with the number *twelve and one-half* $\left(12\frac{1}{2}\right)$, the number *twelve* is a whole number while the *one-half* is a fractional part or piece of the whole. Fractions will be discussed in the next section.

SELF-TEST EVALUATION POINT FOR SECTION 1–1

Now that you have completed this section, you should be able to:

■ **Objective 1.** *Describe the origination of the decimal number system.*

■ **Objective 2.** *Explain the basic process of counting in decimal, and the terms* base *or* radix, positional weight, *and* reset and carry.

Use the following questions to test your understanding of Section 1–1.

1. The decimal number system has a base or radix of _____.

2. Indicate the positional weight of each of the decimal digits in the following number: 35,729.

3. What is the reset and carry action?

4. Which columns will reset and carry when the number 499,997 is advanced or incremented by 5? What will be the final count?

1–2   REPRESENTING FRACTIONS

Fraction

A numerical representation indicating the quotient of two numbers; a piece, fragment, or portion.

Half

Either of two equal parts into which a thing is divisible.

Quarter

One of four equal parts into which something is divisible; a fourth part.

Fourth

See *quarter*.

Another word for **fraction** is "piece"; therefore, if you have a fraction of something, you have a piece of that something. Having only a fraction or piece of something means that the object has been broken up into pieces, and therefore a fraction is something less than the whole thing. For example, if you wanted half of a pizza, a whole pizza would be cut into two pieces and you would be given one of the halves—a **half** is therefore a fraction. If, in another instance, you wanted only a quarter of an apple pie, a whole apple pie would be cut into four pieces and you would be given one of the quarters—a **quarter** is therefore also a fraction. A quarter (or a **fourth**) is a smaller fraction than a half, and therefore we can say that *the more cuts you make, the smaller the fraction.*

Half and *quarter* (or *fourth*) are names that describe a fraction; however, some other system, sign, or symbol is needed to represent or show the fraction. This system will have to show two things:

1. The number of pieces into which the entire object was cut or divided up
2. The number of pieces we have or are concerned with

Numerator

The part of a fraction that is above the line and signifies the number of parts of the denominator taken.

Denominator

The part of a fraction below the line signifying division that functions as the divisor of the numerator.

Vinculum or Fraction Bar

A straight horizontal line placed between two or more numbers of a mathematical expression.

■ EXAMPLE

Write down the proper or written fraction for one-quarter or one-fourth, and label its elements.

■ *Solution:*

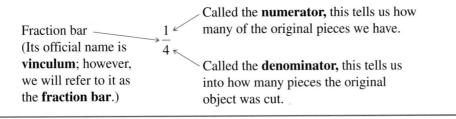

Fraction bar (Its official name is **vinculum**; however, we will refer to it as the **fraction bar**.)

$\dfrac{1}{4}$

Called the **numerator,** this tells us how many of the original pieces we have.

Called the **denominator,** this tells us into how many pieces the original object was cut.

The preceding example fraction represents an amount or quantity. In this example it represents "one-quarter" or "one-fourth" and indicates that the object was cut or divided into four pieces and that we have one of the four pieces.

■ EXAMPLE

First, draw a picture of a pizza cut into six pieces,

a. then show one of the six pieces separately and label it with its fraction
b. then show three of the six pieces separately and label them with their fraction.

■ *Solution:*

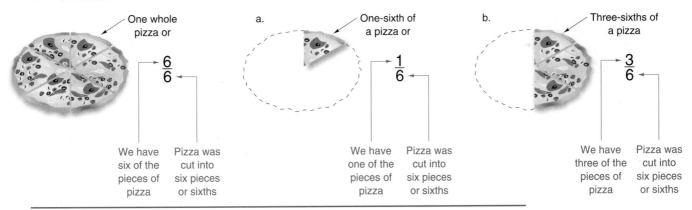

One whole pizza or

$\dfrac{6}{6}$

We have six of the pieces of pizza

Pizza was cut into six pieces or sixths

a.

One-sixth of a pizza or

$\dfrac{1}{6}$

We have one of the pieces of pizza

Pizza was cut into six pieces or sixths

b.

Three-sixths of a pizza

$\dfrac{3}{6}$

We have three of the pieces of pizza

Pizza was cut into six pieces or sixths

In all technical fields you will have to interpret scales, such as the divisions on the straightedge shown next.

One inch

This ruler has one inch divided up into four quarter-inch increments. ($\frac{4}{4}$ = 1 in.)

0    $\frac{1}{4}$ $\frac{2}{4}$ $\frac{3}{4}$ 1 ($\frac{4}{4}$)

Other rulers might have each inch broken up into halves (there would be two halves or $\frac{2}{2}$ in an inch), eighths (there would be eight eighths or $\frac{8}{8}$ in an inch), sixteenths (there would be sixteen sixteenths or $\frac{16}{16}$ in an inch), and so on.

■ INTEGRATED MATH APPLICATION: CONSTRUCTION

As shown in the associated figure, a roof's pitch or steepness is defined as a fraction:

$$\text{Pitch} = \frac{\text{Rise}}{\text{Run}}$$

What fraction would define the roof shown
in the figure?

Roof's rise—Increase in height

Roof's run—Horizontal distance

■ *Solution:*

$$\text{Pitch} = \frac{\text{Rise}}{\text{Run}} = \frac{8\,\text{feet}}{20\,\text{feet}} = \frac{8}{20}$$

■ INTEGRATED MATH APPLICATION: CONSTRUCTION

Use the ruler to measure the length of the screws shown in the following figure.

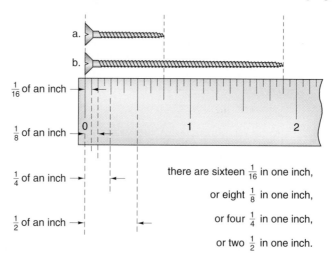

$\frac{1}{16}$ of an inch

$\frac{1}{8}$ of an inch

$\frac{1}{4}$ of an inch

$\frac{1}{2}$ of an inch

there are sixteen $\frac{1}{16}$ in one inch,

or eight $\frac{1}{8}$ in one inch,

or four $\frac{1}{4}$ in one inch,

or two $\frac{1}{2}$ in one inch.

■ *Solution:*

a. The smaller screw measures three-fourths of an inch ($\frac{3}{4}$ in.), or six-eighths of an inch ($\frac{6}{8}$ in.), or twelve-sixteenths ($\frac{12}{16}$ in.) of an inch.

b. The second screw is greater than 1 inch long but less than 2 inches long, which means that it is 1 inch plus a fraction of an inch long. Since its tip falls alongside an $\frac{1}{8}$ inch mark on the ruler, we will use the $\frac{1}{8}$ inch scale. The longer screw measures one and seven-eighths of an inch ($1\frac{7}{8}$ in.) long. We could also say that screw b measures $\frac{15}{8}$ of an inch, since the $\frac{8}{8}$ in the first inch, plus the $\frac{7}{8}$ in the second inch, equals a total of $\frac{15}{8}$. When we use the sixteenths ruler marks, screw b measures $1\frac{14}{16}$ or $\frac{30}{16}$ of an inch long.

■ INTEGRATED MATH APPLICATION: ELECTRICAL

Connecting an ammeter to a robotic arm, an electrician monitors the current drawn at different stages of the manufacturing process.

What are these ampere readings shown on the scale, at points a through e?

■ Solution:

a. $2\frac{1}{2}$ amps   **b.** $4\frac{1}{4}$ amps   **c.** $5\frac{3}{4}$ amps   **d.** $6\frac{7}{8}$ amps   **e.** $8\frac{3}{8}$ amps

---

SELF-TEST EVALUATION POINT FOR SECTION 1–2

Now that you have completed this section, you should be able to:

■ **Objective 3.** *Describe what a fraction is, and name its elements.*

■ **Objective 4.** *Demonstrate how fractions are represented.*

Use the following questions to test your understanding of Section 1–2.

1. What fraction of each of the circular objects is shown shaded?
2. What fraction of each of the circular objects is shown unshaded?

a.   b.   c.

1–3   ADDING FRACTIONS

The *denominator* (number below the fraction bar) determines the fraction type. For example, $\frac{3}{9}$ tells us that the object was divided up into nine parts and that we have three of those nine parts. The number 9 therefore tells us the type of fraction we are dealing with. When fractions are of the same type (have the same denominators), they are added as simply as adding 5 apples and 3 apples.

■ EXAMPLE

Add the following fractions: $\dfrac{5}{9} + \dfrac{3}{9}$

■ Solution:

Since we are dealing with two fractions that are of the same type (ninths), the result is simply the sum or total of the two *numerators* (numbers above the fraction bar).

$$\text{Numerators} \rightarrow \frac{5}{9} + \frac{3}{9} = \frac{8}{9} \leftarrow \text{Denominators}$$

Notice that the numerators were added, whereas the denominators remained the same. This is because the type of fraction remained the same; only the number of pieces changed. To use an analogy, we could say:

$$\frac{5}{\text{apples}} + \frac{3}{\text{apples}} = \frac{8}{\text{apples}} \begin{array}{l} \leftarrow \text{Sum} \\ \leftarrow \text{Type} \end{array}$$

The next step is to address how we can add different types of fractions. For example, how would we add $\frac{1}{2}$ plus $\frac{1}{4}$? In this instance we have two different types of fractions, and therefore they are as different as apples and oranges.

Using pictures, we know that $\frac{1}{2}$ plus $\frac{1}{4}$ equal $\frac{3}{4}$.

The question is: What mathematical procedure do we follow when the answer is not this obvious? If you study the previous answer, you see that the answer was in fourths. Therefore, what we actually did was convert the one-half into two-fourths ($\frac{1}{2} = \frac{2}{4}$, one-half = two-fourths) and then add the other fourth to get three-fourths.

The written fractions would appear as follows:

Since one-half equals two-fourths, we can subsitute $\frac{2}{4}$ (two-fourths) for $\frac{1}{2}$ (one-half).

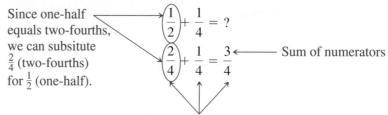

Now the fractions are the same type; therefore the denominator answer indicates the same fraction type.

Common Denominator

A common multiple of the denominators of a number of fractions.

What we have noticed from this example is that to add fractions that are of different types, we need to find a *denominator that is common to both fractions.* There are two methods for finding a **common denominator** so that we can add different types of fractions.

1–3–1   *Method 1 for Finding a Common Denominator*

Probably the easiest way to find a common denominator is to multiply one denominator by another. For example, let us assume that we want to add $\frac{1}{2}$ plus $\frac{1}{3}$.

$$\frac{1}{2} + \frac{1}{3} = ?$$

Our common denominator can be obtained by multiplying one denominator by the other; therefore, $2 \times 3 = 6$.

Now that a common denominator has been found, how many sixths $\left(\frac{1}{6}\right)$ are in one-half $\left(\frac{1}{2}\right)$ and one-third $\left(\frac{1}{3}\right)$? This is determined in the following way.

$$\frac{1}{2} + \frac{1}{3} = \frac{?}{6}$$

(One-half plus one-third equals how many sixths?)

$$\frac{1}{2} + \frac{1}{3} = \frac{3}{6}$$

Step a: 2 into 6 = 3.
Step b: 3 × 1 = 3.
Step c: Place result in answer's numerator.

$$\frac{1}{2} + \frac{1}{3} = \frac{3 + 2}{6}$$

Step a: 3 into 6 = 2.
Step b: 2 × 1 = 2.
Step c: Place result in answer's numerator.

$$\frac{1}{2} + \frac{1}{3} = \frac{3 + 2}{6}$$

The last step is to add the two answers for the numerator ($3 + 2 = 5$).

$$\frac{1}{2} + \frac{1}{3} = \frac{5}{6}$$

Add $\dfrac{1}{4}$ and $\dfrac{2}{3}$

■ *Solution:*

Common denominator = denominator $\times$ denominator

$$= \quad 4 \quad \times \quad 3$$

$$= \quad 12$$

$$\dfrac{1}{4} + \dfrac{2}{3} = \dfrac{3+8}{12} = \dfrac{11}{12}$$

Steps:

4 into 12 = 3,
3 $\times$ 1 = 3,
3 into 12 = 4,
4 $\times$ 2 = 8,
3 + 8 = 11.

1–3–2   *Method 2 for Finding a Common Denominator*

The problem with method 1 is that it often yields a common denominator that is larger than necessary. For example, what is the lowest-value common denominator for the following three-fraction addition?

$$\frac{3}{9} + \frac{1}{4} = \frac{1}{6} = \frac{}{?}$$

Using method 1, we would obtain the answer

$$9 \times 4 \times 6 = 216$$

Even though 216 will work as a common denominator, it is a very large number and there-fore the fraction addition will be cumbersome.

The steps in the following method will always yield the lowest **common denominator.** Using the previous example, we first place all three denominators in an upside-down di-vision box, as shown.

$$\underline{9 \qquad 4 \qquad 6}$$

The next step is to begin with 2 and see if it will divide exactly into any of the three denom-inators. Once 2 can no longer be used as a divisor, increase to 3, and then 4, and then 5, and so on. Let us follow through these steps in more detail.

2 does not go into 9 exactly.
2 goes into 4, giving 2,
and 2 goes into 6, giving 3.
Using 2 again as a divisor,
we can further reduce because 2
goes into 2, giving 1.
Because 2 will no longer reduce
any of the denominators, we increase
the divisor to 3. 3 goes into
9, giving 3, and into 3, giving 1.
Using 3 again as a divisor, we
can reduce because 3 goes into
3, giving 1.

Divisors to be
used as "factors."

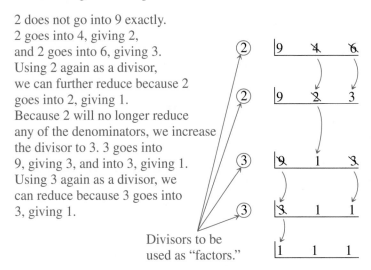

> **Lowest Common Denominator**
> The lowest-value common denominator.

Looking at the previous process, you can see that the point is to keep dividing the denominators until they are all reduced to 1. Because no number is changed when it is multiplied or divided by 1 (for example, $22 \times 1 = 22$, $115 \times 1 = 115$, $17 \div 1 = 17$, $63 \div 1 = 63$), we can ignore the remaining 1s.

The divisors used in this process are called *factors*. A **factor** *is any number that is multiplied by another number and contributes to the result.* For example, consider 3×8. In this multiplication example, 3 is a factor and 8 is a factor that will contribute to the result 24. Let us now take another example: What factors could contribute to a result of 12? In this case we must determine which small numbers, when multiplied together, will produce 12. The answer is

$$2 \times 6 = 12 \quad \text{or} \quad 3 \times 4 = 12$$

Are these, however, the smallest numbers that when multiplied together will produce 12? The answer is no because the 6 can be broken down into 2×3, and the 4 can be broken down into 2×2. Therefore, $2 \times 3 \times 2$ or $3 \times 2 \times 2$ shows that 12 has the three factors 2, 2, and 3.

Some numbers cannot be reduced or broken down into factors. For example, there are no smaller numbers that when multiplied together will produce 5. This number is called a **prime number,** and there are many, such as 3, 5, 7, 11, 13, and 17.

What we were doing, therefore, in the previous three-fraction addition example was extracting the factors from the denominators.

Factor

Any of the numbers or symbols in mathematics that when multiplied together form a product.

Prime Number

Any integer other than 0 or 1 that is not divisible without remainder.

Factoring

Resolving into factors; a reducing process that extracts the common factors.

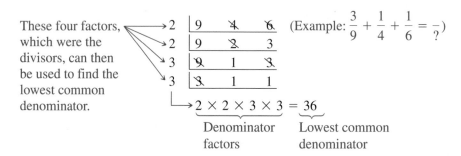

These four factors, which were the divisors, can then be used to find the lowest common denominator.

(Example: $\frac{3}{9} + \frac{1}{4} + \frac{1}{6} = \frac{}{?}$)

$2 \times 2 \times 3 \times 3 = 36$

Denominator factors — Lowest common denominator

Now that we have completed the *factoring* step and determined the lowest common denominator, we can proceed with the example, which was to add $\frac{3}{9} + \frac{1}{4} + \frac{1}{6}$:

$$\frac{3}{9} + \frac{1}{4} + \frac{1}{6} = \frac{}{36} \longleftarrow \text{Lowest common denominator}$$

$$\left(\frac{3}{9}\right) + \frac{1}{4} + \frac{1}{6} = \frac{12 +}{36} \qquad \textit{Steps:} \quad \begin{array}{l} 9 \text{ into } 36 = 4, \\ 4 \times 3 = 12. \end{array}$$

$$\frac{3}{9} + \left(\frac{1}{4}\right) + \frac{1}{6} = \frac{12 + 9 +}{36} \qquad \begin{array}{l} 4 \text{ into } 36 = 9, \\ 9 \times 1 = 9. \end{array}$$

$$\frac{3}{9} + \frac{1}{4} + \left(\frac{1}{6}\right) = \frac{12 + 9 + 6}{36} \qquad \begin{array}{l} 6 \text{ into } 36 = 6, \\ 6 \times 1 = 6. \end{array}$$

$$= \frac{12 + 9 + 6}{36} = \frac{27}{36}$$

■ **EXAMPLE**

Add $\dfrac{1}{4} + \dfrac{3}{5} + \dfrac{1}{8}$

CHAPTER 1 / WHOLE NUMBERS AND FRACTIONS

■ *Solution:*

The first step is to determine the factors in each denominator.

		4	5	8
Divisors or factors 3 and 4 were ignored because 5 was a prime factor.	2	2	5	4
	2	1	5	2
	5	1	5	1
		1	1	1

$2 \times 2 \times 2 \times 5 = 40$ (Lowest common denominator)

$$\frac{1}{4} + \frac{3}{5} + \frac{1}{8} = \frac{10 + 24 + 5}{40} = \frac{39}{40}$$

1-3-3   *Improper Fractions and Mixed Numbers*

All the fractions discussed so far have a numerator that is smaller than the denominator. These are called **proper fractions.** For example:

Proper fractions: $\dfrac{1}{2}$ $\dfrac{3}{4}$ $\dfrac{5}{9}$ $\dfrac{7}{32}$ $\dfrac{15}{27}$ ← Smaller numerator
← Larger denominator

On some occasions, an addition of two or more proper fractions will produce a fraction with a numerator that is larger than the denominator. To explain this, let us use an example.

■ **EXAMPLE**

Add $\dfrac{3}{4} + \dfrac{2}{4}$

■ *Solution:*

$$\frac{3}{4} + \frac{2}{4} = \frac{3 + 2}{4} = \frac{5}{4} \leftarrow \text{Numerator is larger than denominator.}$$

To analyze this answer, $\frac{5}{4}$, we would say that the denominator 4 indicates that the original whole unit was divided into four pieces (quarters) and the numerator indicates that we have five of these pieces. Therefore, with four of the quarters (or fourths) we have one whole unit, leaving one extra fourth.

$$\frac{5}{4} = 1\frac{1}{4}$$

five-fourths = one whole unit and one-fourth

A fraction with a numerator that is larger than the denominator (for example, $\frac{5}{4}$) is called an **improper fraction** and needs to be converted to a **mixed number.** By definition, a mixed fraction or mixed number is a whole number and a fraction (for example, $1\frac{1}{4}$). Since the fraction bar actually indicates the arithmetic operation of division (for example, $\frac{5}{4}$ indicates 5 divided by 4), all we do is perform this division to convert an improper fraction to a mixed number. For example, 4 goes into 5 once with 1 remaining:

$$\frac{5}{4} = 1\frac{1}{4}$$

four into five $=$ one, with one-fourth remaining

■ EXAMPLE

Add $\dfrac{2}{3} + \dfrac{4}{6}$

■ *Solution:*

$$\frac{2}{3} + \frac{4}{6} = \frac{4 + 4}{6} = \frac{8}{6}$$

Lowest common denominator:

2	3	6
3	3	3

$\longrightarrow 2 \times 3 = 6$

Because $\frac{8}{6}$ is an improper fraction, the next step is to convert it to a mixed fraction.

$$\frac{8}{6} = 1\frac{2}{6}$$

six into eight $=$ one, with two-sixths remaining

■ INTEGRATED MATH APPLICATION: MACHINING

What is the total length of the machine part shown in the figure?

■ *Solution:*

$$\frac{1}{8} + \frac{7}{16} + \frac{3}{8} = \frac{15}{16} \text{ inch}$$

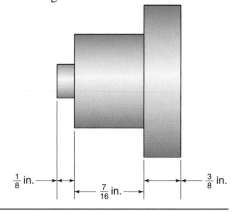

$\frac{1}{8}$ in. $\frac{7}{16}$ in. $\frac{3}{8}$ in.

1–3–4   *Adding Mixed Numbers*

Now that we understand what a mixed number is, let us see how we would add two mixed fractions.

■ EXAMPLE

Add $5\dfrac{1}{4} + 6\dfrac{3}{8}$

■ *Solution:*

The easiest way to add mixed numbers is to deal with the whole numbers and fractions separately.

Step 1: Separate whole number and fractions.

Step 2: Add fractions.

Step 3: Add whole numbers.

Step 4: Combine whole-number result and fraction result.

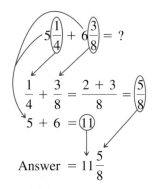

$$5\frac{1}{4} + 6\frac{3}{8} = ?$$

$$\frac{1}{4} + \frac{3}{8} = \frac{2 + 3}{8} = \frac{5}{8}$$

$$5 + 6 = 11$$

$$\text{Answer} = 11\frac{5}{8}$$

■ INTEGRATED MATH APPLICATION: CARPENTRY

If two wood screws measure $6\frac{3}{8}$ inches and $2\frac{1}{4}$ inches, what is the total length of both?

■ *Solution:*

$$6\frac{3}{8} + 2\frac{1}{4} = 8\frac{3 + 2}{8} = 8\frac{5}{8} \text{ inches}$$

■ EXAMPLE

Add $4\frac{15}{20} + 5\frac{4}{5}$

■ *Solution:*

$$4\frac{15}{20} + 5\frac{4}{5} = 9\frac{15 + 16}{20} = 9\frac{31}{20}$$

Because the fraction $\frac{31}{20}$ is improper, we must convert it by dividing 20 into 31.

$$9\frac{31}{20} = 9 + \left(1\frac{11}{20}\right) \longleftarrow 31 \div 20 = 1\frac{11}{20}$$

$$= 10\frac{11}{20} \longleftarrow \begin{array}{l} 9 \text{ whole objects plus} \\ 1 \text{ whole object} = 10 \end{array}$$

■ INTEGRATED MATH APPLICATION: PLUMBING

Calculate the total length of the water pipe shown.

■ *Solution:*

$$14\frac{1}{4} + 7\frac{1}{3} + 6\frac{1}{8} = 27\frac{17}{24} \text{ feet}$$

1–3–5   *Reducing Fractions*

To obtain half of a pizza politely, you could ask for two small quarters because two-quarters equals one-half.

$$\frac{2}{4} = \frac{1}{2}$$

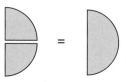

Both fractions therefore indicate the same part of a whole object. In most instances when we have two fractions that describe the same portion, we would use the fraction that has the smaller denominator. For example, the denominator 2 in $\frac{1}{2}$ is smaller than the denominator 4 in $\frac{2}{4}$, and therefore we would use $\frac{1}{2}$ to represent the fraction. The next question is: How do we **reduce** a fraction so that it has the lowest possible denominator? The answer is that we use an algebraic rule that states: *If the same number is divided into both the numerator and denominator of a fraction, the value of the fraction remains the same, only reduced.*

For example, let us divide 2 into both the numerator and denominator of $\frac{2}{4}$:

$$\frac{2 \div 2}{4 \div 2} = \frac{1}{2}$$

To reduce a fraction we must therefore find the largest number that will divide evenly into both its numerator and denominator. This is best achieved by starting with 2, then trying 3, then 4, then 5, and so on. To demonstrate this, let us do the following example.

Reducing

The process of changing to an equivalent but more fundamental expression.

■   **EXAMPLE**

Reduce the following fractions if possible.

    a. $\dfrac{3}{6}$ **b.** $\dfrac{4}{16}$ **c.** $\dfrac{5}{45}$ **d.** $\dfrac{4}{9}$ **e.** $\dfrac{18}{90}$

■   *Solution:*

a. $\dfrac{\left(3\right) \div 3}{\left(6\right) \div 3} = \dfrac{1}{2}$ (three-sixths = one-half)

    └── Will both divide by 2? No. Will both divide by 3? Yes.

b. $\dfrac{\left(4\right) \div 4}{\left(16\right) \div 4} = \dfrac{1}{4}$ (four-sixteenths = one-fourth)

    └── Will both divide by 2? Yes. Will both divide by 3? No. Will both divide by 4? Yes. We will use 4 because dividing by a larger number will give us a smaller denominator in fewer steps. If we had used 2, we would have ended up with the same answer but would have had to reduce in two steps:

$$\frac{\left(4\right) \div 2}{\left(16\right) \div 2} = \frac{\left(2\right) \div 2}{\left(8\right) \div 2} = \frac{1}{4}$$

c. $\left(\dfrac{5}{45}\right) \dfrac{\div\ 5}{\div\ 5} = \dfrac{1}{9}$

⎿—Will both divide by 2? No. Will both divide by 3? No. Will both divide by 4? No. Will both divide by 5? Yes.

d. $\left(\dfrac{4}{9}\right)$

⎿—Will both divide by 2? No. Will both divide by 3? No. Will both divide by 4? No. This fraction cannot be reduced any further.

e. $\left(\dfrac{18}{90}\right) \dfrac{\div\ 18}{\div\ 18} = \dfrac{1}{5}$

⎿—Divide by 2? Yes. Divide by 3? Yes. Divide by 4? No. Divide by 5? No. Divide by 6? Yes. Divide by 7? No. Divide by 8? No. Divide by 9? Yes. . . . Divide by 18? Yes.

■ **INTEGRATED MATH APPLICATION: AUTOMOTIVE**

An automotive technician replaces $2\frac{1}{2}$ quarts of oil, $4\frac{1}{3}$ quarts of coolant, $\frac{1}{4}$ quart of brake fluid, and $3\frac{1}{6}$ quarts of washer fluid for the 15,000-mile routine service. On the maintenance sheet, what should be written alongside "Total number of quarts used"?

■ *Solution:*

$$2\frac{1}{2} + 4\frac{1}{3} + \frac{1}{4} + 3\frac{1}{6} = 10\frac{1}{4} \text{ quarts}$$

SELF-TEST EVALUATION POINT FOR SECTION 1–3

Now that you have completed this section, you should be able to:

■ **Objective 5.** *Describe how fractions are added.*

■ **Objective 6.** *Show two methods used for calculating the lowest common denominator.*

■ **Objective 7.** *Define a proper fraction, an improper fraction, and a mixed fraction.*

■ **Objective 8.** *Show how mixed fractions are added.*

■ **Objective 9.** *Describe how to reduce fractions.*

Use the following questions to test your understanding of Section 1–3.

Add the following fractions. Convert to proper fractions and reduce if necessary.

1. $\dfrac{4}{6} + \dfrac{1}{6} = ?$

2. $\dfrac{3}{64} + \dfrac{2}{64} + \dfrac{25}{64} = ?$

3. $\dfrac{3}{9} + \dfrac{6}{18} = ?$

4. $\dfrac{1}{3} + \dfrac{1}{4} + \dfrac{3}{15} = ?$

5. $\dfrac{3}{4} + \dfrac{4}{5} = ?$

6. $\dfrac{9}{12} + \dfrac{4}{24} + \dfrac{3}{4} = ?$

7. $\dfrac{7}{9} + \dfrac{5}{9} + \dfrac{4}{18} = ?$

Convert to proper fractions and/or reduce if necessary.

8. $\dfrac{15}{6} = ?$

9. $\dfrac{4}{24} = ?$

10. $\dfrac{25}{100} = ?$

1–4 SUBTRACTING FRACTIONS

To subtract one fraction from another, we follow the same steps as in addition except that we subtract the numerators. Let us examine a few examples, starting with a subtraction that involves two fractions with the same denominator.

■ **EXAMPLE**

What is $\dfrac{9}{16} - \dfrac{4}{16}$?

■ *Solution:*

$$\frac{9}{16} - \frac{4}{16} = \frac{9-4}{16} = \frac{5}{16}$$

If the two fractions involved in the subtraction are of different types, we will need to determine the lowest common denominator so that we are subtracting one type of fraction from the same type of fraction.

■ **EXAMPLE**

What is the result of $\dfrac{6}{10} - \dfrac{4}{20}$?

■ *Solution:*

$$\frac{6}{10} - \frac{4}{20} = \frac{12-4}{20} = \frac{8}{20}$$
$$= \frac{8 \div 4}{20 \div 4} = \frac{2}{5}$$

Steps:
Find lowest common denominator (20).
10 into 20 = 2,
2 × 6 = 12.
20 into 20 = 1,
1 × 4 = 4.
12 − 4 = 8.
Reduce result.

■ **INTEGRATED MATH APPLICATION: MACHINING**

The diameter of a steel cylinder is reduced $\frac{14}{1000}$ inch from its original $\frac{750}{1000}$ inch. What is the cylinder's new diameter after being machined?

■ *Solution:*

$$\frac{750}{1000} - \frac{14}{1000} = \frac{736}{1000} \text{ inch}$$

1–4–1 *Subtracting Mixed Fractions*

As with addition, the way to subtract one mixed fraction from another is to deal with the whole numbers and fractions separately.

■ **EXAMPLE**

Subtract $2\dfrac{1}{3}$ from $5\dfrac{4}{5}$

■ *Solution:*

$$5\frac{4}{5} - 2\frac{1}{3} = 3\frac{12-5}{15}$$

$$= 3\frac{7}{15}$$

Steps:
Subtract whole numbers: $5 - 2 = 3$.
Determine the lowest common denominator.

$$\begin{array}{c|cc} 3 & 5 & 3 \\ \hline 5 & 5 & 1 \end{array}$$

$$\longrightarrow 3 \times 5 = 15$$

5 into $15 = 3, 3 \times 4 = 12$.
3 into $15 = 5, 5 \times 1 = 5$.
$12 - 5 = 7$.

■ **INTEGRATED MATH APPLICATION: ELECTRICAL**

What is the power rating difference between a $\frac{1}{4}$ watt resistor and a $2\frac{1}{2}$ watt resistor?

■ *Solution:*

$$2\frac{1}{2} - \frac{1}{4} = 2\frac{1}{4}\text{ watts}$$

1–4–2 *Borrowing*

As is normal in subtraction, we sometimes have problems in which we need to subtract a large number from a small number. In these cases we will need to **borrow**, as shown in the following example.

Borrowing
The process of taking from the next higher order digit so that a subtraction can take place.

■ **EXAMPLE**

What is $6\frac{1}{3} - 2\frac{2}{3}$?

■ *Solution:*

Borrow 1
from 6 and
convert it to $\frac{3}{3}$.

$$6\frac{1}{3} - 2\frac{2}{3} = 4\frac{1-2}{3}?$$

$$6\frac{1}{3} = 5 + \left(\frac{3}{3}\right) + \frac{1}{3} = 5\frac{4}{3}$$

$$5\frac{4}{3} - 2\frac{2}{3} = 3\frac{4-2}{3} = 3\frac{2}{3}$$

Steps:
Subtract whole numbers; $6 - 2 = 4$.
Because fractions have the same
denominator, subtract numerators.
You cannot subtract 2 from 1.
Borrow whole number from 6,
and convert it to a fraction $\left(1 = \frac{3}{3}\right)$.
Add 3 pieces to 1 piece to get 4 pieces,
or 4 thirds.
Perform subtraction.

■ **INTEGRATED MATH APPLICATION: BANKING**

The interest rate on a car loan has been reduced $2\frac{1}{4}$ percent from $6\frac{3}{16}$ percent. What is the new interest rate?

■ *Solution:*

$$6\frac{3}{16} - 2\frac{1}{4} = 4\frac{3-4}{16} \longleftarrow \text{You cannot subtract 4 from 3.}$$

$$5 + \frac{16}{16} + \frac{3}{16} - 2\frac{1}{4} =$$

$$5\frac{19}{16} - 2\frac{1}{4} = 3\frac{19-4}{16} = 3\frac{15}{16}$$

SELF-TEST EVALUATION POINT FOR SECTION 1–4

Now that you have completed this section, you should be able to:

■ **Objective 10.** *Show how to subtract proper and mixed fractions.*

Use the following questions to test your understanding of Section 1–4.

1. $\dfrac{4}{32} - \dfrac{2}{32} = ?$

2. $\dfrac{4}{5} - \dfrac{3}{10} = ?$

3. $\dfrac{6}{12} - \dfrac{4}{16} = ?$

4. $15\dfrac{1}{3} - 2\dfrac{4}{12} = ?$

5. $5\dfrac{15}{36} - 2\dfrac{31}{36} = ?$

1–5   MULTIPLYING FRACTIONS

One of the best ways to explain something new is to use an example that you can imagine easily.

■ **EXAMPLE**

What is half of a half?

■ *Solution:*

$$\frac{1}{2} \text{ of } \bigcirc = \text{◢}$$

half of a half $=$ a fourth

$$\frac{1}{2} \times \frac{1}{2} = \frac{1}{4}$$

It is not surprising that the product (multiplication result) of two fractions has a denominator that is smaller than the denominators of the two fractions, because multiplication is calculating the fractional part of a fraction. Let us look at another example.

■ **EXAMPLE**

What is half of a fourth?

■ *Solution:*

$$\frac{1}{2}\text{ of } = $$

half of one-fourth $=$ one-eighth

$$\frac{1}{2} \times \frac{1}{4} = \frac{1}{8}$$

From the last two examples you can probably see that to obtain the product of two fractions, simply multiply the numerators and multiply the denominators, as shown.

$$\frac{1}{2} \times \frac{1}{2} = \frac{1 \times 1}{2 \times 2} = \frac{1}{4}$$ (one-half of one-half equals one-fourth)

$$\frac{1}{2} \times \frac{1}{4} = \frac{1 \times 1}{2 \times 4} = \frac{1}{8}$$ (one-half of one-fourth equals one-eighth)

■ **EXAMPLE**

What is $\frac{3}{4}$ of $\frac{3}{5}$?

■ *Solution:*

$$\frac{3}{4} \times \frac{3}{5} = \frac{3 \times 3}{4 \times 5} = \frac{9}{20}$$

■ **INTEGRATED MATH APPLICATION: MACHINING**

A bolt has an 18-NF (National Fine) thread, which means that it will advance $\frac{1}{18}$ inch for every complete turn. How far will the bolt advance if a nut driver is used to turn it five complete revolutions?

■ *Solution:*

Pitch $= \frac{1}{18}$ in.

$$5 \times \frac{1}{18}\text{ inch} = \frac{5}{18}\text{ inch}$$

1–5–1 *Multiplying Mixed Fractions*

When adding or subtracting mixed fractions, we deal with the whole numbers and fractions separately. When multiplying mixed fractions, we must first convert the mixed fractions to improper fractions and then perform the multiplication. For example, what is half of one and one-third?

$$\frac{1}{2} \times 1\frac{1}{3} = ?$$

If we convert $1\frac{1}{3}$ to an improper fraction, we will be dealing only with thirds. Because there are three thirds in one whole $\left(\frac{3}{3} = 1\right)$, these three thirds plus the additional one-third piece equals four-thirds.

$$1\frac{1}{3} = \frac{3}{3} + \frac{1}{3} = \frac{4}{3}$$

Now that the mixed fraction is an improper fraction, we can complete the original problem, which is: What is half of one and one-third, or four-thirds?

$$\frac{1}{2} \times \frac{4}{3} = \frac{4}{6} \begin{array}{c} \div 2 = 2 \\ \div 2 = 3 \end{array}$$

As expected, half of four-thirds equals two-thirds. As you can see, it was necessary to reduce the first result, $\frac{4}{6}$, to $\frac{2}{3}$. This often occurs owing to the larger numbers in the improper fractions.

Before doing another example, let's develop a procedure for converting a mixed fraction to an improper fraction. As an example, we will convert the mixed fraction $2\frac{1}{2}$ to an improper fraction. Because the result will be all halves $\left(2\frac{1}{2} = \frac{?}{2}\right)$, we simply have to determine how many halves are in 2 whole units and then add this value to the additional half. In summary:

$$2\frac{1}{2} = \frac{4}{2} + \frac{1}{2} = \frac{5}{2}$$

| Two and | one-half | = | four-halves | + | one-half | = | five-halves |

To change a mixed fraction to an improper fraction, therefore, you *multiply the whole number by the fraction's denominator and then add the fraction's numerator.*

$$2\frac{1}{2} = \frac{5}{2}$$

Steps:
Multiply whole number by fraction's denominator: $2 \times 2 = 4$.
Add result to fraction's numerator: $4 + 1 = 5$.

Now that we understand how to convert a mixed fraction to an improper fraction, we can do a multiplication example involving mixed fractions.

■ EXAMPLE

What is $\frac{4}{9}$ of $5\frac{3}{4}$?

■ *Solution:*

Note that the term *of* implies multiplication. To determine the result, we must first convert the mixed fraction, $5\frac{3}{4}$, to an improper fraction.

$$5\frac{3}{4} = \frac{23}{4}$$

Steps:
$5 \times 4 = 20$.
$20 + 3 = 23$.

The next step is to perform the multiplication using the improper fraction.

$$\frac{4}{9} \times \frac{23}{4} = \frac{92}{36} \begin{array}{c} \div 4 = 23 \\ \div 4 = 9 \end{array}$$

Reduce

Once reduced, the answer may be an improper fraction and therefore will have to be converted back into a mixed fraction. With $\frac{23}{9}$, the denominator 9 indicates that the original whole unit was divided into 9 pieces, and the numerator 23 indicates that we have 23 of these pieces. Dividing 9 into 23 will determine how many whole units we can obtain and how many ninths remain.

$$\frac{23}{9} = 2\frac{5}{9}$$

Steps:
9 into 23 = 2 remainder 5

■ INTEGRATED MATH APPLICATION: AUTOMOTIVE

A hybrid (gasoline–electric) vehicle can average $32\frac{2}{5}$ miles per gallon. How many miles can it travel on $2\frac{1}{4}$ gallons?

■ *Solution:*

$$2\frac{1}{4} \text{ gallons} \times 32\frac{2}{5} \text{ miles per gallon} = 72\frac{9}{10} \text{ miles}$$

■ INTEGRATED MATH APPLICATION: METEOROLOGY

The formula for converting a temperature in degrees Celsius (°C) to degrees Fahrenheit (°F) is:

$$°F = \left(\frac{9}{5} \times °C\right) + 32$$

If you were traveling in Europe and the temperature was 37 °C, what would that temperature be in degrees Fahrenheit?

$$°F = \left(\frac{9}{5} \times °C\right) + 32$$

$$= \left(\frac{9}{5} \times 37°C\right) + 32$$

$$= \left(\frac{9}{5} \times \frac{37}{1}\right) + 32$$

$$= \frac{333}{5} + 32$$

$$= 66\frac{3}{5} + 32$$

$$= 98\frac{3}{5}°F$$

SELF-TEST EVALUATION POINT FOR SECTION 1–5

Now that you have completed this section, you should be able to:

■ *Objective 11.* *Show how to multiply proper and mixed fractions.*

Use the following questions to test your understanding of Section 1–5.

1. $\frac{1}{3} \times \frac{2}{3} = ?$

2. $\frac{4}{8} \times \frac{2}{9} = ?$

3. $\frac{4}{17} \times \frac{3}{34} = ?$

4. $\frac{1}{16} \times 5\frac{3}{7} = ?$

5. $2\frac{1}{7} \times 7\frac{2}{3} = ?$

1–6   DIVIDING FRACTIONS

Once again let us examine what we are doing by using a simple example: How many quarters are in three-quarters? Stated mathematically, this problem appears as follows:

$$\frac{3}{4} \div \frac{1}{4} = ? \qquad \div \qquad = ?$$

The answer is straightforward and requires no mathematical processes or steps because we already know that there are three quarters in three-quarters.

$$\frac{3}{4} \div \frac{1}{4} = 3$$

Since most problems are never this simple, we need to follow a mathematical process to divide fractions. Because addition is the opposite of subtraction, and multiplication is the opposite of division, we can just invert the second fraction in a division problem and then multiply both fractions to obtain the result. Let us try this process on the previous example:

$$\frac{3}{4} \div \frac{1}{4} = \frac{3 \times 4}{4 \times 1} = \frac{12 \div 4}{4 \div 4} = \frac{3}{1} = 3$$

Invert and multiply   Reduce   1 into 3 = 3

■ **EXAMPLE**

How many twelfths are in two-thirds $\left(\frac{2}{3} \div \frac{1}{12}\right)$?

■ *Solution:*

$$\frac{2}{3} \div \left(\frac{1}{12}\right) = \frac{2 \times 12}{3 \times 1} = \frac{\left(24\right) \div 3}{3 \div 3} = \frac{8}{1} = 8$$

Invert and multiply   Reduce

1–6–1   *Dividing Mixed Fractions*

As when we multiply mixed fractions, to divide mixed fractions we must first convert all mixed fractions to improper fractions. Once we have done this, we can divide as usual, which means: Invert the second fraction and multiply. To start, let us use a simple example and test the procedure to see how many halves are in $2\frac{1}{2}$. As you have already determined, the result should indicate that there are five halves in two and one-half.

Steps
Convert mixed fraction to improper fraction:

$2 \times 2 = 4, 4 + 1 = 5.$ $\longrightarrow$ $\left(2\frac{1}{2}\right) \div \frac{1}{2} = ?$

$\frac{5}{2} \div \left(\frac{1}{2}\right) = ?$

Invert second fraction and multiply. $\longrightarrow$ $\frac{5 \times 2}{2 \times 1} = \frac{10}{2}$

Reduce. $\longrightarrow$ $\frac{\left(10\right) \div 2}{2 \div 2} = \frac{5}{1} = 5$

■ **EXAMPLE**

$6\frac{2}{3} \div 1\frac{1}{3} = ?$

■ *Solution:*

$$6\frac{2}{3} \div 1\frac{1}{3} = ?$$

Steps:
Convert mixed fractions to improper fractions:
$6 \times 3 = 18; 18 + 2 = 20.$

$$\frac{20}{3} \div \frac{4}{3} = ?$$

$1 \times 3 = 3; 3 + 1 = 4.$

$$\frac{20 \times 3}{3 \times 4} = \frac{60}{12}$$

Invert second fraction and multiply.

$$\frac{60 \div 12}{12 \div 12} = \frac{5}{1} = 5$$

Reduce.

Now let's do an example that does not work out so neatly.

■ EXAMPLE

How many times will $3\frac{5}{9}$ go into $6\frac{1}{2}$?

■ *Solution:*

Note that the term *go into* implies division and is asking how many times the first number will go into the second number.

$$6\frac{1}{2} \div 3\frac{5}{9} = ?$$

Steps:

$$\frac{13}{2} \div \frac{32}{9} = ?$$

Convert mixed fractions to improper fractions.

$$\frac{13 \times 9}{2 \times 32} = \frac{117}{64}$$

Invert second fraction and multiply.

$$\frac{117}{64} = 1\frac{53}{64}$$

Convert improper fraction to mixed fraction.
$117 \div 64 = 1$ remainder 53.

■ INTEGRATED MATH APPLICATION: ELECTRICAL

How many $34\frac{1}{3}$ watt motors can be run on a 400 watt circuit?

■ *Solution:*

$$400 \text{ watts} \div 34\frac{1}{3} \text{ watts per motor} = 11\frac{67}{103} \text{ motors}$$

■ INTEGRATED MATH APPLICATION: PERSONAL FINANCE

If you are paying an annual percentage rate of $15\frac{1}{8}$ percent on a credit card, what is the monthly percentage rate?

■ *Solution:*

$$15\frac{1}{8} \div 12 = ?$$

$$\frac{120}{8} \div \frac{12}{1} = ?$$

$$\frac{120 \times 1}{8 \times 12} = \frac{120}{96} = 1\frac{24}{96} = 1\frac{1}{4}\%$$

Now that you have completed this section, you should be able to:

■ *Objective 12.* *Show how to divide proper and mixed fractions.*

Use the following questions to test your understanding of Section 1–6.

1. $\dfrac{2}{3} \div \dfrac{1}{12} = ?$

2. $3\dfrac{1}{4} \div \dfrac{3}{4} = ?$

3. $8\dfrac{16}{20} \div 4\dfrac{4}{20} = ?$

4. $4\dfrac{3}{8} \div 2\dfrac{5}{16} = ?$

1–7   CANCELING FRACTIONS

Canceling

The process of removing a common divisor from numerator and denominator.

Canceling is an operation that can be performed on fractions only when a multiplication symbol exists between the two. Its advantage is a reduction in the size of the numbers in the problem and therefore in time needed to solve, because fewer steps are needed. Let us demonstrate the process with a simple example.

$$\frac{5}{12} \times \frac{4}{20} = ?$$

$$\frac{5}{12} \times \frac{4}{20} = \frac{20 \div 20}{240 \div 20} = \frac{1}{12}$$

As you can see from this example, multiplying large numbers generates large results that generally need to be reduced. By canceling before we perform the multiplication, we start with smaller fractions and thus simplify the process.

$$\frac{5}{12} \times \frac{4}{20} \quad = \quad \frac{5}{3} \times \frac{1}{20} \quad = \quad \frac{1}{3} \times \frac{1}{4} = \frac{1}{12}$$

Four divides evenly into 12 and 4.
$12 \div 4 = 3$,
$4 \div 4 = 1$.

Five divides evenly into 5 and 20.
$5 \div 5 = 1$,
$20 \div 5 = 4$.

Multiplication is now easier with no need for reducing.

Up to this point we have "reduced" fractions only by finding a number that will divide evenly into the top and bottom part of the same fraction. In this example you can see that we could not reduce $\frac{5}{12}$ because there was no number that would evenly divide into 5 and 12; however, we could reduce the numerator of one fraction with the denominator of another fraction. Reducing in this crisscross manner is called cancellation and can be done only when a multiplication sign is present between the two fractions.

■ **EXAMPLE**

Calculate the following: $\dfrac{4}{9} \div \dfrac{8}{27}$

■ *Solution:*

$$\frac{4}{9} \div \frac{8}{27} = ?$$

Invert second fraction and multiply.

Now that a multiplication symbol exists between the two fractions, we can see if it is possible to crisscross cancel.

$$\frac{④}{9} \times \frac{27}{⑧} = ?$$

Cancellation: $4 \div 4 = 1, 8 \div 4 = 2$.

$$\frac{1}{⑨} \times \frac{㉗}{2} =$$

Cancellation: $9 \div 9 = 1, 27 \div 9 = 3$.

$$\frac{1}{1} \times \frac{3}{2} = \frac{3}{2} = 1\frac{1}{2}$$

Convert improper fraction to mixed fraction.

■ **EXAMPLE**

Calculate the following: $3\frac{5}{7} \times 2\frac{2}{13}$

■ *Solution:*

$$3\frac{5}{7} \times 2\frac{2}{13} = ?$$

$$\frac{26}{7} \times \frac{28}{13} = \frac{2}{7} \times \frac{28}{1} = \frac{2}{1} \times \frac{4}{1}$$

$$\frac{2}{1} \times \frac{4}{1} = \frac{8}{1} = 8$$

Steps:
Convert mixed fractions to improper fractions:

$3 \times 7 = 21, 21 + 5 = 26$.
$2 \times 13 = 26, 26 + 2 = 28$

Cancellation: $13 \div 13 = 1, 26 \div 13 = 2$.
Cancellation: $28 \div 7 = 4, 7 \div 7 = 1$.

■ **INTEGRATED MATH APPLICATION: CONSTRUCTION**

As discussed previously, a roof's pitch can be expressed as a fraction:

$$\text{Pitch} = \frac{\text{Rise}}{\text{Run}} = \frac{12 \text{ feet}}{36 \text{ feet}}$$

Roofers will normally describe pitch as a single figure that is equal to the amount of rise for every 12 feet (denominator must be 12). Reduce the preceding pitch fraction to get an equivalent fraction with a denominator of 12.

■ *Solution:*

$$\text{Pitch} = \frac{\text{Rise}}{\text{Run}} = \frac{12 \text{ feet} \div 3}{36 \text{ feet} \div 3} = \frac{4}{12}, \text{Pitch} = 4$$

■ **INTEGRATED MATH APPLICATION: ELECTRICAL**

An electrician is given two tasks on one jobsite:

a. A 24 gauge copper wire measures $12\frac{6}{8}$ feet long. What is its length in lowest terms?
b. A web of five fiber-optic cable runs connects cameras to a control center. Add the following lengths to determine the total, and give it in lowest terms.

Cable length in feet: $8\frac{1}{2}, 40\frac{3}{8}, 24\frac{1}{4}, 18\frac{5}{16}, 16\frac{3}{4}$

■ *Solution:*

a. $12\frac{6}{8}$ feet $= 12\frac{3}{4}$ feet

b. $8\frac{1}{2} + 40\frac{3}{8} + 24\frac{1}{4} + 18\frac{6}{16} + 16\frac{3}{4} = 108\frac{4}{16} = 108\frac{1}{4}$ feet

---

SELF-TEST EVALUATION POINT FOR SECTION 1–7

Now that you have completed this section, you should be able to:

■ *Objective 13.* Demonstrate how to use cancellation to simplify fraction multiplication or division.

Use the following questions to test your understanding of Section 1–7.

1. $\dfrac{16}{27} \times \dfrac{27}{240} = ?$

2. $\dfrac{8}{17} \times \dfrac{3}{4} = ?$

3. $\dfrac{15}{32} \div \dfrac{30}{4} = ?$

4. $6\dfrac{2}{3} \times 2\dfrac{7}{10} = ?$

5. $6\dfrac{1}{2} \div \dfrac{9}{32} = ?$

REVIEW QUESTIONS

Multiple Choice Questions

1. What is the base or radix of the decimal number system?
   - **a.** 100
   - **b.** 10
   - **c.** 0 through 9
   - **d.** 10, 100, 1000, 10,000

2. Because each column in the decimal number system increases by _____ times as you move left, two columns to the left will be _____ times greater.
   - **a.** 100, 10
   - **b.** 10, 1000
   - **c.** 1000, 10
   - **d.** 10, 100

3. A fraction is a piece; therefore if you have a fraction, you have something _____.
   - **a.** Between 1 and 10
   - **b.** Greater than 1
   - **c.** Less than 1
   - **d.** Between 1 and 100

4. The more cuts you make in the whole unit, the _____ the fraction.
   - **a.** Larger
   - **b.** Smaller

5. The number above the fraction bar is called the _____, while the number below the fraction bar is called the _____.
   - **a.** Denominator, factor
   - **b.** Numerator, denominator
   - **c.** Denominator, numerator
   - **d.** Numerator, factor

6. How many ninths are in a whole unit?
   - **a.** 9
   - **b.** Depends on the value of the numerator
   - **c.** 16
   - **d.** All the above

7. If you had $6\frac{3}{4}$ apples in one pocket and $2\frac{1}{2}$ apples in the other, what would you have?
   - **a.** Big pockets
   - **b.** $8\frac{1}{4}$ apples
   - **c.** $9\frac{3}{4}$ apples
   - **d.** $9\frac{1}{4}$ apples

8. What is the lowest common denominator of 3 and 15?
   - **a.** 45
   - **b.** 15
   - **c.** 5
   - **d.** 90

9. A(n) _____ fraction has a numerator that is smaller than the denominator.
   - **a.** Mixed
   - **b.** Improper
   - **c.** Proper
   - **d.** All the above

10. A(n) fraction contains both a whole number and a fraction.
    - **a.** Mixed
    - **b.** Proper
    - **c.** Improper
    - **d.** All the above

11. A(n) _____ fraction has a numerator that is larger than the denominator.
    - **a.** Mixed
    - **b.** Proper
    - **c.** Improper
    - **d.** All the above

12. Reduce the following fraction to its lowest terms: $\frac{32}{144}$.
    - **a.** $\dfrac{4}{18}$
    - **b.** $\dfrac{2}{9}$
    - **c.** $\dfrac{1}{6}$
    - **d.** $\dfrac{8}{32}$

Communication Skill Questions

13. Describe the base, positional weight, and reset and carry action of the decimal number system. (1–1)

14. What is a fraction, and what do the two numbers above and below the fraction bar line indicate? (1–2)

---

15. Describe the best method for calculating the lowest-value common denominator.   (1–3)

16. What is a prime factor?   (1–3)

17. What is an improper fraction?   (1–3)

18. What is a mixed fraction?   (1–3)

19. Arbitrarily choose values, and describe the steps used to:

    a. Add fractions.   (1–3)
    b. Subtract fractions.   (1–4)
    c. Multiply fractions.   (1–5)
    d. Divide fractions.   (1–6)

20. How can fractions be reduced?   (1–3)

21. Why do you sometimes need to borrow when subtracting fractions?   (1–4)

22. Arbitrarily choose values, and describe the steps used to:

    a. Add mixed fractions.   (1–3)
    b. Subtract mixed fractions.   (1–4)
    c. Multiply mixed fractions.   (1–5)
    d. Divide mixed fractions.   (1–6)

23. How are fractions canceled?   (1–7)

Practice Problems

24. Indicate the positional weight of each of the following digits: 96,237.

25. Write each of the following values as a number.

    a. Two hundred and seventy-six
    b. Eight thousand and seven tens
    c. Twenty thousand, four tens, and nine units

26. Which columns will reset and carry when each of the following values is advanced or incremented by 1?

    a. 199   **c.** 29,909
    b. 409   **d.** 49,999

27. Write the whole number and fraction for each of the following quantities.

a.

b.

28. Consider the value $5\frac{7}{9}$. State which of the numbers represents the:

    a. Whole number
    b. Denominator
    c. Numerator
    d. Number of original pieces
    e. Number of fractional pieces we have

29. Add the following fractions.

    a. $\dfrac{5}{8} + \dfrac{2}{8} = ?$

    b. $\dfrac{47}{76} + \dfrac{15}{76} + \dfrac{1}{76} = ?$

30. Calculate the lowest common denominator and then add the following fractions.

    a. $\dfrac{4}{9} + \dfrac{1}{3} = ?$ **c.** $\dfrac{2}{14} + \dfrac{1}{8} + \dfrac{1}{4} = ?$

    b. $\dfrac{4}{7} + \dfrac{2}{21} = ?$ **d.** $\dfrac{4}{10} + \dfrac{1}{5} + \dfrac{2}{15} = ?$

31. Convert the following improper fractions to mixed fractions.

    a. $\dfrac{5}{2} = ?$ **c.** $\dfrac{25}{16} = ?$

    b. $\dfrac{17}{4} = ?$ **d.** $\dfrac{37}{3} = ?$

32. Add the following mixed fractions.

    a. $1\dfrac{3}{4} + 2\dfrac{1}{2} = ?$ **b.** $9\dfrac{4}{10} + 6\dfrac{7}{8} = ?$

33. Reduce the following fractions.

    a. $\dfrac{4}{16} = ?$ **c.** $\dfrac{74}{148} = ?$

    b. $\dfrac{16}{18} = ?$ **d.** $\dfrac{28}{45} = ?$

34. Perform the following subtraction of fractions.

    a. $\dfrac{8}{14} - \dfrac{2}{7} = ?$ **c.** $4\dfrac{1}{16} - 2\dfrac{31}{32} = ?$

    b. $3\dfrac{4}{9} - 1\dfrac{1}{3} = ?$ **d.** $5\dfrac{3}{5} - 2\dfrac{4}{15} = ?$

35. Multiply the following. Cancel and reduce if possible.

    a. $\dfrac{1}{9} \times 4 = ?$ **c.** $\dfrac{1}{3} \times 4\dfrac{1}{2} = ?$

    b. $\dfrac{3}{6} \times \dfrac{4}{5} = ?$ **d.** $2\dfrac{3}{4} \times 4\dfrac{4}{11} = ?$

36. Divide the following. Cancel and reduce if possible.

    a. $\dfrac{1}{4} \div 8 = ?$ **c.** $3\dfrac{5}{7} \div 1\dfrac{6}{7} = ?$

    b. $\dfrac{3}{5} \div \dfrac{5}{9} = ?$ **d.** $14\dfrac{8}{9} \div 2\dfrac{4}{5} = ?$

Web Site Questions

Go to the Web site http://www.prenhall.com/cook, select the textbook *Introductory Mathematics,* select this chapter, and then follow the instructions when answering the multiple-choice practice problems.

Decimal Calculation

An Apple a Day

Electronics is the branch of science concerned with the use of electricity and electronic devices to manage information signals. As a science, it has grown out of electrical engineering, which is concerned with the generation, control, distribution, storage, and applications of electrical power. Today, electronics is a separate science involving numerous scientific disciplines, including physics, chemistry, thermodynamics, quantum mechanics, and mathematics.

One of the early pioneers who laid the foundations of many branches of science was Isaac Newton. He was born in a small farmhouse near Woolsthorpe in Lincolnshire, England, on Christmas Day in 1642. He was an extremely small, premature baby, which worried the midwives who went off to get medicine and didn't expect to find him alive when they came back. Luckily for science, however, he did survive.

Newton's father was an illiterate farmer who died three months before he was born. His mother married the local vicar soon after Newton's birth and left him in the care of his grandmother. This parental absence while he was growing up had a traumatic effect on him and throughout his life affected his relationships with people.

At school in the nearby town of Grantham, Newton showed no interest in classical studies but rather in making working models and studying the world around him. When he was in his early teens, his stepfather died and Newton had to return to the farm to help his mother. Newton proved to be a hopeless farmer; in fact, on one occasion when he was tending sheep he became so engrossed with a stream that he followed it for miles and was missing for hours. Luckily, a schoolteacher recognized Newton's single-minded powers of concentration and convinced his mother to let him return to school, where he performed better and later went off to Cambridge University.

In 1665, in his graduation year, Newton left Cambridge to return home to escape an epidemic of bubonic plague that had spread throughout London. During this time, Newton reflected on his years of seclusion at his mother's cottage and called them the most significant time in his life. It was here on a warm summer's day that Newton saw the apple fall to the ground, leading him to develop his laws of motion and gravitation. It was here that he wondered about the nature of light and later built a prism and proved that white light contains all the colors in a rainbow.

Later, when Newton returned to Cambridge, he demonstrated many of his discoveries but was reluctant to publish the details and did so finally only at the insistence of others. Newton went on to build the first working reflecting astronomical telescope and wrote a paper on optics that was fiercely challenged by the physicist Robert Hooke. Hooke quarreled bitterly with Newton over the years, and there were also heated debates about whether Newton or the German mathematician Gottfried Leibniz invented calculus.

The truth is that many of Newton's discoveries roamed around with him in the English countryside, and even though many of these would, before long, have been put forward by others, it was Newton's genius and skill (and long walks) that tied together all the loose ends.

Outline and Objectives

VIGNETTE: AN APPLE A DAY

INTRODUCTION

2–1 THE DECIMAL POINT

Objective 1: Define the function of the decimal point.

2–1–1 Positional Weight

Objective 2: Describe the positional weight of decimal fractions.

2–1–2 No Decimal Point and Extra Zeros

2–1–3 Converting Written Fractions to Decimal Fractions

Objective 3: Demonstrate how to:
    a. Convert written fractions to decimal fractions.
    b. Calculate reciprocals.
    c. Convert decimal fractions to written fractions.

2–1–4 Reciprocals

2–1–5 Converting Decimal Fractions to Written Fractions

2–2 CALCULATING IN DECIMAL

Objective 4: Describe the relationships among the four basic arithmetic operations: addition, subtraction, multiplication, and division.

2–2–1 Decimal Addition

Objective 5: Demonstrate how to perform the following arithmetic operations.
    a. Decimal addition
    b. Decimal subtraction
    c. Decimal multiplication
    d. Decimal division

2–2–2 Decimal Subtraction

2–2–3 Decimal Multiplication

2–2–4 Decimal Division

2–2–5 Ratios

Objective 6: Describe the following mathematical terms and their associated operations.
    a. Ratios
    b. Rounding off
    c. Significant places
    d. Percentages
    e. Averages

2–2–6 Rounding Off

2–2–7 Significant Places

2–2–8 Percentages

2–2–9 Averages

Introduction

The *decimal system of counting and calculating* was developed by Hindu mathematicians in India around A.D. 400. The system involved using fingers and thumbs, so it was natural that it would have 10 numerals or *digits*, which is a word meaning "finger." The use of this 10-digit or 10-finger system became widespread and by A.D. 800 was used extensively by Arab cultures. The 10 digits were represented by combinations of the 10 arabic numerals: *0, 1, 2, 3, 4, 5, 6, 7, 8,* and *9*. This is still called the *arabic number system*. The system eventually found its way to England, where it was adopted by nearly all the European countries by A.D. 1200. The Europeans called it the *decimal number system* after the Latin word *decimus*, which means "ten."

In this chapter we will discuss the decimal system of counting and calculating. **Decimal** is a topic with which you probably already feel extremely comfortable because we make use of it every day in countless applications. For example, our money is based on the decimal number system (there are 100 cents in 1 dollar, or 10 dimes in 1 dollar), and the metric system of measurement is also based on the decimal system (there are 10 millimeters in 1 centimeter, and 100 centimeters in 1 meter). Basic as it seems, however, you should still review this chapter because it will discuss all the terminology and procedures regarding how the decimal system is used, and this will serve as our foundation for all subsequent chapters.

Decimal

Numbered or proceeding by tens; based on the number 10.

2–1   THE DECIMAL POINT

Decimal Point

A period or centered dot between the parts of a decimal mixed number, separating the whole number from the fraction.

Decimal Fraction

A decimal value that is less than one.

The **decimal point (dp),** *symbolized by* ".", *is used to separate a whole decimal number from a* **decimal fraction.** Using a monetary example, we ask: What does the amount $12.75 actually mean? The number to the left of the decimal point, which is 12 in this example, indicates how many whole or complete dollars we have. On the other hand, the number to the right of the decimal point indicates the fraction, or the amount that is less than one dollar. Because there are 100 cents in $1, and this example indicates that we have 75 cents, this fraction is actually three-fourths of a dollar. In summary:

Decimal point

The numbers to the left of the decimal point indicate the amount of whole or complete units we have. In this monetary example, we have 12 complete or whole dollars.

$12.75

The numbers to the right of the decimal point indicate the *decimal fraction*, or amount that is less than one. In this monetary example, we have 75 cents or $\frac{75}{100}$ (which reduces to $\frac{3}{4}$).

The decimal point (dp) separates the whole numbers on the left from the fraction on the right. In this monetary example, the decimal point separates the complete dollars on the left from the fraction of a dollar or cents on the right.

Most calculators have a key specifically for entering the decimal point. It operates as follows.

CALCULATOR KEYS

Name: Decimal-point key

Function: Used in conjunction with the digit keys 0 through 9 to enter a decimal point. The position of the decimal point separates the whole number on the left from the decimal fraction on the right.

Example: Enter the number 12.75.

Press keys: ☐1 ☐2 ☐. ☐7 ☐5

Display shows: 12.75

2–1–1 *Positional Weight*

Most of us know that $0.50 is half a dollar, or 50 cents; $0.25 is a fourth or a quarter of a dollar, or 25 cents; and $0.75 is three-fourths or three-quarters of a dollar, or 75 cents; however, when it comes to explaining why $\frac{1}{2}$ = $0.50, $\frac{1}{4}$ = $0.25, and $\frac{3}{4}$ = $0.75, many people have difficulty. Why are these decimal fractions of 0.50, 0.25, and 0.75 equivalent to $\frac{1}{2}, \frac{1}{4}$, and $\frac{3}{4}$? The answer is in the position that each digit occupies in relation to the decimal point. To explain this further, Figure 2–1 repeats our previous example of $12.75, or twelve dollars and seventy-five cents. Once again, each place or position carries its own weight or value, and these positions are all relative to the decimal point. For instance, the digit in the first column to the left of the decimal point indicates how many units or ones are in the value, whereas the digit in the second column to the left of the decimal point indicates how many tens are in the value. Digits to the left of the decimal point all indicate the number of whole numbers in the value. On the other hand, digits to the right of the decimal point indicate the fraction or piece of a whole that is in the value. A value such as 3.5, therefore, indicates that we have three whole units and five-tenths $\left(\frac{5}{10}\right)$. Remembering our proper fractions from Chapter 1, we know that $\frac{5}{10}$ means that a whole unit was divided into ten parts and we have five of those parts. Because 5 is half of 10, we have half of a whole unit ($\frac{5}{10}$ can be reduced to $\frac{1}{2}$).

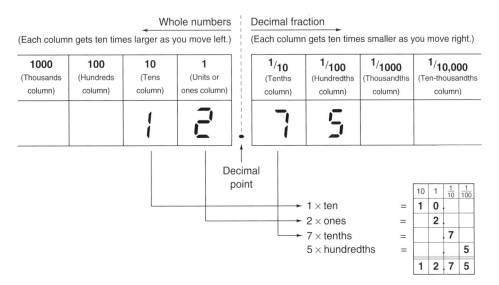

FIGURE 2–1   The Positional Weight of Decimal Fractions.

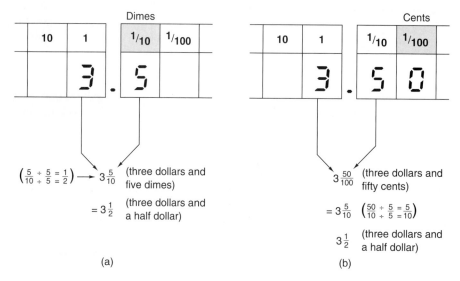

$\left(\frac{5}{10} \div \frac{5}{5} = \frac{1}{2}\right) \longrightarrow 3\frac{5}{10}$ (three dollars and five dimes)

$= 3\frac{1}{2}$ (three dollars and a half dollar)

$3\frac{50}{100}$ (three dollars and fifty cents)

$= 3\frac{5}{10}$ $\left(\frac{50}{10} \div \frac{5}{5} = \frac{5}{10}\right)$

$3\frac{1}{2}$ (three dollars and a half dollar)

(a)　　　　　　　　　　　　(b)

FIGURE 2–2　Interpreting the Decimal Fraction.

Is there a difference, therefore, between \$3.5 and \$3.50? The answer is no, which can be explained by looking at Figure 2–2. In Figure 2-2(a) we have interpreted the value as 3 dollars and $\frac{5}{10}$ of a dollar. Because $\frac{5}{10} = \frac{1}{2}$, the value in Figure 2-2(a) is 3 dollars and a half dollar. In Figure 2-2(b) we can interpret the decimal fraction as $\frac{5}{10}$ and $\frac{0}{100}$, or $\frac{50}{100}$. Because $\frac{50}{100} = \frac{5}{10} = \frac{1}{2}$, the value in Figure 2-2(b) is also 3 dollars and a half dollar. As far as money is concerned, we tend to refer to the fraction of a dollar in $\frac{1}{100}$ parts because a dollar is divided into 100 parts, and each of these parts is called a cent (there are 100 fractional parts or cents in a dollar).

Looking at this another way, we could label the $\frac{1}{10}$ column as dimes (10¢) and the $\frac{1}{100}$ column as cents (1¢). A value of 3.5 would then be interpreted as 3 dollars and 5 dimes, and a value of 3.50 would be interpreted as 3 dollars and 50 cents, as shown in Figure 2-2(a) and (b).

■　**EXAMPLE**

State the individual digit weights of the following number: 7.25.

■　*Solution:*

The first step is to give each column its respective weight or value.

1	$\frac{1}{10}$	$\frac{1}{100}$
7	2	5

It is now easier to see that we have 7 whole units, $\frac{2}{10}$ (two-tenths), and $\frac{5}{100}$ (five-hundredths). Said another way, we have 7 whole units and $\frac{25}{100}$ (twenty-five hundredths). Because $\frac{25}{100}$ can be reduced to $\frac{1}{4}$,

$$\frac{25 \div 25 = 1}{100 \div 25 = 4}$$

we could also interpret this value as seven and one-fourth $\left(7\frac{1}{4}\right)$.

2–1–2 No Decimal Point and Extra Zeros

What happens if a value such as

$$76$$

does not have a decimal point? The answer is: If you do not see a decimal point, always assume that it is to the right of the last digit in the number. In this example, therefore, the decimal point will follow the 6 in the number 76, as follows:

$$76.$$

Another point to remember: Always ignore extra zeros that are added to the left or right of a number. For example, to emphasize that a number such as .33 is a fraction, many people write this number as 0.33. The extra zero in this example is used to frame the decimal point so that it is not missed. You may have noticed that your calculator will add this zero if you press the keys $\boxed{.}\,\boxed{3}\,\boxed{3}$ and will display 0.33. Sometimes, zeros are added to the left of decimal whole numbers as well as to the right of decimal fractions. For example:

These extra zeros are used to align the digits of each number in the appropriate column. Without them, digits may be placed in the wrong column, as follows:

$$1367.5$$

$$0293.3$$

$$6.67$$

$$49.24$$

In other instances, extra zeros are placed to the right of a decimal fraction to indicate how to interpret the fraction. For example, three dollars and fifty cents would be written as 3.50 instead of 3.5 even though the extra zero after the five has no bearing on the value. This is so the value will be interpreted as three dollars and fifty cents $\left(\$3 + \frac{50}{100}\right)$ instead of three dollars and five dimes $\left(\$3 + \frac{5}{10}\right)$.

In all instances, extra zeros that are outside the number (to the left of the decimal whole number or to the right of the decimal fraction) do not change the value of the number.

2–1–3 Converting Written Fractions to Decimal Fractions

To convert proper fractions (such as $\frac{1}{2}, \frac{2}{6}, \frac{2}{3}, \frac{9}{16}$) to decimal fractions (such as 0.5, 0.3, 0.666, 0.562) we simply perform the operation that is indicated by the fraction bar. For example, when a fraction bar separates a numerator of 1 from a denominator of $2\left(\frac{1}{2}\right)$, the arithmetic operation described by the fraction bar is a division of 2 into 1. Performing this operation, you will convert the proper fraction $\frac{1}{2}$ to its decimal fraction equivalent.

$$\frac{1}{2} = 1 \div 2 = 0.5 \qquad \textit{Calculator sequence:}$$

$$\boxed{1}\,\div\,\boxed{2}\,\boxed{=}$$

Answer: 0.5

■ **EXAMPLE**

Convert the following proper fractions to decimal fractions.

a. $\dfrac{3}{4}$ **b.** $\dfrac{1}{4}$

c. $\dfrac{2}{3}$ **e.** $\dfrac{3}{9}$

d. $\dfrac{8}{16}$ **f.** $\dfrac{1}{8}$

■ *Solution:*

a. $\dfrac{3}{4} = 3 \div 4 = 0.75$

b. $\dfrac{1}{4} = 1 \div 4 = 0.25$

c. $\dfrac{2}{3} = 2 \div 3 = 0.666$

d. $\dfrac{8}{16} = 8 \div 16 = 0.5$ $\left(\dfrac{8 \div 8}{16 \div 8} = \dfrac{1}{2} = 0.5 \right)$

e. $\dfrac{3}{9} = 3 \div 9 = 0.333$ $\left(\dfrac{3 \div 3}{9 \div 3} = \dfrac{1}{3} = 0.3 \right)$

f. $\dfrac{1}{8} = 1 \div 8 = 0.125$

■ **INTEGRATED MATH APPLICATION: MACHINING**

Convert the following drill bit diameters to decimal equivalent values:

a. $\dfrac{3}{4}$ **b.** $\dfrac{5}{16}$ **c.** $\dfrac{3}{8}$ **d.** $\dfrac{5}{32}$

■ *Solution:*

a. $\dfrac{3}{4} = 0.75$

b. $\dfrac{5}{16} = 0.3125$

c. $\dfrac{3}{8} = 0.375$

d. $\dfrac{5}{32} = 0.15625$

The next question is: How do we convert a mixed fraction such as $3\frac{1}{2}$ to its decimal equivalent? The answer is to do nothing with the whole number except place it in the whole-number decimal columns and then convert the proper fraction as before. Let us do the following example.

■ **EXAMPLE**

Convert the following mixed fractions to their decimal equivalents.

a. $3\dfrac{1}{2}$ **c.** $27\dfrac{4}{9}$

b. $486\dfrac{1}{4}$ **d.** $31{,}348\dfrac{8}{9}$

■ *Solution:*

a. $3\dfrac{1}{2} = 3.5$ $(1 \div 2 = 0.5)$

b. $486\dfrac{1}{4} = 486.25$ $(1 \div 4 = 0.25)$

c. $27\dfrac{4}{9} = 27.444$ $(4 \div 9 = 0.444)$

d. $31{,}348\dfrac{8}{9} = 31{,}348.888$ $(8 \div 9 = 0.888)$

Finally, let us convert an improper fraction such as $\frac{4}{2}$ (four halves) to its decimal equivalent. This is the easiest operation to perform, as shown in the following example, because all you do is perform the division on your calculator to obtain a decimal whole number and fraction.

■ **EXAMPLE**

Convert the following improper fractions to their decimal equivalents.

a. $\dfrac{4}{2}$

b. $\dfrac{7}{3}$

c. $6\dfrac{7}{2}$

d. $\dfrac{18}{16}$

■ *Solution:*

a. $\dfrac{4}{2} = 4 \div 2 = 2.0$

b. $\dfrac{7}{3} = 7 \div 3 = 2.333$

c. $6\dfrac{7}{2} = 6 + (7 \div 2) = 6 + 3.5 = 9.5$ $\left(\dfrac{7}{2} = 3\dfrac{1}{2} = 3.5\right)$

d. $\dfrac{18}{16} = 18 \div 16 = 1.125$ $\left(\dfrac{18 \div 2}{16 \div 2} = \dfrac{9}{8} = 1\dfrac{1}{8}\right)$

2–1–4 *Reciprocals*

Whenever you divide a number into 1, you get the **reciprocal** *of that number.* The word *reciprocal* means "inverse" or "opposite relation." This term can best be described by determining the reciprocal of two numbers, 2 and 2000. As just mentioned, to get the reciprocal of these numbers, we simply divide these numbers into 1, as shown:

Reciprocals of original numbers

$$\dfrac{1}{2} = 0.5 \qquad \dfrac{1}{2000} = 0.0005$$

Original numbers

Comparing these two answers, you can see why the reciprocals, or result of dividing the numbers into 1, are inversely related to the original numbers. For instance, the reciprocal of a small number ($\downarrow$), such as 2, results in a large fraction ($\uparrow$), 0.5, whereas the reciprocal of a large number ($\uparrow$), such as 2000, results in a small fraction ($\downarrow$), 0.0005. This inverse relationship is even more evident when you consider that 2000 is one thousand

Reciprocal
Inversely related.

times greater than 2, yet the reciprocal of 2000 is one thousand times smaller than the reciprocal of 2. To summarize:

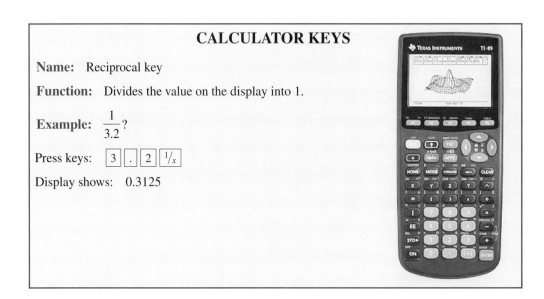

Original number						Reciprocal of number			
1000	100	10	1			$\frac{1}{10}$	$\frac{1}{100}$	$\frac{1}{1000}$	$\frac{1}{10,000}$
		2	0	$\frac{1}{2} =$	0	5			
2	0	0	0 . 0	$\frac{1}{2000} =$	0 . 0	0	0	0	5

2000 is three columns or 1000 times greater than 2.

0.0005 is three columns or 1000 times smaller than 0.5.

Most calculators have a key specifically for calculating the reciprocal of any number. It is called the reciprocal key and operates as follows.

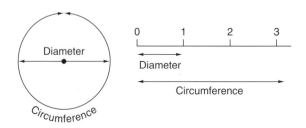

CALCULATOR KEYS

Name: Reciprocal key

Function: Divides the value on the display into 1.

Example: $\frac{1}{3.2}$?

Press keys: $\boxed{3}$ $\boxed{.}$ $\boxed{2}$ $\boxed{1/x}$

Display shows: 0.3125

Pi

The sixteenth letter of the Greek alphabet; the symbol π denoting the ratio of the circumference of a circle to its diameter; a transcendental number having a value to eight decimal places of 3.14159265.

The reciprocal of π is used in many different formulas. The Greek lowercase letter pi, which is pronounced "pie" and symbolized by π, is a constant value that describes how much bigger a circle's circumference is than its diameter.

Diameter

Circumference

0   1   2   3

Diameter

Circumference

The approximate value of this constant is 3.142, which means that the circumference of any circle is approximately 3 times bigger than its diameter.

Most calculators have a key for quickly obtaining the value of pi. It is called the pi (π) key and operates as follows.

CALCULATOR KEYS

Name: Pi key

Function: Enters the value of pi correct to eight decimal digits.

$\pi = 3.1415927$

Example: $2 \times \pi = ?$

Press keys: $\boxed{2}\ \boxed{\times}\ \boxed{\pi}\ \boxed{=}$

Display shows: 6.2831853

The factor π is used frequently in many formulas involved with the analysis of circular motion. The reciprocal $\frac{1}{2}\pi$ (1 over $2 \times \pi$) is frequently used, so we should do a few problems involving the reciprocal and pi keys on the calculator.

■ **EXAMPLE**

Calculate the reciprocals of the following numbers.

    a. 5   **b.** π **c.** 60   **d.** $2 \times \pi$

■ *Solution:*

  a. The reciprocal of $5 = \dfrac{1}{5} = 1 \div 5 = 0.2$.

      Calculator sequence: $\boxed{5}\ \boxed{^1/_x}$; display shows: 0.2

  b. The reciprocal of $\pi = \dfrac{1}{\pi} = \dfrac{1}{3.142} = 0.318$.

      Calculator sequence: $\boxed{\pi}\ \boxed{^1/_x}$; display shows: 0.3183098

  c. The reciprocal of $60 = \dfrac{1}{60} = 0.0166$.

      Calculator sequence: $\boxed{6}\ \boxed{0}\ \boxed{^1/_x}$; display shows: 0.0166667

  d. The reciprocal of $2 \times \pi = \dfrac{1}{2 \times \pi} = \dfrac{1}{6.28} = 0.159$.

      Calculator sequence: $\boxed{2}\ \boxed{\times}\ \boxed{\pi}\ \boxed{=}\ \boxed{^1/_x}$; display shows: 0.1591549.

2–1–5   *Converting Decimal Fractions to Written Fractions*

First, let us discuss why we would want to convert a decimal fraction, such as 0.5, into a written fraction, such as $\frac{1}{2}$. The answer is so that we can convert from a scale of 10 (decimal) to a scale that is something other than 10. For example, imagine that you have a standard ruler with 1 inch divided into 16 parts (there are sixteen $\frac{1}{16}$ in 1 inch). Next, imagine that you have to measure a length of 0.125 of an inch. Since the scale of the ruler is dealing with

16 parts and your decimal value of 0.125 is based on 10 parts, the two are not compatible and a conversion is needed. The first step is to convert $\frac{1}{16}$ into its decimal equivalent:

$$\frac{1}{16} = 1 \div 16 = 0.0625$$

This answer tells us that one-sixteenth of an inch is equivalent to 0.0625 of an inch in decimal. This can be proved by multiplying 0.0625 by 16 to see if we have 16 of these parts in 1 inch.

$$16 \times 0.0625 = 1 \text{ inch}$$

Now that we know that the value 0.0625 inch is equal to $\frac{1}{16}$ inch, the next step is to find out how many 0.0625 parts (or $\frac{1}{16}$) are in our decimal measurement of 0.125.

$$\frac{0.125}{0.0625} = 0.125 \div 0.0625 = 2$$

The answer 2 tells us that there are two 0.0625 parts (or sixteenths) in 0.125. Therefore, to measure off a length of 0.125 of an inch using a $\frac{1}{16}$ inch scale, simply measure a length of two-sixteenths $\left(\frac{2}{16}\right)$.

$$0.125 \text{ in.} = \frac{2}{16} \text{ in.}$$

■ **EXAMPLE**

Convert the following decimal fractions to proper fractions.

a. $0.25 = \dfrac{?}{8}$

b. $0.666 = \dfrac{?}{3}$

c. $0.59375 = \dfrac{?}{32}$

■ *Solution:*

a. 0.25 equals how many eighths? $\frac{1}{8}$ has a decimal equivalent of $0.125(1 \div 8 = 0.125)$. How many 0.125 parts are in 0.25?

$$\frac{0.25}{0.125} = 0.25 \div 0.125 = 2$$

$$0.25 = \frac{2}{8} = \frac{1}{4}$$

b. $\dfrac{1}{3} = 0.333$

$$\frac{0.666}{0.333} = 2$$

$$0.666 = \frac{2}{3}$$

c. $\dfrac{1}{32} = 0.03125$

$$\frac{0.59375}{0.03125} = 19$$

$$0.59375 = \frac{19}{32}$$

■ **INTEGRATED MATH APPLICATION: METALWORKING**

Convert the thicknesses of the following pieces of sheet metal to their fractional equivalents in 64ths:

    a. 0.046875 inch   **b.** 0.859375 inch

■ *Solution:*

    a. $\frac{3}{64}$ inch   **b.** 55.64 inch

■ **INTEGRATED MATH APPLICATION: AUTOMOTIVE**

A main cylinder measures $2\frac{9}{16}$ inches in diameter before it is rebored to increase its size by 0.05 inch. What is the new size of the cylinder as a decimal value?

■ *Solution:*

$$2\frac{9}{16} = \quad 2.5625 \text{ inches}$$
$$\underline{+ \ 0.05 \text{ inch}}$$
$$2.6125 \text{ inches}$$

CALCULATOR KEYS

Name: Conversion Functions
Display the answer as a fraction (▶ Frac).
Display the answer as a decimal (▶ Dec).

Function:
Value ▶ Frac (display as a fraction) displays an answer as its written fraction equivalent.
Value ▶ Dec (display as a decimal) displays an answer as its decimal equivalent.

Example: $\frac{1}{2} + \frac{1}{3}$ ▶ Frac

                   $\frac{5}{6}$

        Ans ▶ Dec
           .8333333333

SELF-TEST EVALUATION POINT FOR SECTION 2–1

Now that you have completed this section, you should be able to:

■ **Objective 3.** *Define the function of the decimal point.*

■ **Objective 4.** *Describe the positional weight of decimal fractions.*

■ **Objective 5.** *Demonstrate how to convert written fractions to decimal fractions, calculate reciprocals, and convert decimal fractions to written fractions.*

Use the following questions to test your understanding of Section 2–1.

1. What are the individual digit weights of the number 178.649?

2. Convert the following written fractions to decimal fractions.
    a. $1\frac{5}{2}$ **b.** $192\frac{3}{4}$ **c.** $67\frac{6}{9}$

3. Calculate the reciprocal of the following numbers.
    a. 2500   **b.** 0.25   **c.** 0.000025

4. Convert the following decimal fractions to written fractions.
    a. $7.25 = \frac{?}{16}$ **b.** $156.90625 = \frac{?}{32}$

In this section we first review the basic principles of decimal addition, subtraction, multiplication, and division. As you will discover, there are really only *two basic operations: addition and subtraction.* If you can perform an addition, you can perform a multiplication because *multiplication is simply repeated addition.* For example, a problem such as 4×2 (4 multiplied by 2) is asking you to calculate the sum of four 2s ($2 + 2 + 2 + 2$). Similarly, *division is simply repeated subtraction.* For example, a problem such as $9 \div 3$ (9 divided by 3) is asking you to calculate how many times 3 can be subtracted from 9 ($9 - 3 = 6$, $6 - 3 = 3$, $3 - 3 = 0$; therefore, there are three 3s in 9). In summary, therefore, the two basic mathematical operations (addition and subtraction) and their associated mathematical operations (multiplication and division) are as follows:

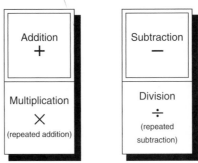

There is another relationship that needs to be discussed, and that is the inverse or opposite nature between addition and subtraction and between multiplication and division. For example, *subtraction is the opposite arithmetic operation of addition.* This can be proved by starting with a number such as 10, then adding 5, then subtracting 5.

Similarly, *multiplication is the opposite arithmetic operation of division.* This can be proved by starting with a number such as 4, then multiplying by 3, then dividing by 3.

In summary, therefore, addition is the opposite mathematical operation of subtraction, and multiplication is the opposite mathematical operation of division. This opposite relationship will be made use of in a subsequent chapter.

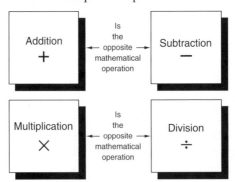

Every calculator has keys for performing these four basic arithmetic operations. The name, function, and procedure for using these keys are as follows.

CALCULATOR KEYS

Name: Decimal point key $\boxed{.}$ (discussed previously)

Name: Equals key $\boxed{=}$

Function: Combines all previously entered numbers and operations. Used to obtain intermediate and final results.

Example: $3 \times 4 + 6 = ?$

Press keys: $\boxed{3}\ \boxed{\times}\ \boxed{4}\ \boxed{=}\ \boxed{+}\ \boxed{6}\ \boxed{=}$

Display shows: 18.0

Name: Add key $\boxed{+}$

Function: Instructs calculator to add the next entered quantity to the displayed number.

Example: Add $31.6 + 2.9 = ?$

Press keys: $\boxed{3}\ \boxed{1}\ \boxed{.}\ \boxed{6}\ \boxed{+}\ \boxed{2}\ \boxed{.}\ \boxed{9}\ \boxed{=}$

Displays shows: 34.5

Name: Subtract key $\boxed{-}$

Function: Instructs calculator to subtract the next entered quantity from the displayed number.

Example: $266.3 - 5.35 = ?$

Press keys: $\boxed{2}\ \boxed{6}\ \boxed{6}\ \boxed{.}\ \boxed{3}\ \boxed{-}\ \boxed{5}\ \boxed{.}\ \boxed{3}\ \boxed{5}\ \boxed{=}$

Display shows: 260.95

Name: Multiply key $\boxed{\times}$

Function: Instructs calculator to multiply the displayed number by the next entered quantity.

Example: $3.85 \times 2.9 = ?$

Press keys: $\boxed{3}\ \boxed{.}\ \boxed{8}\ \boxed{5}\ \boxed{\times}\ \boxed{2}\ \boxed{.}\ \boxed{9}\ \boxed{=}$

Display shows: 11.165

Name: Divide key $\boxed{\div}$

Function: Instructs calculator to divide the displayed number by the next entered quantity.

Example: $28.98 \div 2.3 = ?$

Press keys: $\boxed{2}\ \boxed{8}\ \boxed{.}\ \boxed{9}\ \boxed{8}\ \boxed{\div}\ \boxed{2}\ \boxed{.}\ \boxed{3}\ \boxed{=}$

Display shows: 12.6

2–2–1   *Decimal Addition*

Addition is the process of gathering together two or more numbers into one total sum.

■ **EXAMPLE**

Add 3 and 2.

■ *Solution:*

$$3 + 2 = 5$$

You have probably noticed that the order which numbers are added has no effect on the final result or sum. To use the values in the previous example, whether you add $3 + 2$ or $2 + 3$, the result will always be 5. This property of addition is formally known as the *commutative law of addition.*

$$a + b = b + a$$

Commutative Law of Addition: The order in which you add two numbers will have no effect on the sum.

When three numbers are added together, we generally add two numbers first to obtain a partial sum, and then add this to the remaining number.

■ **EXAMPLE**

Add 3 and 7 and 4.

■ *Solution:*

	Steps:
$3 + 7 + 4 =$	$3 + 7 = 10$
$10 + 4 =$	$10 + 4 = 14$
14	

From the previous example, notice that whichever two numbers you group and add first has no effect on the final sum. For example,

Parentheses are used to indicate which group is to be added first.

$(3 + 7) + 4 =$	$3 + (7 + 4) =$
$10 + 4 =$	$3 + 11 =$
14	14

This property is known as the *associative law of addition.*

$$(a + b) + c = a + (b + c)$$

Associative Law of Addition: The grouping of the numbers you add first will have no effect on the final sum.

To review decimal **addition,** let us examine the steps involved in adding the following numbers.

$$66.332 + 285 + 0.002 + 182.788 = ?$$

First, list the numbers vertically, making sure that the decimal points are lined up as follows.

Addition process:

$(100)(10)(1).\left(\frac{1}{10}\right)\left(\frac{1}{100}\right)\left(\frac{1}{1000}\right)$

```
Carry →   2  1 | 1  1  1
          0  6 | 6.  3  3  2
          2  8 | 5.  0  0  0
          0  0 | 0.  0  0  2
         +1  8 | 2.  7  8  8
Sum →     5  3   4.  1  2  2
```

The extra zeros are included to keep the digits in their proper columns.

Steps:

Starting at the right column $\left(\frac{1}{1000}\right)$, add vertically.

$2 + 0 + 2 + 8 = 12, 2$ carry 1.
$1 + 3 + 0 + 0 + 8 = 12, 2$ carry 1.
$1 + 3 + 0 + 0 + 7 = 11, 1$ carry 1.
$1 + 6 + 5 + 0 + 2 = 14, 4$ carry 1.
$1 + 6 + 8 + 0 + 8 = 23, 3$ carry 2.
$2 + 0 + 2 + 0 + 1 = 5.$

Calculator sequence:

| 6 | 6 | . | 3 | 3 | 2 | + | 2 | 8 | 5 | + | . | 0 | 0 | 2 | + | 1 | 8 | 2 | . | 7 | 8 | 8 | = |

Answer: 534.122

There are a few points to note regarding this previous addition. First, the number 285 did not have a decimal point indicated; therefore, it was assumed to be to the right of the last digit (285.0). Second, what is really happening when we **carry** or overflow into the next-left column? The answer is best explained by examining the addition of the units column in the previous example. In this column we had a result of $1 + 6 + 5 + 0 + 2 = 14$. As you know from our discussion on positional weight, 14 can be broken down into 1 ten (1×10) and 4 ones (4×1). This is why the 1 ten is carried over to the tens column, and the 4 ones or units are placed in the units sum column. Explained another way, whenever the sum of all the digits in any column exceeds the maximum single decimal digit 9, a reset and carry action occurs. In the case of the units column with a sum of 14, we exceeded 9 and therefore the units sum was reset to 0 and then advanced to 4, and a 1 was carried into the tens column to make 14.

In summary, therefore, because each column to the left is 10 times larger, a reset and carry action will occur every time the sum of any column is incremented or advanced by one, from 9 to 10.

Carrying

The process of transferring from one column to another during an addition process.

Sum

The result obtained when adding numbers.

■ **EXAMPLE**

How much is $976.73 + $998.28?

■ *Solution:*

Carry:	1111 1
Augend:	0976.73
Addend:	+ 0998.28
Sum:	$1975.01

Steps:

$3 + 8 = 11$, 1 carry 1.
$1 + 7 + 2 = 10$, 0 carry 1.
$1 + 6 + 8 = 15$, 5 carry 1.
$1 + 7 + 9 = 17$, 7 carry 1.
$1 + 9 + 9 = 19$, 9 carry 1.
$1 + 0 + 0 = 1$, 1 carry 0.

Augend

The number to which a new quantity, called the addend, is to be added.

Addend

The number or quantity to be added to an existing quantity, called the augend.

Calculator sequence: [9] [7] [6] [.] [7] [3] [+]
[9] [9] [8] [.] [2] [8] [=]

Answer: 1975.01

■ **INTEGRATED MATH APPLICATION: ACCOUNTING**

Calculate the total cost of the following service station repair invoice:

Parts:	$ 64.95
Labor:	$226.35
Coolant:	$ 4.25
Tax:	$ 22.17

■ *Solution:*

```
     64.95
    226.35
      4.25
  +  22.17
  $317.72
```

■ **INTEGRATED MATH APPLICATION: PERSONAL FINANCE**

How many hours did you work in the previous week if your time card shows the following:

Monday—6 hours
Tuesday—4.5 hours
Wednesday—7.5 hours
Thursday—5 hours
Friday—8 hours

Solution:

$$6 + 4.5 + 7.5 + 5 + 8 = 31 \text{ hours}$$

2–2–2 *Decimal Subtraction*

Subtraction is the process of finding the difference between two numbers by taking away the lesser number from the greater number.

■ EXAMPLE

Are all of the following subtractions?

   a. Subtract 4 from 7.
   b. What is the difference between 4 and 7?
   c. How much greater is 7 than 4?
   d. How much less is 4 than 7?

■ *Solution:*

Yes, these are all different ways of describing subtraction, and in all cases:

$$7 - 4 = 3$$

Subtraction

The operation of deducting one number from another.

To review decimal **subtraction**, let us examine the steps involved in the following example:

$$\$2005.23 - \$192.41 = ?$$

To begin, list the numbers vertically, making sure that the decimal points are lined up as follows:

Subtraction process:

			Dimes	Cents
(1000)(100)(10)(1)			$(\frac{1}{10})$	$(\frac{1}{100})$

Borrow: 1 9   4

Minuend: 2 ^{1}0 ^{1}0 5. ^{1}2 3
Subtrahend: 0 1 9 2. 4 1
Difference: 1 8 1 2. 8 2

Minuend

A number from which the subtrahend is to be subtracted.

Subtrahend

A number that is to be subtracted from a minuend.

Difference

The amount by which things differ in quantity or measure.

Calculator sequence:

2	0	0	5	.	2	3	−
1	9	2	.	4	1	=	

Answer: 1812.82
        or $1812.82

Steps:
Starting at the right column ($\frac{1}{100}$ or cent) subtract vertically:
$3 - 1 = 2$
$2 - 4$: impossible to subtract 4 from the smaller number 2. Therefore, we will *borrow* \$1 from the \$5 in the units column. This will leave \$4 in the units column; because \$1 = 10 dimes, we will be adding 10 dimes to the 2 dimes already present, giving a total of 12 dimes.
$12 - 4 = 8$
 $4 - 2 = 2$
 $0 - 9$: impossible to subtract 9 from nothing or zero.
Because we cannot borrow from the tens or hundreds columns, we will have to borrow from the thousands column.
Borrowing $1000(10 \times 100)$ gives 10 hundreds; borrowing 1 hundred from 10 hundreds gives 9 hundreds. Borrowing $100(10 \times 10)$ gives 10 tens.
$10 - 9 = 1$
 $9 - 1 = 8$
 $1 - 0 = 1$

46   CHAPTER 2 / DECIMAL CALCULATION

As you can see from this example, *borrowing* is the exact opposite of carrying. With addition we sometimes end up with a sum of 10 or more, which results in an overflow or carry into the next-higher or next-left column, whereas with subtraction we sometimes end up with a larger number in the subtrahend than the minuend, and it becomes necessary to borrow a 1 from the next-higher place to the left in the minuend.

■ EXAMPLE

Perform the following subtraction:

$$27,060.0143 - 963.86$$

■ *Solution:*

(10,000)	(1000)	(100)	(10)	(1)	$\left(\frac{1}{10}\right)$	$\left(\frac{1}{100}\right)$	$\left(\frac{1}{1000}\right)$	$\left(\frac{1}{10,000}\right)$	
	6	9	¹5	9.	9				*Steps:*
2	~~7~~	~~10~~	6	~~10.~~	~~10~~	¹1	4	3	3 − 0 = 3
−0	0	9	6	3.	8	6	0	0	4 − 0 = 4
Difference: 2	6	0	9	6.	1	5	4	3	1 − 6: can't do; borrow.

Calculator sequence:

| 2 | 7 | 0 | 6 | 0 | . | 0 | 1 | 4 | 3 | − |

| 9 | 6 | 3 | . | 8 | 6 | = |

Answer: 26,096.1543

Borrowing 1 ten from 6 tens gives 5 tens and 10 ones. Borrowing 1 one from 10 ones gives 9 ones and 10 tenths.
Borrowing 1 tenth from 10 tenths gives 9 tenths and 10 hundredths.
10 hundredths + the original 1 hundredth = 11 hundredths:
11 − 6 = 5
9 − 8 = 1
9 − 3 = 6
5 − 6: can't do; borrow.
Borrowing 1 thousand from 7 thousands gives 6 thousands and 10 hundreds.
Borrowing 1 hundred from 10 hundreds gives 9 hundreds and 10 tens. 10 tens + the original 5 tens = 15 tens:
15 − 6 = 9
9 − 9 = 0
6 − 0 = 6
2 − 0 = 2

■ INTEGRATED MATH APPLICATION: RETAIL

If the regular price of a television is $856.99 and the sale price is $699.99, how much will you save?

■ *Solution:*

$$
\begin{array}{r}
\$859.99 \\
-\$699.99 \\
\hline
\$160.00
\end{array}
$$

A savings of $160.00.

■ INTEGRATED MATH APPLICATION: OFFICE MANAGEMENT

If an employee's gross weekly salary is $695.00 and her deductions total $212.65, what is her take-home pay?

■ *Solution:*

$$
\begin{array}{r}
\$695.00 \\
-\$212.65 \\
\hline
\$482.35
\end{array}
$$

Her weekly take-home pay is $482.35.

■ INTEGRATED MATH APPLICATION: ELECTRICAL

The National Fire Protection Association developed a set of standards known as the American Wire Gauge (AWG) for all copper conductors, which lists their diameter, resistance, and maximum safe current in amperes. The table lists some frequently used residential wires.

a. Measured with a micrometer, the diameter of a wire is 0.092 inch. What is the AWG?

b. What AWG wire size would you choose for a circuit that can draw a maximum of 21 amps?

c. What is the diameter difference between an AWG #10 and a #12 wire?

AWG #	DIAMETER	MAX. CURRENT
10	0.102 in.	30 A
11	0.091 in.	25 A
12	0.081 in.	20 A
14	0.064 in.	15 A
15	0.051 in.	6 A

■ *Solution:*

a. AWG #11

b. AWG #11

c.
$$
\begin{array}{rl}
\text{AWG \#10} = & 0.102 \text{ inch} \\
\text{AWG \#12} = & -0.081 \text{ inch} \\
\hline
& 0.021 \text{ inch}
\end{array}
$$

2–2–3   *Decimal Multiplication*

Multiplication is a process in which a number is added to itself a given number of times. The given number is called the *multiplicand*, the number that specifies how many times the number is to be added to itself is called the *multiplier*, and the result is called the *product*.

■ EXAMPLE

What is the product of 3 times 2?

■ *Solution:*

$$3 \times 2 = 6$$

Two threes $(3 + 3)$ are six.

$$
\begin{array}{rl}
3 & \text{Multiplicand} \\
\times\, 2 & \text{Multiplier} \\
\hline
6 & \text{Product}
\end{array}
$$

The commutative law also applies to multiplication, in that the order in which you multiply two numbers will have no effect on the result. To use the values in the previous example, whether you multiply 2 × 3 or 3 × 2, the result will in both cases be 6. This property of multiplication is known as the *commutative law of multiplication.*

$$a \times b = b \times a$$

Commutative Law of Multiplication: The order in which you multiply two numbers will have no effect on the product.

When three numbers are to be multiplied, we generally multiply one number by another to obtain a partial product, and then multiply this result by the final number.

■ EXAMPLE

What is the result of 2 times 3 times 10?

■ *Solution:*

$$2 \times 3 \times 10 = ?$$

No matter which pair is multiplied first, the result is still the same.

$2 \times (3 \times 10) =$	$(2 \times 3) \times 10 =$
$2 \times 30 =$	$6 \times 10 =$
60	60

The previous example confirms the *associative law of multiplication.*

$$a \times (b \times c) = (a \times b) \times c$$

Associative Law of Multiplication: The grouping of the numbers you multiply first will have no effect on the final product.

To review decimal **multiplication,** let us examine the steps involved in the following example:

$$27.66 \times 2.4 = ?$$

Unlike in the previous mathematical operations, we can ignore the alignment of the decimal point and simply make sure that the two numbers are flush on the right side.

Multiplicand: 27.66
Multiplier: ×   2.4

Two numbers are listed vertically so that they are flush on the right side.

We can now use the following multiplication steps:

Multiplication process:

Second carry:	11 1
First carry:	132 2
Multiplicand:	27.6 6
Multiplier:	× ②④
	↓
First partial product:	110 6 4
	↓
Second partial product:	553 2 0
Final product:	663 8 4

Steps:

To begin, multiply each digit in the multiplicand by the first digit of the multiplier (which is 4) to obtain the first partial product. Line up the rightmost digit of the first partial product under the multiplier digit being used (see arrow).

Multiplication

The operation of adding an integer to itself a specified number of times.

Multiplicand

The number that is to be multiplied by another.

Multiplier

One that multiplies; a number by which another number is multiplied.

$4 \times 6 = 24$, 4 carry 2.
$4 \times 6 = 24$, $24 +$ previous carry of $2 = 26$, 6 carry 2.
$4 \times 7 = 28$, $28 +$ carry of $2 = 30$, 0 carry 3.
$4 \times 2 = 8$, $8 +$ carry of $3 = 11$, 1 carry 1; bring 1 down.

The next step is to multiply each digit in the multiplicand by the second digit of the multiplier (which is 2) to obtain the second partial product. Line up the rightmost digit of the second partial product under the multiplier digit being used (see arrow).

$2 \times 6 = 12$, 2 carry 1.
$2 \times 6 = 12$, $12 +$ carry of $1 = 13$, 3 carry 1.
$2 \times 7 = 14$, $14 +$ carry of $1 = 15$, 5 carry 1.
$2 \times 2 = 4$, $4 +$ carry of $1 = 5$.

Calculator sequence:

| 2 | 7 | . | 6 | 6 | × | 2 | . | 4 | = |

Answer: 66.384

The next step is to add the first and second partial products to obtain the final product.
$4 + 0 = 4$
$6 + 2 = 8$
$0 + 3 = 3$
$1 + 5 = 6$
$1 + 5 = 6$

27.6 6̂ The multiplicand has two digits to the right of the decimal point.
2.4 The multiplier has one digit to the right of the decimal point.
$2 + 1 = 3$ places.

← The final step is to determine the position of the decimal point in the answer by counting the number of digits to the right of the decimal point in the multiplier and multiplicand. There are three in this example, so the final product will have three digits to the right of the decimal point.

Answer is 66.384

66.3 8 4

■ **EXAMPLE**

How much is 0.035 multiplied by 0.005?

■ *Solution:*

In this example we will take a few shortcuts. Since the decimal points do not make a difference at this stage, let us drop them completely. This will produce:

$$\begin{array}{r} 0035 \\ \times\ 0005 \\ \hline \end{array}$$

Second, we learned earlier that extra zeros appearing to the left or right of a number are not needed in the multiplication process, so let us remove these zeros:

$$\begin{array}{r} 35 \\ \times\ 5 \\ \hline \end{array}$$

The multiplication is now a lot easier because it is simply / 35 × 5. Performing the multiplication, we obtain the product:

$$
\begin{array}{r}
\overset{2}{35} \\
\times\ \ 5 \\
\hline
175
\end{array}
$$

The final step is to determine the position of the decimal point. The original multiplicand had three digits to the right of the decimal point (0.035), and the original multiplier also had three digits to the right of the decimal point (0.005), so the final product should have six digits to the right (3 + 3) of the decimal point.

$$.000175 \quad \text{or} \quad 0.000175$$

■ INTEGRATED MATH APPLICATION: AUTOMOTIVE

If your car averages 26.3 miles per gallon, how far can you travel if your gas tank is filled with 12 gallons of gasoline?

■ *Solution:*

$$26.3 \text{ miles per gallon} \times 12 \text{ gallons} = 315.6 \text{ miles}$$

■ INTEGRATED MATH APPLICATION: SURVEYING

If an inch on a street map is equivalent to 0.5 mile, how many miles will you have to travel if the distance on the map measures 14.5 inches?

■ *Solution:*

$$14.5 \text{ inches} \times 0.5 \text{ mile per inch} = 7.25 \text{ miles}$$

■ INTEGRATED MATH APPLICATION: METALWORKING

The accompanying table lists a section of the U.S. Standard sheet metal gauge.

a. What is the difference in thickness between gauge numbers 5–0 and 2?
b. What would be the thickness if seven gauge number 4–0 sheets were stacked together?
c. Which gauge number is twice as thick as gauge #3?

GAUGE #	THICKNESS
7–0	0.5 in.
6–0	0.469 in.
5–0	0.438 in.
4–0	0.406 in.
3–0	0.375 in.
2–0	0.344 in.
0	0.313 in.
1	0.281 in.
2	0.266 in.
3	0.25 in.

■ *Solution:*

a.
$$
\begin{array}{ll}
\text{Gauge \#5–0} = & 0.438 \text{ in.} \\
\text{Gauge \#2} \quad = & -0.266 \text{ inch} \\
\hline
& 0.172 \text{ inch}
\end{array}
$$

b. 7 × 0.406 inch = 2.842 inches

c. 7–0

2–2–4 *Decimal Division*

Division
The operation of separating a number into equal parts.

Quotient
The number resulting from the division of one number by another.

Divisor
The number by which a dividend is divided.

Dividend
A number to be divided.

Remainder
The final undivided part after division that is less or of lower degree than the divisor.

To review decimal **division** (or *long division*, as it is sometimes called) let us step through the following division example: 1362 ÷ 12. The procedure is as follows.

Long-division process:

Steps:

The first step is to see how many times the divisor (12) can be subtracted from the first digit of the dividend (1). Because the dividend is smaller, the answer or quotient is zero (1 − 12 cannot be done). We place this zero above the dividend 1.

Next, we see how many times the divisor (12) can be subtracted from the first two digits of the dividend (13). 13 − 12 = 1, remainder 1. The answer or quotient (1) is placed above the next digit of the dividend (3), and the remainder of the subtraction (1) is dropped below the 13 − 12 line.

We carry the next digit of the dividend (6) down to the remainder (1) to address the next part of the dividend, which is 16. Now we repeat the process by seeing how many times the divisor, 12, can be subtracted from these two digits, 16. The answer is a quotient of 1 (which is placed above the 6 dividend) with a remainder of 4 (which is dropped below the 16 − 12 line).

Calculator sequence:

Answer: (113).(5)

The whole decimal number is telling you how many times 12 will go into 1362. Therefore,

1362 ÷ 12 = (113)

The decimal fraction indicates how much of a fraction of the divisor is remaining. Therefore,

0.5 × 12 = 6

Answer to 1362 ÷ 12 is 113, remainder 6.

As another example, 13 ÷ 4 = ?

Answer: (3).(25)

4 goes into 13 three times.

13 ÷ 4 = (3)

Remainder is $\frac{1}{4}$ or 0.25 of 4; therefore,

0.25 × 4 = (1)

Answer to 13 ÷ 4 is 3, remainder 1.

Bringing down the next dividend digit, 2, and combining it with the previous remainder, 4, we now have the final part of the dividend, which is 42. Because 12 can be subtracted from 42 three times, a quotient of 3 is placed above the final digit of the dividend (2), and the remainder of the subtraction, 6, is dropped below the 42 − 36 line.

Therefore,

1362 ÷ 12 = 113, remainder 6. Said another way, 12 can be subtracted from 1362 113 times, with 6 remaining.

■ EXAMPLE

Calculate the result of dividing 19,205.628 by 1.56.

■ *Solution:*

$19{,}205.628 \div 1.56 = 12{,}311.3$

Calculator sequence: ⌑1⌑9⌑2⌑0⌑5⌑.⌑6⌑2⌑8⌑÷⌑1⌑.⌑5⌑6⌑=

Answer: 12,311.3

■ INTEGRATED MATH APPLICATION: PERSONAL FINANCE

If your round-trip commute to work for five days totals 87.5 miles, how many miles are you driving per day?

■ *Solution:*

$$87.5 \text{ miles} \div 5 \text{ days} = 17.5 \text{ miles per day}$$

■ INTEGRATED MATH APPLICATION: PERSONAL FINANCE

If your gross pay for a week is $536.50 and you worked 37 hours, what is your hourly rate of pay?

■ *Solution:*

$$\$536.50 \div 37 \text{ hours} = \$14.50 \text{ per hour}$$

2–2–5 *Ratios*

A **ratio** is a comparison of one number with another number. For example, a ratio such as $7 : 4$ ("seven to four") is comparing the number 7 with the number 4. This ratio could be written in one of the following three ways:

$\dfrac{7}{4}$ ← Ratio is written as a fraction.

$7 : 4$ ← Ratio is written using the ratio sign (:).

$1.75 \text{ to } 1$ ← Ratio is written as a decimal.

To express the ratio as a decimal, the number 7 was divided by the number 4, giving a result of 1.75 ($7 \div 4 = 1.75$). This result indicates that the number 7 is 1.75 times (or one and three-quarter times) larger than the number 4.

Ratio

The relationship in quantity, amount, or size between two or more things.

■ INTEGRATED MATH APPLICATION: RETAIL

If one store has 360 cell phones and another store has 100, express the ratio of 360 to 100:

a. Using the divide sign
b. Using the ratio sign
c. As a decimal

■ *Solution:*

a. $\dfrac{360}{100} = \dfrac{360 \div 20}{100 \div 20} = \dfrac{18}{5}$ ← 360 to 100 and 18 to 5 are equivalent ratios.

b. $360 : 100 = 18 : 5$
c. 3.6 ← The first store has 3.6 times as many items as the other store ($18 \div 5 = 3.6$).

■ **EXAMPLE**

Calculate the ratio of the following two like quantities. If necessary, reduce the ratio to its lowest terms.

    a. $0.8 \div 0.2 = ?$
    b. $0.0008 \div 0.0002 = ?$

■ *Solution:*

    a. How many 0.2s are in 0.8? The answer is the same as saying, How many 2s are in 8?

$$0.8 \div 0.2 = 4 \qquad \frac{8 \div 2}{2 \div 2} = \frac{4}{1} \quad \text{(a ratio of 4 to 1)}$$

    b. How many 0.0002s are in 0.0008?

$$0.0008 \div 0.0002 = 4 : 1$$

The ratio in example (a) is the same as in (b). Another way to describe this is to say that because the size of each decimal fraction compared with the other decimal fraction (or the *ratio*) remained the same, the answer or quotient remained the same.

    8 is four times larger than 2, or 2 is four times smaller than 8.

    0.8 is four times larger than 0.2, or 0.2 is four times smaller than 0.8.

    0.0008 is four times larger than 0.0002, or 0.0002 is four times smaller than 0.0008. Therefore, all have the same $4 : 1$ ratio.

■ **INTEGRATED MATH APPLICATION: ADMINISTRATION**

In some instances, a ratio will be used to describe how a quantity is to be divided. For example, if 600 textbooks are to be divided between two schools in the ratio of 1 to 2 ($1 : 2$), how many textbooks will each school get? To calculate this, you should first add the terms of the ratio:

■ *Solution:*

    Step 1:   The ratio is $1 : 2$, so $1 + 2 = 3$ (this means that the total is to be divided into 3 parts).

Next, you should multiply the total by each fractional part:

    Step 2:   The first school will receive $\frac{1}{3}$ of the total 600 textbooks, which is 200 texts $\left(\frac{1}{3} \times 600 = 200\right)$. The second school will receive $\frac{2}{3}$ of the total 600 textbooks, which is 400 texts $\left(\frac{2}{3} \times 600 = 400\right)$.

■ **INTEGRATED MATH APPLICATION: FOOD SERVICE**

If three restaurants are shipped 216 hamburgers in the ratio of $6 : 14 : 7$, how many hamburgers will each restaurant get?

■ *Solution:*

The first step is to add the terms of the ratio:

    Step 1:   The ratio is $6 : 14 : 7$, so $6 + 14 + 7 = 27$ (this means that the total is to be divided into 27 parts).

Next, you should multiply the total by each fractional part:

    Step 2:   The first restaurant will receive $\frac{6}{27}$ of the total 216 hamburgers, which is 48 $\left(\frac{6}{27} \times 216 = 48\right)$. The second restaurant will receive $\frac{14}{27}$ of the total 216 hamburgers, which is 112 $\left(\frac{14}{27} \times 216 = 112\right)$. The third restaurant will receive $\frac{7}{27}$ of the total 216 hamburgers, which is 56 $\left(\frac{7}{27} \times 216 = 56\right)$.

CHAPTER 2 / DECIMAL CALCULATION

2–2–6   *Rounding Off*

Rounding Off

An operation in which a value is abbreviated by applying the following rule: When the first digit to be dropped is 6 or more, or 5 followed by a digit that is more than zero, increase the previous digit by 1. When the first digit to be dropped is 4 or less, or 5 followed by a zero, do not change the previous digit.

In many cases we will **round off** *decimal numbers because we do not need the accuracy indicated by a large number of decimal digits.* For example, it is usually unnecessary to have so many decimal digits in a value such as 74.139896428. If we were to round off this value to the nearest hundredths place, we would include two digits after the decimal point, which is 74.13. This is not accurate, however, since the digit following 74.13 was a 9 and therefore one count away from causing a reset and carry action into the hundredths column. To take into account the digit that is to be dropped when rounding off, therefore, we follow this basic 3-step procedure:

Step 1:   To round a decimal to a particular decimal place, first locate the digit in that place and call it the *rounding digit.*

Step 2:   Look at the *test digit* to the right of the rounding digit.

Step 3:   If the test digit is 5 or greater, add 1 to the rounding digit and drop all digits to the right of the rounding digit. If the test digit is less than 5, do not change the rounding digit, and drop all digits to the right of the rounding digit.

Therefore, the value 74.139896428, rounded off to the nearest hundredths place, would equal:

Rounding digit: hundredths place

74.13 | 9896428 = 74.14

Test digit: First digit to be dropped is greater than 5,
so rounding digit should be increased by 1.

■ **EXAMPLE**

Round off the value 74.139896428 to the nearest ten, whole number, tenth, hundredth, thousandth, and ten-thousandth.

■ *Solution:*

7 | 4.139896428 rounded off to the nearest ten = 70
   └─ Test digit to be dropped is less than 5; therefore, do not change rounding digit.

74. | 139896428 rounded off to the nearest whole number = 74
    └─ Test digit to be dropped is less than 5; therefore, do not change rounding digit.

74.1 | 39896428 rounded off to the nearest tenth = 74.1
     └─ Test digit to be dropped is less than 5; therefore, do not change rounding digit.

74.13 | 9896428 rounded off to the nearest hundredth = 74.14
      └─ Test digit to be dropped is 5 or greater; therefore, increase rounding digit by 1
      (3 to 4).

74.139 | 896428 rounded off to the nearest thousandth = 74.140
       └─ Test digit to be dropped is 5 or greater; therefore, increase rounding digit by 1.
       Since rounding digits is 9, allow reset and carry action to occur. The steps are:

   74.1398 ← 8 carries 1 into thousandths column.
   74.130 ← 9 resets to 0, and carries 1 into hundredths column.
   74.140 ← Hundredths digit is increased by 1.

74.1398 | 96428 rounded off to the nearest ten-thousandth = 74.1399
        └─ Test digit to be dropped is 5 or greater; therefore, increase rounding digit by 1.

CALCULATOR KEYS

Name: Round function
round (*value*[,*#decimals*])

Function: Value is rounded to the number of decimals specified.

Example:
round (π, 4)
  3.1416

2-2-7   Significant Places

The number of significant places describes how many digits are in the value and how many digits in the value are accurate after rounding off. For example, a number such as 347.63 is a five-significant-place (or significant-figure) value because it has five digits in five columns. If we were to round off this value to a whole number, we would get 348.00. This value would still be a five-significant-place number; however, it would now be accurate to only three significant places. Let us examine a few problems to practice using the terms *significant places* and *accurate to significant places.*

■ EXAMPLE

On a calculator, the value of π will come up as 3.141592654.

  a. Write the number π to six significant places.
  b. Give π to five significant places, and also round off π to ten-thousandths.
  c. 3.14159000 is the value of π to _____ significant places; however, it is accurate to only _____ significant places.

■ *Solution:*

  a. The value π to six significant places or figures is $\boxed{3.14159}$ 27

$$\underbrace{3.14159}$$
          ⌐—— Uses six digits or columns.

  b. The value π to five significant places.
  $\boxed{3.1415}$ 92 rounded off to ten-thousandths = 3.1416
  c. 3.14159000 is the value of π to nine significant places (has nine digits); however, it is accurate to only six significant places (because the zeros to the right are just extra, and the value 3.14159 has only six digits).

2–2–8   *Percentages*

The percent sign (%) means hundredths, which as a proper fraction is $\frac{1}{100}$ or as a decimal fraction is 0.01. For example, 50% means 50 hundredths $\left(\frac{50}{100} \frac{+50}{+50} = \frac{1}{2}\right)$, which is one-half. In decimal, 50% means 50 hundredths ($50 \times 0.01 = 0.5$), which is also one-half. The following shows how some of the more frequently used percentages can be expressed in decimal.

Percent

In the hundred; of each hundred.

$$1\% = 1 \times 0.01 = 0.01$$

(For example, 1% is expressed mathematically in decimal as 0.01.)

$$5\% = 5 \times 0.01 = 0.05$$
$$10\% = 10 \times 0.01 = 0.10$$
$$25\% = 25 \times 0.01 = 0.25$$
$$50\% = 50 \times 0.01 = 0.50$$
$$75\% = 75 \times 0.01 = 0.75$$

Most calculators have a percent key that automatically makes this conversion to decimal hundredths. It operates as follows.

CALCULATOR KEYS

Name: Percent key

Function: Converts the displayed number from a percentage to a decimal fraction.

Example: 18.6% = ?

Press keys: | 1 | 8 | . | 6 | % |

Display shows: 0.186

Now that we understand that the percent sign stands for hundredths, what does the following question actually mean: What is 50% of 12? We now know that 50% is 50 hundredths $\left(\frac{50}{100}\right)$, which is one-half. This question is actually asking: What is half of 12? Expressed mathematically as both proper fractions and decimal fractions, the question would appear as follows:

Proper Fractions	**Decimal Fractions**
50% of 12 = ?	50% of 12 = ?
$= \frac{50}{100} \times 12$	$= (50 \times 0.01) \times 12$
$= \frac{1}{2} \times 12$	$= 0.5 \times 12$
$= 6$	$= 6$

Explaining this another way, we could say that percentages are fractions in hundredths. In the preceding example, therefore, the question is actually asking us to divide 12 into 100 parts and then to determine what value we would have if we had 50 of those 100 parts. Therefore, if we were to split 12 into 100 parts, each part would have a value of 0.12 $(12 \div 100 = 0.12)$. Having 50 of these 100 parts would give us a total of $50 \times 0.12 = 6$.

■ **INTEGRATED MATH APPLICATION: RETAIL**

If sales tax is 8% and the merchandise price is $256.00, what is the tax on the value, and what will be the total price?

■ *Solution:*

First, determine the amount of tax to be paid.

$$8\% \text{ of } \$256 =$$
$$0.08 \times 256 = \$20.48$$

Calculator sequence: 8 % × 2 5 6 =

Answer: 20.48
The total price paid will therefore equal

$$\$256.00 + \$20.48 = \$276.48$$

■ **INTEGRATED MATH APPLICATION: PERSONAL FINANCE**

Which is the better buy—a $599.95 refrigerator at 25% off or a $499.95 refrigerator with a $50 manufacturer's rebate?

■ *Solution:*

$$25\% \text{ of } \$599.95 = \$150$$
$$\text{Refrigerator A: } \$599.95 - \$150.00 = \$449.95$$
$$\text{Refrigerator B: } \$499.95 - \$50.00 = \$449.95$$

The refrigerators are the same price.

■ **INTEGRATED MATH APPLICATION: ELECTRONICS**

A 330 ohm (330 Ω) resistor is listed as having a 10% (plus or minus ten percent) tolerance, which means the manufacturer guarantees the value will not be less than 330 ohms plus 10%, or less than 330 ohms minus 10%. List the resistor's possible range of values.

■ *Solution:*

$$330 \ \Omega \times 10\% = 33 \ \Omega$$
$$330 \ \Omega + 33 \ \Omega = 363 \ \Omega$$
$$330 \ \Omega - 33 \ \Omega = 297 \Omega$$

2–2–9 *Averages*

Average

A single value that summarizes or represents the general significance of a set of unequal values.

A *mean* **average** *is a value that summarizes a set of unequal values.* This value is equal to the sum of all the values divided by the number of values. For example, if a team has scores of 5, 10, and 15, what is their average score? The answer is obtained by adding all the scores $(5 + 10 + 15 = 30)$ and then dividing the result by the number of scores $(30 \div 3 = 10)$.

■ INTEGRATED MATH APPLICATION: ELECTRICAL

The voltage at the wall outlet in your home was measured at different times in the day and equaled 108.6, 110.4, 115.5, and 123.6 volts (V). Find the average voltage.

■ *Solution:*

Add all the voltages.

$$108.6 + 110.4 + 115.5 + 123.6 = 458.1 \text{ V}$$

Divide the result by the number of readings.

$$\frac{458.1 \text{ V}}{4} = 114.525 \text{ V}$$

The average voltage present at your home was therefore 114.525 volts.

Statistics are generally averages and can misrepresent the actual conditions. For example, if five people earn an average salary of $204,600 a year, one would believe that all five are very well paid. Studying the individual figures, however, we find out that their yearly earnings are $8000, $6000, $4000, $20,000, and $985,000 per year. Therefore,

$$8000 + 6000 + 4000 + 20,000 + 985,000 = 1,023,000$$

$$\frac{1,023,000}{5} = 204,600$$

If one value is very different from the others in a group, this value will heavily influence the result, creating an average that misrepresents the actual situation.

Statistics

A branch of mathematics dealing with the collection, analysis, interpretation, and presentation of masses of numerical data.

■ INTEGRATED MATH APPLICATION: SPORTS

A football player's "average number of yards gained per carry" ranking is calculated by dividing the total number of yards gained by the total number of times the player carried the football. Which of the following players had the higher average number of yards gained per carry?

Player	Total Yards	Total Carries
Payton	16,726	3,838
Brown	12,312	2,359

■ *Solution:*

$$\text{Payton: } \frac{16,726}{3,838} = 4.4 \text{ yards per carry average}$$

$$\text{Brown: } \frac{12,312}{2,359} = 5.2 \text{ yards per carry average}$$

On average, Brown gained a higher number of yards per carry.

SELF-TEST EVALUATION POINT FOR SECTION 2–2

Now that you have completed this section, you should be able to:

■ *Objective 6.* *Describe the relationships among the four basic arithmetic operations: addition, subtraction, multiplication, and division.*

■ *Objective 7.* *Demonstrate how to perform the following arithmetic operations: decimal addition, decimal subtraction, decimal multiplication, and decimal division.*

■ **Objective 8.** *Describe the following mathematical terms and their associated operations: ratios, rounding off, significant places, percentages, and averages.*

Use the following questions to test your understanding of Section 2–2.

1. The two basic arithmetic operations are _____ and _____.

2. Give the opposite of the following arithmetic operations.
   a. Addition
   c. Multiplication
   b. Subtraction
   d. Division

3. Perform the following arithmetic operations by hand, and then confirm your answers with a calculator.
   a. 26.443 + 197.1 + 2.1103 + 0.004 = ?
   b. 19,637.224 − 866.43 = ?
   c. 894.357 × 8.6 = ?
   d. 1.3397 ÷ 0.015 = ?

4. Express a 176 meter to 8 meter ratio as a fraction, using the ratio sign and as a decimal.

5. Round off the following values to the nearest hundredth.
   a. 86.43760
   b. 12,263,415.00510
   c. 0.176600

6. Referring to the values in Question 5, describe:
   a. Their number of significant places
   b. To how many significant places they are accurate

7. Calculate the following.
   a. 15% of 0.5 = ?
   b. 22% of 1000 = ?
   c. 2.35% of 10 = ?
   d. 96% of 20 = ?

8. Calculate the average of the following values.
   a. 20, 30, 40, and 50
   b. 4000, 4010, 4008, and 3998

REVIEW QUESTIONS

Multiple Choice Questions

1. The decimal point is used to:
   a. Separate the whole numbers from the fraction
   b. Separate the tens column from the units column
   c. Identify the decimal number system
   d. All the above

2. What is the positional weight of the second column to the right of the decimal point?
   a. Tenths
   c. Units
   b. Tens
   d. Hundredths

3. The Greek letter pi represents a constant value that describes how much bigger a circle's _____ is compared with its _____, and it is frequently used in many formulas involved with the analysis of circular motion.
   a. Circumference, radius
   b. Circumference, diameter
   c. Diameter, radius
   d. Diameter, circumference

4. _____ is the opposite arithmetic operation of _____, and _____ is the opposite arithmetic operation of _____.

   a. Multiplication, division, subtraction, addition
   b. Division, addition, multiplication, subtraction
   c. +, ÷, −, ×
   d. ×, −, ÷, +

5. Is π a ratio?
   a. Yes
   b. No

6. Round off the following number to the nearest hundredth: 74.8552.
   a. 74.85
   c. 74.84
   b. 74.86
   d. 74.855

7. The number 27.0003 is a _____ significant-place number and is accurate to _____ significant places.
   a. 6, 2
   c. 6, 6
   b. 2, 6
   d. 4, 6

8. Express 29.63% mathematically as a decimal.
   a. 29.63
   c. 0.02963
   b. 2.963
   d. 0.2963

Communication Skill Questions

9. How are whole numbers and fractions represented in decimal? (2–1)

10. Describe how written fractions are converted to decimal fractions and, conversely, how decimal fractions are converted to written fractions. (2–1)

11. How do you obtain the reciprocal of a number, and what is its relationship to the original number? (2–1)

12. Arbitrarily choose values, and describe the steps involved in: (2–1)
   a. Decimal addition
   c. Decimal multiplication
   b. Decimal subtraction
   d. Decimal division

13. What is a ratio, and how is it expressed? (2–2)

14. Briefly describe the rules for rounding off. (2–2)

15. What is a percentage? (2–2)

16. How is an average calculated? (2–2)

17. What is the approximate value of the constant pi (π) and what does it describe? (2–1)

Practice Problems

18. What are the individual digit weights of 2.7463?

19. Write each of the following values as a number.

    a. Two units and three-tenths

    b. Five-tenths and seven-thousandths

    c. Nine thousand, three tens, and four-hundredths

20. Indicate which zeros can be dropped in the following values (excluding the zero in the units place).

    a. 02.017

    b. 375.011020

    c. 0.0102000

21. Convert the following written fractions to their decimal equivalents.

    a. $\dfrac{16}{32}$ **c.** $4\dfrac{9}{8}$

    b. $3\dfrac{8}{9}$ **d.** $195\dfrac{7}{3}$

22. Calculate the reciprocal of the following values.

    a. π

    b. 0.707

    c. 0.3183

23. Convert the following decimal fractions to written fractions.

    a. $0.777 = \dfrac{?}{9}$

    b. $0.6149069 = \dfrac{?}{161}$

    c. $43.125 = \dfrac{?}{8}$

24. Perform the following arithmetic operations by hand, and then confirm your answers with a calculator.

    a. $764 + 37 + 2199 = ?$

    b. $22{,}763 + 4 + 56{,}003 = ?$

    c. $451 + 19 + 17 = ?$

    d. $5555 + 3333 + 9999 = ?$

    e. $83.57 + 4 + 0.663 = ?$

    f. $898 - 457 = ?$

    g. $7{,}660{,}232 - 23{,}000 = ?$

    h. $4.9 - 3.7 = ?$

    i. $22 - 14.7 = ?$

    j. $2.01 - 1.1 = ?$

    k. $4 \times 16 = ?$

    l. $488 \times 14.0 = ?$

    m. $0.3 \times 99 = ?$

    n. $22{,}000 \times 14{,}000 = ?$

    o. $0.5 \times 10{,}000 = ?$

    p. $488 \div 4 = ?$

    q. $64.8 \div 8.1 = ?$

    r. $99{,}000 \div 10 = ?$

    s. $123 \div 17.5 = ?$

    t. $0.662 \div 0.002 = ?$

25. Express the ratio of the following in their lowest terms. (Remember to compare like quantities.)

    a. The ratio of 20 feet to 5 feet

    b. The ratio of $2\frac{1}{2}$ minutes to 30 seconds

26. Round off as indicated.

    a. 48.36 to the nearest whole number

    b. 156.3625 to the nearest tenth

    c. 0.9254 to the nearest hundredth

27. Light travels in a vacuum at a speed of 186,282.3970 miles per second. This _____-significant-place value is accurate to _____ significant places.

28. Calculate the following percentages.

    a. 23% of 50 = ?

    b. 78% of 10 = ?

    c. 3% of 1.5 = ?

    d. 20% of 3300 = ?

    e. 5% of 10,000 = ?

29. Calculate the average value of the following groups of values (include units).

    a. 4, 5, 6, 7, and 8

    b. 15 seconds, 20 seconds, 18 seconds, and 16 seconds

    c. 110 volts, 115 volts, 121 volts, 117 volts, 128 volts, and 105 volts

    d. 33 ohms, 17 ohms, 1000 ohms, and 973 ohms

    e. 150 meters, 160 meters, and 155 meters

Web Site Questions

Go to the Web site http://www.prenhall.com/cook, select the textbook *Introductory Mathematics,* select this chapter, and then follow the instructions when answering the multiple-choice practice problems.

Positive and Negative Numbers

Blaise of Genius

Blaise Pascal, the son of a regional tax official, was born in France in 1623. His father realized early that his son was a genius when his son's understanding of mathematics at the age of 8 exceeded his own. By the age of 16, Pascal had published several mathematical papers on conic sections and hydraulics. While still a youth he invented, among other things, the syringe, the hydraulic press, and the first public bus system in Paris.

At the age of 19 he began working on an adding machine in an attempt to reduce the computational drudgery of his father's job. Pascal's machine, which became known as the *Pascaline*, became the rage of European nobility and is recognized today as one of the first computers. The operator loaded a number into a machine by setting a series of wheels. Each wheel, marked with the digits zero through nine, stood for each of the decimal columns: 1s, 10s, 100s, and so on. A carry was achieved when a total greater than nine caused a complete revolution of a wheel, which advanced the wheel to the left by one digit.

Although widely praised, the Pascaline, whose adding principle of operation would remain the standard for the next 300 years, did not make Pascal rich. Like many geniuses, he was a troubled person whose ideas and problems would plague him until he could find solutions. At the age of 37, he dropped out of society to join a religious monastery that refrained from all scientific pursuits. He died just two years later at the age of 39.

In his lifetime he became famous as a mathematician, physicist, writer, and philosopher, and in honor of his achievements one of today's computer programming languages has been named after him—Pascal.

Outline and Objectives

VIGNETTE: BLAISE OF GENIUS

INTRODUCTION

Objective 1: Define the difference between a positive number and a negative number.

Objective 2: Describe some applications in which positive and negative numbers are used.

3–1 EXPRESSING POSITIVE AND NEGATIVE NUMBERS

Objective 3: Explain how to express positive and negative numbers.

3–2 ADDING POSITIVE AND NEGATIVE NUMBERS

Objective 4: Describe the three basic rules of positive and negative numbers as they relate to:
- a. Addition
- b. Subtraction
- c. Multiplication
- d. Division

3–3 SUBTRACTING POSITIVE AND NEGATIVE NUMBERS

3–4 MULTIPLYING POSITIVE AND NEGATIVE NUMBERS

3–5 ORDER OF OPERATIONS

3–6 DEALING WITH STRINGS OF POSITIVE AND NEGATIVE NUMBERS

Objective 5: Explain how to combine strings of positive and negative numbers.

Introduction

Positive Number
A value that is greater than zero.

Negative Number
A value that is less than zero.

Until now we have been dealing only with **positive numbers.** In fact, all the values we have discussed so far have been positive. For example, even though 12 does not have a positive (+) sign in front of it, it is still a positive number and therefore could be written as +12. We do not normally include the positive sign in front of a positive value unless we have other values that are negative. In instances when a **negative number** is present, we include the positive sign (+) in front of positive numbers to distinguish them from negative numbers that have the negative sign (−) in front of them. For example, if a list contained both positive and negative values, we would have to include their signs, as follows:

$$-4 \qquad +12 \qquad -6 \qquad +156 \qquad -198{,}765 \qquad +1{,}000{,}000$$

On the other hand, if we were to list only the positive values, there would be no need to include the positive sign because all values are positive.

$$12 \qquad 156 \qquad 1{,}000{,}000 \qquad or \qquad +12 \qquad +156 \qquad +1{,}000{,}000$$

By definition, therefore, *a positive number is any value that is greater than zero, whereas a negative number is any value that is less than zero,* as shown in Figure 3–1.

The fractions discussed so far, which are pieces of a whole, are less than 1 and can be positive values or negative values. For example, some of the following fractions are positive (have a value between zero and positive one), and some are negative (have a value between zero and negative one).

$$-\frac{1}{2} \qquad +0.5 \qquad +\frac{1}{8} \qquad -0.0016$$

The next important question is: *Why do we need both positive and negative numbers? The answer is: to indicate values that are above and below a reference point.* For example, consider the thermometer shown in Figure 3–2, which measures temperature in degrees Celsius (symbolized °C). This scale is used to indicate the amount of heat. On this scale, a reference point of 0 °C (zero degrees Celsius) was assigned to the temperature at which water freezes or ice melts, and this is an easy point to remember. For example, let us assume that the ice cubes inside your freezer are at a temperature of −15 °C (negative 15 degrees Celsius), and you remove several ice cubes and place them in a glass on the kitchen counter. At the warmer room temperature the ice cubes will absorb heat, and their temperature value will increase, or become less negative, as follows: −15 °C, −12 °C, −10 °C, −6 °C, and so on.

As we cross this temperature scale's reference point of 0 °C, the ice cubes change from a solid (ice cubes) to a liquid (water). If we then pour the water into a kettle and heat it even further, its temperature value will increase or become more positive, as follows: −5 °C, +12 °C, +22 °C, and so on.

Negative and positive numbers therefore are used to designate some quantity or value relative to, or compared with, a zero reference point. Temperature is not the only application for these numbers. For example, positive values can be used to express money that you have, whereas negative values can be used to express money that you owe, or have paid out. Similarly, a zero reference point can be used to indicate the average rainfall or snowfall, with positive values indicating the amount of rain or snow above the average, and negative values indicating the amount of rain or snow below the average.

In Chapter 3 we will discuss how to add, subtract, multiply, and divide positive and negative numbers, or signed numbers (numbers with signs).

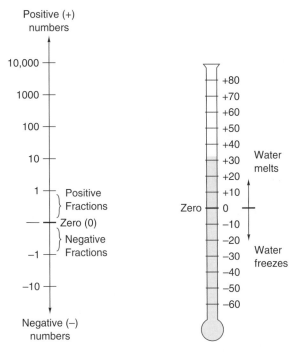

FIGURE 3–1   The
Positive and Negative
Number Line.

FIGURE 3–2   The
Celsius Temperature
Scale.

3–1   EXPRESSING POSITIVE AND NEGATIVE NUMBERS

Let us first discuss how we will mathematically express an addition, subtraction, multiplication, or division involving positive and negative numbers. As an example, let us assume that the rainfall over the last 2 years has been 3 and 9 graduations (marks on the scale) above the average reference point of zero. Because these two values of 3 and 9 are above the zero reference, they are positive numbers and can therefore be written as

$$3 \text{ and } 9 \quad \text{or} \quad +3 \text{ and } +9$$

To calculate the total amount of rainfall over the last 2 years, therefore, we must add these two values, as follows:

$$3 + 9 = 12 \quad \text{or} \quad +3 + (+9) = +12$$

If each graduation or mark on the scale was equivalent to 1 inch, over the past 2 years we would have had 12 inches of rain $(+12)$ above the average of zero. Both preceding statements or expressions are identical, although the positive signs precede the values or numbers in the statement on the right. You can also see how the parentheses were used to separate the *plus sign* (indicating mathematical addition) from the *positive sign* (indicating that 9 is a positive number). If the parentheses had not been included, you might have found the statement confusing, because it would have appeared as follows:

$$+3 + +9 = +12$$

Plus sign ⤴   ⤴ Positive sign

The parentheses are included to isolate the sign of a number from a mathematical operation sign and therefore to prevent confusion.

As another example, let us imagine that last year the rainfall was 7 inches or graduations above the average, and the year before that there was a bit of a drought and the rainfall was 5 graduations or inches below the average. The total amount of rainfall for the last two years was therefore

$$+7 + (-5) = +2$$ ← Expression or meaningful combination of symbols

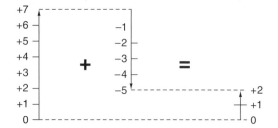

← Graphic representation of expression

Once again, notice how the parentheses were used to separate the addition symbol (plus sign) from the sign of the number (negative 5). This example is probably best understood by referring to the graphic representation below the expression, which shows how the -5 arrow cancels 5 graduations of the $+7$ arrow, resulting in a sum of 2 positive graduations ($+2$).

■ INTEGRATED MATH APPLICATION: ELECTRONICS

The direct current (dc) power supply for a notebook computer converts the 120 volt alternating current (ac) input to three dc voltage outputs.

a. List the dc voltages present at points A, B and C.

b. What is the difference in voltage between points A and B, B and C, A and C?

120 V (AC) —▭— | DC A → +5 V (DC) | POWER B → −7 V (DC) | SUPPLY C → +12 V (DC)

■ *Solution:*

a. A $= +5$ volts
    B $= -7$ volts
    C $= +12$ volts

b. A and B $= 12$ volts
    B and C $= 19$ volts
    A and C $= 7$ volts

Before addressing the details of adding positive and negative numbers, let us discuss the *positive/negative* or *change-sign key* available on most calculators and used to change the sign of an inputted number.

CALCULATOR KEYS

Name: Change-sign, positive/negative, or negation key

Function: Generally used after a number has been entered to change the sign of the number. Numbers with no sign indicated are always positive. Normally you press the negation key $\boxed{(-)}$ before you enter the value.

Example: $\boxed{+/-}$ $+7 + (-5) = ?$

Press keys: $\boxed{7}$ $\boxed{+}$ $\boxed{5}$ $\boxed{+/-}$ $\boxed{=}$

Answer: 2 or +2

Example: $\boxed{(-)}$ $-3 + (-15) = ?$

Press keys: $\boxed{(-)}$ $\boxed{3}$ $\boxed{+}$ $\boxed{(-)}$ $\boxed{1}$ $\boxed{5}$ $\boxed{=}$

Answer: -18

Note: If the subtraction key $\boxed{-}$ is used in place of the negation key $\boxed{(-)}$, or vice versa, an error occurs.

■ INTEGRATED MATH APPLICATION: PLUMBING

To offset daily gasoline and equipment expenses of $65, a plumber needs to exceed this value in order to turn a profit.

a. Express the values earned over the last two weeks as positive and negative numbers relative to $65.

b. What is the average of the positive and negative values for week 1 and week 2?

c. Did the plumber make a profit or stand a loss in week 1 and in week 2?

	WEEK 1	WEEK 2
M	$75	$470
T	$0	$950
W	$195	$14
T	$45	$0
F	$30	$60

■ Solution:

a.

	WEEK 1	WEEK 2
M	+10	+405
T	−65	+885
W	+130	−51
T	−20	−65
F	−35	−5

b. Week 1 = +20   week 2 = +1169

c. Week 1: Profit $20

  Week 2: Profit $1169

SELF-TEST EVALUATION POINT FOR SECTION 3–1

Now that you have completed this section, you should be able to:

■ **Objective 1.** *Define the difference between a positive number and a negative number.*

■ **Objective 2.** *Describe some applications in which positive and negative numbers are used.*

■ **Objective 3.** *Explain how to express positive and negative numbers.*

Use the following questions to test your understanding of Section 3–1.

1. Use parentheses to separate the arithmetic operation sign from the sign of the number, and then determine the result.
   a. $+3 - -4 =$
   b. $12 \times +4 =$
   c. $-5 \div -7 =$
   d. $-0.63 \times +6.4 =$

2. Express the following statements mathematically.
   a. Positive six plus negative seven
   b. Negative zero point seven five multiplied by eighteen

3. Write the calculator key sequences to input the following problems.
   a. $-7.5 + (-4.6) =$
   b. $+5 \times (-2) =$
   c. $+316.629 \div (-1.44) =$

3–2   ADDING POSITIVE AND NEGATIVE NUMBERS

To begin, we will review how to add one positive number $(+)$ to another positive number $(+)$.

> ## RULE 1: $(+) + (+) = (+)$
>
> **Description:**   A positive number $(+)$ plus a positive number $(+)$ equals a positive number $(+)$.

Example:

$$5 + 15 = ?$$
$$+5 + (+15) = +20$$

Explanation:

A positive 5-unit arrow plus a positive 15-unit arrow equals a positive 20-unit arrow.

Calculator sequence: 5 + 1 5 =

Answer: 20 (+20)

Other Examples:

$$+9 + 6 = ?$$
$$+9 + (+6) = +15$$
$$81 + +16 = ?$$
$$+81 + (+16) = +97$$
$$+(15) + +15 = ?$$
$$+15 + (+15) = +30$$

The preceding box helps to collect all the facts about this rule of addition. As you know, this basic rule of addition was covered in Chapter 2.

RULE 2: $(-) + (+)$ or $(+) + (-) =$ (sign of larger number) and difference

Description: A negative number $(-)$ plus a positive number $(+)$ or a positive number $(+)$ plus a negative number $(-)$ equals the sign of the larger number and the difference.

Two Examples:

$$-8 + (+15) = ? \qquad \text{or} \qquad +15 + (-8) = ?$$
$$-8 + (+15) = +7 \qquad\qquad +15 + (-8) = +7$$

Because 15 is the larger number, sign is positive.

Difference between 8 and 15 is 7.

Explanation:

A negative 8-unit arrow plus a positive 15-unit arrow equals a positive 7-unit arrow.

Calculator sequence: $\boxed{8}\ \boxed{+/-}\ \boxed{+}\ \boxed{1}\ \boxed{5}\ \boxed{=}$

Answer: 7(+7)

Other Examples:

$$6 + (-3) = ?$$
$$+6 + (-3) = +3 \qquad \text{(larger number is } +, \text{ difference } = 3)$$
$$-112 + +59 = ?$$
$$-112 + (+59) = -53 \qquad \text{(larger number is } -, \text{ difference } = 53)$$
$$-18 + +12 = ?$$
$$-18 + (+12) = -6 \qquad \text{(larger number is } -, \text{ difference } = 6)$$

■   **INTEGRATED MATH APPLICATION: ELECTRICAL**

The current drawn by an automated manufacturing station varies based on the process being performed.

a. What is the average current drawn compared with the zero-reference optimum load current of 6.6 amperes?

Manufacturing step:	1.	2.	3.	4.	5.	6.
Load current:	−0.3 A	+2.4 A	+3.6 A	0 A	−2.2 A	−1.4 A

b. What is the actual average load current?

■   *Solution:*

a. Average current relative to zero-reference = +2.1 amps
b. Average load current = 6.6 A + 2.1 A = 8.7 amps

RULE 3: (−) + (−) = (−)

Description: A negative number (−) plus a negative number (−) equals a negative number (−).

Example:

$$-16 + (-4) = ?$$
$$-16 + (-4) = -20$$

Explanation:

A negative 16-unit arrow plus a negative 4-unit arrow equals a negative 20-unit arrow.

Calculator sequence: $\boxed{1}\ \boxed{6}\ \boxed{+/-}\ \boxed{+}\ \boxed{4}\ \boxed{+/-}\ \boxed{=}$

Answer: −20

Other Examples:

$$-15 + -25 = ?$$
$$-15 + (-25) = -40$$
$$(-1063) + (-1000) = ?$$
$$-1063 + (-1000) = -2063$$

■ **INTEGRATED MATH APPLICATION: METEOROLOGY**

The average temperature for the second week of summer should be 78 degrees Fahrenheit. The following lists the daily temperature measured compared with the zero-reference point of 78 °F.

Week Day:	2/1	2/2	2/3	2/4	2/5	2/6	2/7
Temperature:	+2	−1	0	+6	−3	−2	−4

  a. What is the average temperature compared with the zero reference of 78 °F?
  b. What is the actual average temperature?

■ *Solution:*

  a. Average temperature relative to zero reference $= -2\,°F$
  b. Average temperature $= 78\,°F - 2 = 76\,°F$

SELF-TEST EVALUATION POINT FOR SECTION 3–2

Now that you have completed this section, you should be able to:

■ *Objective 4.* *Describe the three basic rules of positive and negative numbers as they relate to addition.*

Use the following questions to test your understanding of Section 3–2.
Add the following values.

1. $+15 + (+4) = ?$
2. $-3 + (+45) = ?$
3. $+114 + (-111) = ?$
4. $-357 + (-74) = ?$
5. $17 + 15 = ?$
6. $-8 + (-177) = ?$
7. $+4600 + (-3400) = ?$
8. $-6.25 + (+0.34) = ?$

3–3 SUBTRACTING POSITIVE AND NEGATIVE NUMBERS

In Chapter 2 we performed a basic subtraction in the following way:

$$19 - 12 = ?$$

$$
\begin{array}{r}
19 \\
-12 \\
\hline
07
\end{array}
$$

Steps:
$9 - 2 = 7.$
$1 - 1 = 0.$

Therefore, $19 - 12 = 7$.

  If we wanted to, however, we could perform the same subtraction simply by changing the sign of the second number and then adding the two numbers. Let us apply this method to the previous example of $19 - 12$ and see if it works.

$19 - 12 = ?$ ← Original example
$+19 - (+12) = ?$ ← Values written as signed numbers

Step 1: Change sign of second number.

Step 2: Add the two numbers.

$+19 + (-12) = +7$ ← Same result is achieved as performing the operation $19 - 12$.

Previous rule of addition
$(+) + (-) =$ (sign of larger number) and difference

In summary, therefore, *to subtract one number from another, we can change the sign of the second number and then add the two numbers together.* If you apply this "general rule of subtraction" to all the following positive and negative number subtraction examples, you will find that it simplifies the operation because all you have to do is change the sign of the second number and then apply the rules of addition previously discussed.

RULE 1: $(+) - (+)$ or $(-) - (-) =$ (sign of larger number) and difference

Description: A positive number $(+)$ minus a positive number $(+)$, or a negative number $(-)$ minus a negative number $(-)$, equals the sign of the larger number and the difference.

Two Examples:

$$+15 - (+8) = ?$$
$$+15 + (-8) = ?$$

$$-10 - (-2) = ?$$
$$-10 + (+2) = ?$$

Apply rule of subtraction:
— *Step 1:* Change sign of second number. —
— *Step 2:* Add the two numbers. —

$$+15 + (-8) = +7 \qquad\qquad -10 + (+2) = -8$$

Apply rule of addition:
$(+) + (-)$ or $(-) + (+) =$
(sign of larger number)
and difference

Explanation:

Calculator sequence: | 1 | 5 | − | 8 | = | | 1 | 0 | +/− | − | 2 | +/− | = |

Answer: $7(+7)$ -8

Other Examples:

$$-15 - (+8) = ?$$
$$-15 + (-8) = -23 \qquad \leftarrow \text{(Change sign of second number and then add.)}$$

$$-2 - (-10) = ?$$
$$-2 + (+10) = +8 \qquad \leftarrow \text{(Change sign of second number and then add.)}$$

$$-654 - (-601) = ?$$
$$-654 + (+601) = -53 \qquad \leftarrow \text{(Change sign of second number and then add.)}$$

RULE 2: $(+) - (-) = (+)$

Description: A positive number $(+)$ minus a negative number $(-)$ will always equal a positive number $(+)$.

Example:

$$+6 - (-5) = ? \longleftarrow \text{Apply rule of subtraction:}$$

Step 1: Change sign of second number.

Step 2: Add two numbers.

$$+6 + (+5) = +11 \qquad \text{Apply rule of addition:}$$
$$(+) + (+) = +.$$

Explanation: As you can see from the preceding example, subtracting a negative number is the same as adding a positive number, so the result will always be positive. To explain this further, the minus sign is immediately followed by a negative sign, causing two reversals in direction, and therefore the two numbers are combined.

This *double negative* (minus sign followed by a negative sign) can also be explained by using an English language analogy. If you were to use the expression "I don't have no money," you would be using two negatives (don't and no) or reversals and therefore be saying the opposite. If you don't have "no money," you must have some money.

Calculator sequence: $\boxed{6}\ \boxed{-}\ \boxed{5}\ \boxed{+/-}\ \boxed{=}$

Answer: $11(+11)$

Other Examples:

$$+63 - (-14) = ? \qquad \leftarrow (\text{Change sign of second number and then add.})$$
$$+63 + (+14) = +77$$
$$1 - -2 = ?$$
$$+1 - (-2) = ? \qquad \leftarrow (\text{Change sign of second number and then add.})$$
$$+1 + (+2) = +3$$
$$(18) - -14 = ?$$
$$+(18) - -14 = ? \qquad \leftarrow (\text{Change sign of second number and then add.})$$
$$+18 + (+14) = +32$$

■ INTEGRATED MATH APPLICATION: ELECTRONICS

A quality assurance (QA) technician is checking the output voltage from a dc power supply. The $+5$ V (positive five volt) output has a ±0.75 tolerance, whereas the -12 V (negative 12 volt) output has a ±2.2 V tolerance. Which of the following units should pass QA inspection? (Explain why a unit fails.)

	UNIT A	UNIT B	UNIT C	UNIT D
Voltmeter reading of $+5$ V output	$+5.35$ V	$+4.36$ V	$+5.83$ V	$+4.29$ V
Voltmeter reading of -12 V output	-11.98 V	-12.22 V	-11.75 V	-12.25 V

■ *Solution:*

$$5 \text{ V} \pm 0.75 \text{ V} = 4.25 \text{ V} \text{ to } 5.75 \text{ V}$$
$$-12 \text{ V} \pm 2.2 \text{ V} = -9.8 \text{ V} \text{ to } -14.2 \text{ V}$$

UNITS	A	B	C	D
$+5$ V	✓	✓	✗	✓
-12 V	✓	✓	✓	✓

✓ = Within tolerance
✗ = Outside tolerance

RULE 3: $(-) - (+) = (-)$

Description: A negative number $(-)$ minus a positive number $(+)$ will always equal a negative number $(-)$.

Example:

$$-7 - (+4) = ? \longleftarrow \text{Apply rule of subtraction:}$$

Step 1: Change sign of second number.

Step 2: Add two numbers.

$$-7 + (-4) = -11 \quad \text{Apply rule of addition:}$$
$$(-) + (-) = -.$$

Explanation: A negative number minus a positive number $[(-) - (+) = (-)]$ is the same as a negative number plus a negative number $[(-) + (-) = (-)]$. The result will always be negative because the arithmetic operation is designed to calculate the sum of two negatives.

Calculator sequence: $\boxed{7} \boxed{+/-} \boxed{-} \boxed{4} \boxed{=}$

Answer: $\qquad -11$

Other Examples:

$$-63 - (+15) = ?$$
$$-63 + (-15) = -78 \qquad \leftarrow \text{(Change sign of second number and then add.)}$$
$$-(3) - (+15) = ?$$
$$-3 - (+15) = ?$$
$$-3 + (-15) = -18 \qquad \leftarrow \text{(Change sign of second number and then add.)}$$

■ INTEGRATED MATH APPLICATION: MACHINING

In reference to the drilled hole, point A should be 4.2 centimeters from point x, and point B should be 2.6 centimeters from point y. If both points have a tolerance of ± 0.3 cm:

a. What is their range of acceptance?
b. Which of the following are within or outside tolerance?

PART	1	2	3	4	5
$x \rightarrow A$	3.8	4.2	3.9	4.6	4.2
$y \rightarrow B$	2.4	3.0	2.5	2.6	2.7

■ *Solution:*

a. $4.2 \text{ cm} \pm 0.3 \text{ cm} = 3.9 \text{ cm to } 4.5 \text{ cm}$
$2.6 \text{ cm} \pm 0.3 \text{ cm} = 2.3 \text{ cm to } 2.9 \text{ cm}$

b.

PART	1	2	3	4	5
A	✗	✓	✓	✗	✓
B	✓	✗	✓	✓	✓

✓ = Within tolerance
✗ = Outside tolerance

Now that you have completed this section, you should be able to:

■ **Objective 4.** *Describe the three basic rules of positive and negative numbers as they relate to subtraction.*

Use the following questions to test your understanding of Section 3–3.

Subtract the following values.

1. $+18 - (+7) = ?$
2. $-3.4 - (-5.7) = ?$
3. $19,665 - (-5031) = ?$
4. $-8 - (+5) = ?$
5. $467 - 223 = ?$
6. $-331 - (-2.6) = ?$
7. $8 - (+25) = ?$
8. $-0.64 - (-0.04) = ?$

3–4 MULTIPLYING POSITIVE AND NEGATIVE NUMBERS

Let us begin by reviewing how to multiply one positive number by another positive number. This basic rule of multiplication was covered in Chapter 2.

RULE 1: $(+) \times (+) = (+)$

Description: A positive number $(+)$ multiplied by a positive number $(+)$ will always yield a positive product or result $(+)$.

Example:

$$+6 \times (+3) = ? \qquad \leftarrow (\text{same as } 6 \times 3 = ?)$$
$$+6 \times (+3) = +18$$

Explanation: Multiplication is simply repeated addition, so adding a series of positive numbers will always produce a positive result.

Six positive 3s

Six positive 3-unit arrows equals a positive 18-unit arrow.

Calculator sequence: $\boxed{6}\ \boxed{\times}\ \boxed{3}\ \boxed{=}$

Answer: $18(+18)$

Other Examples:

$$24 \times 2 = ?$$
$$+24 \times (+2) = +48$$
$$+3 \times +7 = ?$$
$$+3 \times (+7) = +21$$
$$+26,397 \times +31 = ?$$
$$+26,397 \times (+31) = +818,307$$

RULE 2: $(+) \times (-)$ OR $(-) \times (+) = (-)$

Description: A positive number $(+)$ multiplied by a negative number $(-)$, or a negative number $(-)$ multiplied by a positive number $(+)$, will always yield a negative result $(-)$.

Example:

$$+6 \times (-4) = ? \qquad \text{or} \qquad -4 \times (+6) = ?$$
$$+6 \times (-4) = -24 \qquad \qquad -4 \times (+6) = -24$$

Explanation: The result will always be negative because multiplication is simply repeated addition, and therefore adding a series of negative numbers will always result in a negative answer.

Calculator sequence: $\boxed{6}\ \boxed{\times}\ \boxed{4}\ \boxed{+/-}\ \boxed{=}$ or $\boxed{4}\ \boxed{+/-}\ \boxed{\times}\ \boxed{6}\ \boxed{=}$

Answer: -24 $\qquad\qquad\qquad\qquad$ -24

Other Examples:

$$18 \times (-2) = ?$$
$$+18 \times (-2) = -36$$
$$-4 \times +3 = ?$$
$$-4 \times (+3) = -12$$
$$+21.3 \times -0.2 = ?$$
$$+21.3 \times (-0.2) = -4.26$$

■ **INTEGRATED MATH APPLICATION: ELECTRONICS**

The positive segment of the waveform shown on the oscilloscope display rises one-third the amplitude of the negative segment. What is the voltage of the positive peak of the waveform?

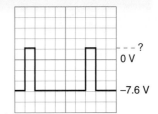

■ *Solution:*

$$\tfrac{1}{3} \times (-7.6\,\text{V}) = -2.53 \text{ volts}$$
The positve peak $= +2.53$ volts

RULE 3: $(-) \times (-) = (+)$

Description: A negative number $(-)$ multiplied by a negative number $(-)$ will always equal a positive number $(+)$.

Example:

$$-3 \times (-5) = ?$$
$$-3 \times (-5) = +15$$

Explanation: The explanation for this is once again the double negative. After a series of negative numbers are added repeatedly, the negative answer is reversed to a positive. In summary, the reversals of direction in the operation coupled with the two negative numbers reverse the result to a positive value.

Calculator sequence: $\boxed{3}\ \boxed{+/-}\ \boxed{\times}\ \boxed{5}\ \boxed{+/-}\ \boxed{=}$

Answer: $15\ (+15)$

Other Examples:

$$-30 \times (-2) = ?$$
$$-30 \times (-2) = +60$$
$$-(2) \times -3 = ?$$
$$-2 \times (-3) = +6$$
$$-76.44 \times -1.3 = ?$$
$$-76.44 \times (-1.3) = +99.37$$

To convert a Fahrenheit temperature to Celsius, subtract 32 from the °F reading and multiply the result by $\frac{5}{9}$. Convert the following °F readings to °C.

    a. 16 °F   **b.** 76 °F   **c.** 4 °F   **d.** 98 °F

■ *Solution:*

$$°C = \frac{5}{9} \times (°F - 32)$$

a. $\frac{5}{9} \times (16 - 32) = \frac{5}{9} \times (-16) = -8\frac{8}{9} = -8.89 \,°C$

b. $\frac{5}{9} \times (76 - 32) = 24\frac{4}{9} = +24.44 \,°C$

c. $\frac{5}{9} \times (4 - 32) = -15\frac{5}{9} = -15.55 \,°C$

d. $\frac{5}{9} \times (98 - 32) = 36\frac{2}{3} = 36.66 \,°C$

SELF-TEST EVALUATION POINT FOR SECTION 3–4

Now that you have completed this section, you should be able to:

■ ***Objective 4.*** *Describe the three basic rules of positive and negative numbers as they relate to multiplication.*

Use the following questions to test your understanding of Section 3–4.

Multiply the following values.

1. $4 \times (+3) = ?$
2. $+17 \times (-2) = ?$
3. $-8 \times 16 = ?$
4. $-8 \times (-5) = ?$
5. $+12.6 \times 15 = ?$
6. $-3.3 \times +1.4 = ?$
7. $0.3 \times (-4) = ?$
8. $-4.6 \times -3.3 = ?$

3–5   DIVIDING POSITIVE AND NEGATIVE NUMBERS

Let us begin by reviewing how to divide one positive number by another positive number. This basic rule of division was covered in Chapter 2.

RULE 1: (+) ÷ (+) = (+)

Description: A positive number (+) divided by a positive number (+) will always result in a positive number (+).

Example:

$$+12 \div (+3) = ? \longleftarrow \text{(same as } 12 \div 3)$$

$$\text{or} \quad \frac{+12}{+3} = ?$$

$$+12 \div (+3) = +4$$

Explanation: Because division is simply repeated subtraction, subtracting a series of positive numbers from a positive number will yield a positive answer.

Calculator sequence: ☐1☐ ☐2☐ ☐÷☐ ☐3☐ ☐=☐

Answer: 4 (+4)

Other Examples:

$$+36 \div (+2) = \,?$$
$$+36 \div (+2) = +18$$
$$\frac{+504}{(+8)} = \,?$$
$$+504 \div (+8) = +63$$
$$+22.7 \div +3.6 = \,?$$
$$+22.7 \div (+3.6) = +6.31$$

RULE 2: $(+) \div (-)$ OR $(-) \div (+) = (-)$

Description: A positive number $(+)$ divided by a negative number $(-)$ or a negative number $(-)$ divided by a positive number $(+)$ will always yield a negative result $(-)$.

Example:

Dividend is larger than divisor:

$$+16 \div (-4) = \,? \qquad \text{or} \qquad -16 \div (+4) = \,?$$
$$+16 \div (-4) = -4 \qquad \text{or} \qquad -16 \div (+4) = -4$$

Dividends (16) are larger than divisors (4).

Dividend is smaller than divisor:

$$+4 \div (-16) = -0.25 \qquad \text{or} \qquad -4 \div (+16) = -0.25$$

Even if dividends (4) are smaller than divisors (16), the answer is still negative.

Explanation: The result will always be negative because division is simply repeated subtraction, and therefore subtracting a series of negative numbers will always result in a negative answer.

Calculator sequence:

(Dividend is larger) → ☐1☐ ☐6☐ ☐÷☐ ☐4☐ ☐+/−☐ ☐=☐ or ☐1☐ ☐6☐ ☐+/−☐ ☐÷☐ ☐4☐ ☐=☐

Answer: −4   −4

(Divisor is larger) → ☐4☐ ☐÷☐ ☐1☐ ☐6☐ ☐+/−☐ ☐=☐ or ☐4☐ ☐+/−☐ ☐÷☐ ☐1☐ ☐6☐ ☐=☐

Answer: −0.25   −0.25

Other Examples:

$$+48 \div (-4) = \,?$$
$$+48 \div (-4) = -12$$
$$-4 \div +12 = \,?$$
$$-4 \div (+12) = -0.333$$
$$\frac{+677.25}{-3.66} = \,?$$
$$+677.25 \div (-3.66) = -185.04$$

A -180 volt supply voltage for an oscilloscope is applied to the voltage divider shown. What will be the voltage at points A and B?

■ *Solution:*

$$A = -180\,\text{V} \div 3 = -60\,\text{volts}$$
$$B = -180\,\text{V} \div 8 = -22.5\,\text{volts}$$

RULE 3: $(-) \div (-) = (+)$

Description: A negative number $(-)$ divided by a negative number $(-)$ will always equal a positive number $(+)$.

Example:

$$-15 \div (-5) = ?$$
$$-15 \div (-5) = +3$$

Explanation: The explanation is once again the double negative. After a series of negative numbers are repeatedly subtracted, the negative answer is reversed to a positive. In summary, the reversals of direction in the operation coupled with the two negative numbers reverse the result to a positive value.

Calculator sequence: | 1 | 5 | +/− | ÷ | 5 | +/− | = |

Answer: 3 $(+3)$

Other Examples:

$$-4 \div (-2) = ?$$
$$-4 \div (-2) = +2$$
$$\frac{-96}{-3} = ?$$
$$-96 \div (-3) = +32$$
$$-2.2 \div -11.0 = ?$$
$$-2.2 \div (-11.0) = +0.2$$

■ INTEGRATED MATH APPLICATION: METEOROLOGY

To convert a Celsius temperature to Fahrenheit, use the following formula:

$$°\text{F} = \frac{9 \times °\text{C}}{5} + 32$$

Convert the following:

a. $-26\,°\text{C}$ **b.** $-8\,°\text{C}$

■ *Solution:*

a. $°\text{F} = \dfrac{9 \times °\text{C}}{5} + 32 = \dfrac{9 \times -26}{5} + 32 = -14.8\,°\text{F}$

b. $°\text{F} = \dfrac{9 \times °\text{C}}{5} + 32 = \dfrac{9 \times -8}{5} + 32 = +17.6\,°\text{F}$

Now that you have completed this section, you should be able to:

■ *Objective 4. Describe the three basic rules of positive and negative numbers as they relate to division.*

Use the following questions to test your understanding of Section 3–5.

Divide the following values.

1. $\dfrac{+16.7}{+2.3} = ?$

2. $18 \div (+6) = ?$

3. $-6 \div (+2) = ?$

4. $+18 \div (-4) = ?$

5. $2 \div (-8) = ?$

6. $-8 \div +5 = ?$

7. $\dfrac{-15}{-5} = ?$

8. $0.664 \div (-0.2) = ?$

3–6 ORDER OF OPERATIONS

For easy access, Figure 3–3 summarizes all the rules for signed number addition, subtraction, multiplication, and division. Remember that it is more important to understand these rules rather than just to memorize them. Returning to the detailed in-chapter explanation for each of the end-of-chapter problems is not a waste of time but an investment, because you will be developing logic and reasoning skills that can be applied to all aspects of science and technology.

When a string of more than two positive and negative numbers has to be combined by some arithmetic operation, the order in which the mathematical operations should be combined is as follows:

ORDER OF OPERATIONS	MEMORY AID
First: **P**arentheses	**P**lease
Second: **E**xponents	**E**xcuse
Third: **M**ultiplication	**M**y
Fourth: **D**ivision	**D**ear
Fifth: **A**ddition	**A**unt
Sixth: **S**ubtraction	**S**ally

Order of Operations

The sequence in which a string of two or more signed numbers should be combined.

This **order of operations** can more easily be remembered with the phrase "Please Excuse My Dear Aunt Sally."

Addition	Subtraction
$(+) + (+) = (+)$	$(+) - (+)$ or $(-) - (-) =$ (sign of larger number) and difference
$(-) + (+)$ or $(+) + (-) =$ (sign of larger number) and difference	$(+) - (-) = (+)$
$(-) + (-) = (-)$	$(-) - (+) = (-)$
Multiplication	**Division**
$(+) \times (+) = (+)$	$(+) \div (+) = (+)$
$(+) \times (-)$ or $(-) \times (+) = (-)$	$(+) \div (-)$ or $(-) \div (+) = (-)$
$(-) \times (-) = (+)$	$(-) \div (-) = (+)$

FIGURE 3–3 **Positive and Negative Number Rules of Addition, Subtraction, Multiplication, and Division.**

■ EXAMPLE

Calculate the following:

    a. $3 + 6 \cdot 4 = ?$

    b. $4(4 + 3) = ?$

    c. $8^2 - 5 = ?$

■ *Solution:*

    a. Problem has multiplication and addition operations. According to the order of operations (PEMDAS), multiplication should be performed first and then addition second.

$$3 + 6 \cdot 4 = ?$$
$$3 + 24 = \qquad\qquad (6 \cdot 4 = 24)$$
$$27 \qquad\qquad (3 + 24 = 27)$$

    b. $4(4 + 3) = ?$ Operations in parentheses should be performed first, multiplication second.

$$4(4 + 3) = ?$$
$$4(7) =$$
$$4 \cdot 7 =$$
$$28$$

    c. $8^2 - 5 = ?$ Exponents first, subtraction second.

$$8^2 - 5 = ?$$
$$64 - 5 =$$
$$59$$

■ INTEGRATED MATH APPLICATION: CONSTRUCTION

The length of the rope in feet on the drum shown is determined with the formula:

$$L = x \times 1.05 \times y \times (x + z)$$

Calculate the length of the rope following the order of operations if:

$$x = 8 \text{ inches}$$
$$y = 36 \text{ inches}$$
$$z = 30 \text{ inches}$$

■ *Solution:*

$$L = x \times y \times 1.05 \times (x + z)$$
$$= 8 \times 36 \times 1.05 \times (8 + 30)$$

(For $\frac{1}{2}$ inch rope use 1.05 in formula)

PEMDAS: **1.** Parentheses
                2. Multiplication

$$L = 8 \times 36 \times 1.05 \times (38) = 11{,}491 \text{ feet}$$

CALCULATOR KEYS

Name: Order of evaluation key

Function: Most calculators define the order in which functions in expressions are entered and evaluated. This order is as follows:

ORDER NUMBER	FUNCTION
1	Functions that precede the argument, such as $\sqrt{\ }($, **sin(**, or **log(**
2	Functions that are entered after the argument, such as ^{2}, $^{-1}$, $\circ$, ^{r}, and conversions
3	Powers and roots, such as **2^5** or **5$^x\sqrt{32}$**
4	Permutations (**nPr**) and combinations (**nCr**)
5	Multiplication, implied multiplication, and division
6	Addition and subtraction
7	Relational functions, such as $>$ or $\leq$
8	Logic operator **and**
9	Logic operators **or** and **xor**

Example: Calculate $3.76 \div (-7.9 + \sqrt{5}) + 2\log 45$.

Press keys: $\boxed{3}$ $\boxed{.}$ $\boxed{76}$ $\boxed{\div}$ $\boxed{(}$ $\boxed{(-)}$ $\boxed{7}$ $\boxed{.}$ $\boxed{9}$ $\boxed{+}$ $\boxed{\text{2nd}}$ $[\sqrt{\ }]$

$\boxed{5}$ $\boxed{)}$ $\boxed{+}$ $\boxed{2}$ $\boxed{\times}$ $\boxed{\text{LOG}}$ $\boxed{(}$ $\boxed{45}$ $\boxed{)}$

$3.76/(-7.9 + \sqrt{5})$
$+2\log(45)$
$\qquad 2.642575252$

$\boxed{\text{ENTER}}$

■ INTEGRATED MATH APPLICATION: AUTOMOTIVE

The length of the fan belt can be calculated with the following formula:

$$L = ((2 \times D) + 1.57) \times (d_1 + d_2) + \frac{(d_1 + d_2)}{4 \times D}$$

$D = 12$ inches
$d_1 = 18$ inches
$d_2 = 2$ inches

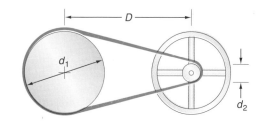

D = Distance between centers
d_1 = Diameter of load
d_2 = Diameter of driver

■ *Solution:*

$$L = ((2 \times D) + 1.57) \times (d_1 + d_2) + \frac{(d_1 + d_2)}{4 \times D}$$

$$= ((2 \times 12 \text{ in.}) + 1.57) \times (18 \text{ in.} + 2 \text{ in.}) + \frac{(18 \text{ in.} + 2 \text{ in.})}{4 \times 12 \text{ in.}}$$

PEMDAS: **1.** Parentheses
2. Multiplication
3. Division
4. Addition

$$L = (24 + 1.57) \times 20 + \frac{20}{4 \times 12} \qquad \text{(P)}$$

$$L = 25.57 \times 20 + \frac{20}{4 \times 12} \qquad \text{(M)}$$

$$L = 45.57 + \frac{20}{48} \qquad \text{(D)}$$

$$L = 45.57 + 0.42 \qquad \text{(A)}$$

$$L = 46 \text{ inches}$$

SELF-TEST EVALUATION POINT FOR SECTION 3–6

Now that you have completed this section, you should be able to:

■ **Objective 5.** *Explain how to combine strings of positive and negative numbers.*

Use the following questions to test your understanding of Section 3–6.

1. $+6 + (+3) + (-7) + (-5) = ?$
2. $+9 - (+2) - (-13) - (-4) = ?$

3. $-6 \times (-4) \times (-5) = ?$
4. $+8 \div (+2) \div (-5) = ?$

or $\dfrac{\left(\dfrac{+8}{+2}\right)}{(-5)} = ?$

5. $-4 \div (-2) \times (+8) = ?$ or $\dfrac{-4}{-2} \times (+8) = ?$
6. $-9 + (+5) - (-7) = ?$

REVIEW QUESTIONS

Multiple Choice Questions

1. A _____ number is any value greater than zero, whereas a _____ number is any value less than zero.
   a. Positive, +   c. +, negative
   b. −, +   d. −, −

2. The fraction 0.3417 is a positive number.
   a. True   b. False

3. Negative and positive numbers are used to designate some value relative to a _____ reference point.
   a. 0   c. 10 (decimal)
   b. 0.5   d. π

4. Which of the following calculator sequences would be used to input the following expression? $+5 - (-15) =$
   a. [5] [+/−] [−] [1] [5] [=]
   b. [+/−] [5] [−] [−] [1] [5] [=]
   c. [5] [−] [−] [1] [5] [=]
   d. [5] [−] [1] [5] [+/−] [=]

5. $(+) + (+) = ?$
   a. $(+)$
   b. $(-)$
   c. (Sign of larger number) and difference
   d. None of the above

6. $(+) + (-) = ?$
   a. $(+)$
   b. $(-)$
   c. (Sign of larger number) and difference
   d. None of the above

7. $(-) + (-) = ?$
   a. $(+)$
   b. $(-)$
   c. (Sign of larger number) and difference
   d. None of the above

8. $(+) - (-)$ is equivalent to?
   a. $(-) + (-)$ c. $(+) + (+)$
   b. $(-) + (+)$ d. $(+) + (-)$

9. $(-) - (-)$ is equivalent to?
   a. $(-) + (+)$ c. $(-) + (-)$
   b. $(+) + (+)$ d. $(+) + (-)$

10. $(-) - (+)$ is equivalent to?
    a. $(-) + (+)$ c. $(+) + (+)$
    b. $(-) + (-)$ d. $(+) + (-)$

Communication Skill Questions

11. Define the following: (Introduction)
    a. Positive number   c. Positive fraction
    b. Negative number   d. Negative fraction

12. What method is used to show whether the number is positive or negative? (3–1)

13. Describe the three rules of adding positive and negative numbers. (3–2)

14. How can you achieve subtraction through addition? (3–3)

15. Describe the three rules of subtracting positive and negative numbers. (3–3)

16. What are the three rules of multiplying positive and negative numbers? (3–4)

17. List and describe the three rules of positive and negative number division. (3–5)

18. Arbitrarily choose values, and then describe the following: (3–2)

    a. $(+) + (-) = ?$

    b. $(-) + (-) = ?$

    c. $(+) + (+) = ?$

19. Arbitrarily choose values, and then describe the following: (3–3)

    a. $(+) - (+) = ?$

    b. $(-) - (+) = ?$

    c. $(+) - (-) = ?$

20. Arbitrarily choose values, and then describe the following: (3–4 and 3–5)

    a. $(+) \times (-) = ?$

    b. $(-) \times (-) = ?$

    c. $(+) \div (-) = ?$

    d. $(-) \div (-) = ?$

Practice Problems

21. Express the following statements mathematically.

    a. Positive eight minus negative five

    b. Negative zero point six times thirteen

    c. Negative twenty-two point three divided by negative seventeen

    d. Positive four divided by negative nine

22. Add the following values.

    a. $+8 + (+9) = ?$

    b. $+9 + (-6) = ?$

    c. $-6 + (+9) = ?$

    d. $-8 + (-9) = ?$

23. Subtract the following values.

    a. $+6 - (+8) = ?$

    b. $+9 - (-6) = ?$

    c. $-75 - (+62) = ?$

    d. $-39 - (-112) = ?$

24. Multiply the following values.

    a. $+18 \times (+4) = ?$

    b. $+8 \times (-2) = ?$

    c. $-20 \times (+3) = ?$

    d. $-16 \times (-10) = ?$

25. Divide the following values.

    a. $+19 \div (+3) = ?$ **c.** $-80 \div (+5) = ?$

    b. $+36 \div (-3) = ?$ **d.** $-44 \div (-2) = ?$

Calculate the answers to the following problems.

26. $+0.75 + (-0.25) - (+0.25) = ?$

27. $\left(\dfrac{+15}{+5}\right) \times (-3.5) = +15 \div (+5)$

    $\times (-3.5) = ?$

28. $\dfrac{+2 \times (-4) \times (-7)}{+3} = +2 \times (-4) \times (-7)$

    $\div (+3) = ?$

29. $-6 \div (-4) \div (-3) \times (-15)$

    $= \dfrac{\left(\dfrac{-6}{-4}\right)}{-3} \times (-15) = ?$

30. $[+15 - (-4) - (+5) - (-10)] \div 4 = ?$

    $\dfrac{+15 - (-4) - (+5) - (-10)}{4} = ?$

Web Site Questions

Go to the Web site http://www.prenhall.com/cook, select the textbook *Introductory Mathematics*, select this chapter, and then follow the instructions when answering the multiple-choice practice problems.

Exponents and the Metric System

Not a Morning Person

René Descartes was born in 1596 in Brittany, France, and at a very early age began to display an astonishing analytical genius. At the age of 8 he was sent to Jesuit College in La Flèche, then one of the most celebrated institutions in Europe, where he studied several subjects, none of them interesting him as much as mathematics. His genius impressed all his professors, who gave him permission, because of his delicate health, to study in bed until midday—a practice that he retained throughout his life. In 1612 he left La Flèche to go to the University of Poitiers, where he graduated in law in 1616, a profession he never practiced.

Wanting to see the world, he joined the army, which made use of his mathematical ability in military engineering. In 1619 he met Dutch philosopher Isaac Beeckman, who convinced Descartes to "turn his mind back to science and worthier occupations." After leaving the army, Descartes traveled to Neuberg, Germany, where he shut himself in a well-heated room for the winter. On the eve of St. Martin's Day (November 10, 1619), Descartes described a vivid dream that determined all his future endeavors. The dream clarified his purpose and showed him that physics and all sciences could be reduced to geometry, and therefore all were interconnected "as by a chain." From this point on, his genius was displayed in his invention of coordinate geometry and in contributions to theoretical physics, methodology, and metaphysics.

In his time he was heralded as an analytical genius, a reputation that has lasted to this day. His fame was so renowned that he was asked frequently to tutor royalty. When in Paris in 1649, Descartes was asked to tutor Queen Christina of Sweden. He did not want the opportunity to, in his words, "live in the land of bears among rock and ice, and lose my independence," but he was persuaded to do so by the French ambassador a month later. The queen chose five o'clock in the morning for her lessons, and on his travels one bitter morning Descartes caught a severe chill and died within two weeks.

Descartes's problem-solving ability was incredible. You may find his four-step method to solving a problem helpful:

1. Never accept anything as true unless it is clear and distinct enough to exclude all doubt from your mind.
2. Divide the problem into as many parts as necessary to reach a solution.
3. Start with the simplest things and proceed step by step toward the complex.
4. Review the solution so completely and generally that you are sure nothing was omitted.

Outline and Objectives

VIGNETTE: NOT A MORNING PERSON

INTRODUCTION

Objective 1: Define the term *exponent*.

4–1 RAISING A BASE NUMBER TO A HIGHER POWER

Objective 2: Describe what is meant by raising a number to a higher power.

4–1–1 Square of a Number

Objective 3: Explain how to find the square and root of a number.

4–1–2 Root of a Number

4–1–3 Powers and Roots in Combination

Objective 4: Calculate the result of problems with powers and roots in combination.

4–2 POWERS OF 10

Objective 5: Explain the powers-of-10 method and how to convert to powers of 10.

4–2–1 Converting to Powers of 10

4–2–2 Powers of 10 in Combination

Objective 6: Calculate the result of problems with powers of 10 in combination.

4–2–3 Scientific and Engineering Notation

Objective 7: Describe the two following floating-point number systems:
      a. Scientific notation
      b. Engineering notation

4–3 WEIGHTS AND MEASURES

4–3–1 Customary System of Weights and Measures

Objective 8: List and describe how to convert between customary units of weight and measure.

4–3–2 Metric System of Weights and Measures

Objective 9: List the metric prefixes and describe the value of each.
    1. Metric Units of Length

Objective 10: Describe the following metric units for:
      a. Length
      b. Area
      c. Weight
      d. Volume
      e. Temperature

Objective 11: Convert U.S. customary units to metric units, and metric units to U.S. customary units.
    2. Metric Units of Area
    3. Metric Units of Weight
    4. Metric Units of Volume
    5. Metric Unit of Temperature
    6. Electrical Units, Prefixes, and Conversions

Objective 12: List and describe many of the more frequently used electrical units and prefixes, and describe how to interconvert prefixes.

4–3–3 Using the Calculator to Convert Weights and Measures

Objective 13: Show how the calculator can be used to convert between the customary and metric systems, and convert prefixes.

4–3–4 Measuring Devices

Objective 14: Describe how the micrometer and vernier calipers are used to measure the size of an object.
    1. Measuring Physical Properties
    2. Measuring Electrical Properties

Introduction

Exponent

A symbol written above and to the right of a mathematical expression to indicate the operation of raising to a power.

Like many terms used in mathematics, the word **exponent** sounds as if it will be complicated; however, once you find out that exponents are simply a sort of "math shorthand," the topic loses its intimidation. Many of the values used in science and technology contain numbers that have exponents. An exponent is a number in a smaller type size that appears to the right of and slightly higher than another number, for example:

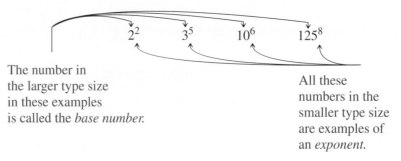

$2^2 \qquad 3^5 \qquad 10^6 \qquad 125^8$

The number in the larger type size in these examples is called the *base number.*

All these numbers in the smaller type size are examples of an *exponent.*

However, what does a term like 2^3 mean? It means that the base 2 is to be used as a factor 3 times; therefore,

$$2^3 = 2 \times 2 \times 2 = 8$$

Similarly, 3^5 means that the base 3 is to be used as a factor 5 times; therefore,

$$3^5 = 3 \times 3 \times 3 \times 3 \times 3 = 243$$

As you can see, it is much easier to write 3^5 than to write $3 \times 3 \times 3 \times 3 \times 3$, and since both mean the same thing and equal the same amount (which is 243), exponents are a quick and easy math shorthand.

In this chapter we will examine the details relating to squares, roots, exponents, scientific and engineering notation, and prefixes.

4–1   RAISING A BASE NUMBER TO A HIGHER POWER

A base number's exponent indicates how many times the base number must be multiplied by itself. This is called *raising a number to a higher power.* For example, $5 \times 5 \times 5 \times 5$ can be written as 5^4, which indicates that the base number 5 is raised to the fourth power by the exponent 4. As another example, 3×3 can be written as 3^2, which indicates that the base number 3 is raised to the second power by the exponent 2. The second power is also called the **square** of the base number, and therefore 3^2 can be called "three squared" or "three to the second power."

Square

The product of a number multiplied by itself.

■ EXAMPLE

What does 10^3 mean?

■ *Solution:*

10^3 indicates that the base number 10 is raised to the third power by the exponent 3. Described another way, it means that the base 10 is to be used as a factor 3 times; therefore,

$$10^3 = 10 \times 10 \times 10 = 1000$$

Thus, instead of writing $10 \times 10 \times 10$, or 1000, you could simply write 10^3 (pronounced "ten to the three," "ten to the third power," or "ten cubed").

Because the square of a base number is used very frequently, let us begin by discussing raising a base number to the second power.

4–1–1   *Square of a Number*

The square of a base number means that the base number is to be multiplied by itself. For example, 4^2, which is pronounced "four squared" or "four to the second power," means 4×4. The squares of the first ten base numbers are used very frequently in numerical problems and are as follows:

$$0^2 = 0 \times 0 = 0$$
$$1^2 = 1 \times 1 = 1$$
$$2^2 = 2 \times 2 = 4$$
$$3^2 = 3 \times 3 = 9$$
$$4^2 = 4 \times 4 = 16$$
$$5^2 = 5 \times 5 = 25$$
$$6^2 = 6 \times 6 = 36$$
$$7^2 = 7 \times 7 = 49$$
$$8^2 = 8 \times 8 = 64$$
$$9^2 = 9 \times 9 = 81$$
$$10^2 = 10 \times 10 = 100$$

Many people get confused with the first three of the squares. Be careful to remember that $0^2 = 0$ because nothing $\times$ nothing equals nothing; $1^2 = 1$ because one times one is still one; and 2^2 means 2×2, not $2 + 2$, even though the answer works out both ways to be 4.

■ **EXAMPLE**

Give the square of the following.

    a. 12^2 **b.** 7^2

■ *Solution:*

    a. $12^2 = 12 \times 12 = 144$
    b. $7^2 = 7 \times 7 = 49$

Finding the square of a base number is done so frequently in science and technology that most calculators have a special key just for that purpose. It is called the *square key* and operates as follows.

CALCULATOR KEYS

Name: Square key

Function: Calculates the square of the number in the display.

Example: $16^2 = ?$

Press keys: $\boxed{1}\ \boxed{6}\ \boxed{x^2}$

Display shows: 256

■ **INTEGRATED MATH APPLICATION: HEALTH CARE**

Body mass index (BMI) is used to determine whether a person is considered overweight. This value is determined with the formula:

$$\text{BMI} = \frac{703 \times \text{Weight (pounds)}}{\text{Height (inches)}^2}$$

20–25: Ideal
25–30: Overweight
Greater than 30: Obese

A person with a BMI of greater than 25 is considered overweight, and with a BMI greater than 30 is considered obese.

What is the BMI of a 5-foot 6-inch patient who weighs 180 pounds?

■ *Solution:*

$$\text{BMI} = \frac{703 \times 180\,\text{lb}}{5\,\text{ft}\,6\,\text{in.}^2} = \frac{703 \times 180}{66^2} = \frac{126{,}540}{4356} = 29$$

The patient is overweight, bordering on obese.

Most calculators also have a key for raising a base number of any value to any power. This is called the *y to the x power key,* and it operates as follows.

CALCULATOR KEYS

Name: *y* to the *x* power key

Function: Raises the displayed value *y* to the *x*th power.

Example: $5^4 = ?$

Press keys: $\boxed{5}\ \boxed{y^x}\ \boxed{4}\ \boxed{=}$

Display shows: 625

4–1–2 *Root of a Number*

What would we do if we had the result 64 and we didn't know the value of the number that was multiplied by itself to get 64? In other words, we wanted to *find the source or root number that was squared to give us the result.* This process, called **square root,** uses a special symbol called a *radical sign* and a smaller number called the *index.*

Square Root

A factor of a number that when squared gives the number.

Smaller number is called the *index.* It indicates how many times a number was multiplied by itself to get the value shown inside the radical sign. A 2 index is called the *square root.*

$$\sqrt[2]{64} = ?$$

Radical sign ($\sqrt{}$) indicates that the value inside is the result of a multiplication of a number two or more times.

In this example the index is 2, which indicates that the number we are trying to find was multiplied by itself two times. Of course, in this example we already know that the answer is 8 because $8 \times 8 = 64$.

If squaring a number takes us forward ($8^2 = 8 \times 8 = 64$), taking the square root of a number must take us backward ($\sqrt[2]{64} = 8$). Almost nobody extracts the squares from square root problems by hand because calculators make this process more efficient. Most calculators have a special key just for determining the square root of a number. Called the *square root key,* it operates as follows.

CALCULATOR KEYS

Name: Square root key

Function: Calculates the square root of the number in the display.

Example: $\sqrt{81} = ?$

Press keys: $\boxed{8}\ \boxed{1}\ \boxed{\sqrt{x}}$

Display shows: 9

Some of the more frequently used *square root* values are as follows:

$$\sqrt{1} = 1$$
$$\sqrt{2} = 1.414$$
$$\sqrt{3} = 1.732$$
$$\sqrt{4} = 2$$
$$\sqrt{9} = 3$$
$$\sqrt{16} = 4$$
$$\sqrt{25} = 5$$
$$\sqrt{36} = 6$$
$$\sqrt{49} = 7$$
$$\sqrt{64} = 8$$
$$\sqrt{81} = 9$$
$$\sqrt{100} = 10$$

Check each of these answers using the $\boxed{\sqrt{x}}$ key on your calculator.

The maximum speed of a car involved in an accident on a wet tarmac road can be determined with the following formula:

$$\text{Maximum speed} = \sqrt{30 \times \text{Road's coefficient of friction} \times \text{Length of skid}}$$

Calculate the maximum speed of a car based on the following data:

Road's coefficient of friction = 38%   (Use 0.38.)

Length of skid = 129 feet

■ *Solution:*

$$\text{Speed (max)} = \sqrt{30 \times 0.38 \times 129} = 38 \text{ miles per hour}$$

In some cases the index may be 3, meaning that some number was multiplied by itself three times to get the value within the radical sign; for example,

$$\sqrt[3]{125} = ?$$

Cube Root

A factor of a number that when multiplied by itself three times gives the number.

Using the index 3 instead of the index 2 is called taking the **cube root.** The answer to this problem is 5 because 5 multiplied by itself three times equals 125.

$$\sqrt[3]{125} = 5$$

Most calculators also have a key for calculating the root of any number with any index. It operates as follows.

CALCULATOR KEYS

Name: *x*th root of *y* key

Function: Calculate the *x*th root of the displayed value *y*.

Example: $\sqrt[3]{512} = ?$

Press keys: ⟨5⟩ ⟨1⟩ ⟨2⟩ ⟨$\sqrt[x]{y}$⟩ ⟨3⟩ ⟨=⟩

Display shows: 8

In most cases a radical sign will not have an index, in which case you can assume that the index is 2, or square root; for example,

$$\sqrt{16} = 4$$

Square root of sixteen = four

4–1–3   *Powers and Roots in Combination*

In some instances you may have several operations to perform before the result can be determined.

■ EXAMPLE

What is the sum of $4^2 + 3^3$ (four squared plus three cubed)?

$$4^2 + 3^3 = ?$$

■ *Solution:*

The first step in a problem like this is to raise the base numbers to the power indicated by the exponent and then add the two values.

$$4^2 + 3^3 =$$
$$16 + 27 = 43$$

Steps:

$4^2 = 4 \times 4 = 16$

$3^3 = 3 \times 3 \times 3 = 27$

$16 + 27 = 43$

Calculator sequence: $\boxed{4}\;\boxed{x^2}\;\boxed{+}\;\boxed{3}\;\boxed{y^x}\;\boxed{3}\;\boxed{=}$

Following are some other examples in which a mathematical operation needs to be performed on numbers with exponents to determine an answer or result.

■ EXAMPLE A

$$(2 + 9)^2 = 11^2 = 121$$

Exponent 2 indicates that the sum within the parentheses is to be squared.

Parentheses are used to group terms $2 + 9$.

Steps:

$2 + 9 = 11$

$11^2 = 11 \times 11 = 121$

Calculator sequence: $\boxed{2}\;\boxed{+}\;\boxed{9}\;\boxed{=}\;\boxed{x^2}$

■ EXAMPLE B

$$(8^2)^2 = 64^2 = 4096$$

Steps:

$8^2 = 8 \times 8 = 64$

$64^2 = 64 \times 64 = 4096$

Calculator sequence: $\boxed{8}\;\boxed{x^2}\;\boxed{x^2}$

■ EXAMPLE C

$$(5^2 \times 3^3)^2 = (25 \times 27)^2$$
$$= 675^2 = 455{,}625$$

Steps:

$5^2 = 5 \times 5 = 25$

$3^3 = 3 \times 3 \times 3 = 27$

$25 \times 27 = 675, 675 \times 675 = 455{,}625$

Calculator sequence:

$\boxed{5}\;\boxed{x^2}\;\boxed{\times}\;\boxed{3}\;\boxed{y^x}\;\boxed{3}\;\boxed{=}\;\boxed{x^2}$

■ **EXAMPLE D**

$$(5^2 - 15)^3 = (25 - 15)^3$$
$$= 10^3 = 1000$$

Steps:

$5^2 = 5 \times 5 = 25$

$25 - 15 = 10$

$10^3 = 10 \times 10 \times 10 = 1000$

Calculator sequence:

| 5 | x^2 | $-$ | 1 | 5 | $=$ | y^x | 3 |

■ **EXAMPLE E**

$$\left(\frac{4^2}{2}\right)^2 + 7^2 \quad \text{or} \quad (4^2 \div 2)^2 + 7^2$$

$$= (16 \div 2)^2 + 7^2$$
$$= 8^2 + 7^2$$
$$= 64 + 49 = 113$$

Steps:

$4^2 = 4 \times 4 = 16$

$16 \div 2 = 8$

$8^2 = 8 \times 8 = 64$

$7^2 = 7 \times 7 = 49$

$64 + 49 = 113$

Calculator sequence: | 4 | x^2 | $\div$ | 2 | $=$ |

| x^2 | $+$ | 7 | x^2 | $=$ |

Now let us examine a few examples involving powers and roots.

■ **EXAMPLE A**

No index means square root.

$$\sqrt{3 + 7 + 6} = \sqrt{16} = 4$$

Radical sign extends over terms $3 + 7 + 6$.

Steps:

$3 + 7 + 6 = 16$

$\sqrt{16} = 4$

Calculator sequence: | 3 | $+$ | 7 | $+$ | 6 | $=$ | $\sqrt{x}$ |

■ **EXAMPLE B**

$$\sqrt{14^2} = \sqrt{196} = 14$$

(Because square is the opposite of square root, the square root of fourteen squared $= 14$.)

Steps:

$14^2 = 14 \times 14 = 196$

$\sqrt{196} = 14$

Calculator sequence: | 1 | 4 | x^2 | $\sqrt{x}$ |

■ **EXAMPLE C**

$$\sqrt[2]{\frac{8}{9}} \quad \text{or} \quad \sqrt[2]{8 \div 9}$$

$$= \sqrt{0.888} = 0.943$$

Steps:

$8 \div 9 = 0.889$

Calculator sequence: | 8 | $\div$ | 9 | $=$ | $\sqrt{x}$ |

■ EXAMPLE D

$\sqrt[3]{5^2 \times 2^2} = \sqrt[3]{25 \times 4}$ *Steps:*

$= \sqrt[3]{100} = 4.6416$

$5^2 = 25$

$2^2 = 4$

$25 \times 4 = 100$

$\sqrt[3]{100} = 4.6416$

Calculator sequence: $\boxed{5}\;\boxed{x^2}\;\boxed{\times}\;\boxed{2}\;\boxed{x^2}\;\boxed{=}\;\boxed{\sqrt[x]{y}}\;\boxed{3}\;\boxed{=}$

■ EXAMPLE E

$\sqrt{6^2 - 5} + \sqrt{121}$ *Steps:*

$= \sqrt{36 - 5} + \sqrt{121}$

$= \sqrt{31} + \sqrt{121}$

$= 5.568 + 11 = 16.568$

$6^2 = 6 \times 6 = 36$

$36 - 5 = 31$

$\sqrt{31} = 5.568$

$\sqrt{121} = 11$

$5.568 + 11 = 16.568$

Calculator sequence: $\boxed{6}\;\boxed{x^2}\;\boxed{-}\;\boxed{5}\;\boxed{=}$

$\boxed{\sqrt{x}}\;\boxed{+}\;\boxed{1}\;\boxed{2}\;\boxed{1}\;\boxed{\sqrt{x}}\;\boxed{=}$

■ INTEGRATED MATH APPLICATION: ELECTRONICS

The total impedance, or opposition, of the series circuit shown is calculated with the formula:

$$Z \text{ (Impedance)} = \sqrt{R^2 + X^2}$$

Determine Z if resistance is 33 ohms ($R = 33\ \Omega$) and reactance is 257.8 ohms ($X = 257.8\ \Omega$).

■ *Solution:*

$$Z = \sqrt{33\ \Omega^2 + 257.8\ \Omega^2}$$
$$= \sqrt{1089 + 66460.8}$$
$$= \sqrt{67549.8} = 260 \text{ ohms}$$

SELF-TEST EVALUATION POINT FOR SECTION 4–1

Now that you have completed this section, you should be able to:

■ *Objective 1.* *Define the term* exponent.

■ *Objective 2.* *Describe what is meant by raising a number to a higher power.*

■ *Objective 3.* *Explain how to find the square and root of a number.*

■ *Objective 4.* *Calculate the result of problems with powers and roots in combination.*

Use the following questions to test your understanding of Section 4–1.

1. Raise the following base numbers to the power indicated by the exponent, and give the answer.
   a. $16^4 = ?$ **d.** $15^6 = ?$
   b. $32^3 = ?$ **e.** $2^3 = ?$
   c. $112^2 = ?$ **f.** $3^{12} = ?$

2. Give the following roots.
   a. $\sqrt[2]{144} = ?$ **c.** $\sqrt{20} = ?$
   b. $\sqrt[3]{3375} = ?$ **d.** $\sqrt[3]{9} = ?$

3. Calculate the following.
   a. $(9^2 + 14^2)^2 - \sqrt[3]{3 \times 7} = ?$
   b. $\sqrt{3^2 \div 2^2} + \dfrac{(151 - 9^2)}{3.5^2} = ?$

Many of the sciences deal with numbers that contain a large number of zeros, for example:

$$14,000$$

$$0.000032$$

By using exponents, we can eliminate the large number of zeros to obtain a shorthand version of the same number. This method is called *powers of 10*.

As an example, let us remove all the zeros from the number 14,000 until we are left with simply 14; however, this number (14) is not equal to the original number (14,000), and therefore simply removing the zeros is not an accurate shorthand. Another number needs to be written with the 14 to indicate what has been taken away—this number is called a *multiplier*. The multiplier must indicate what you have to multiply 14 by to get back to 14,000; therefore,

$$14,000 = 14 \times 1000$$

As you know from our discussion on exponents, we can replace the 1000 with 10^3 because $1000 = 10 \times 10 \times 10$. Therefore, the powers-of-10 notation for 14,000 is 4×10^3. To convert 14×10^3 back to its original form (14,000), simply remember that each time a number is multiplied by 10, the decimal place is moved one position to the right. In this example 14 is multiplied by 10 three times (10^3, or $10 \times 10 \times 10$), and therefore the decimal point will have to be moved three positions to the right.

$$14 \times 10^3 = 14 \times 10 \times 10 \times 10 = 14.000. = 14,000$$

As another example, what is the powers-of-10 notation for the number 0.000032? If we once again remove all the zeros to obtain the number 32, we will again have to include a multiplier with 32 to indicate what 32 has to be multiplied by to return it to its original form. In this case 32 will have to be multiplied by 1/1,000,000 (one millionth) to return it to 0.000032.

$$32 \times \frac{1}{1,000,000} = 0.000032$$

This can be verified because when you divide any number by 10, you move the decimal point one position to the left. Therefore, to divide any number by 1,000,000, you simply move the decimal point six positions to the left.

$$32 \times \frac{1}{1,000,000} = \frac{32}{1,000,000} = \frac{32}{10 \times 10 \times 10 \times 10 \times 10 \times 10} = 0.0000032.$$

Once again an exponential expression can be used in place of the 1/1,000,000 multiplier, namely,

$$\frac{1}{1,000,000} = \frac{1}{10^6} = 0.000001 = 10^{-6}$$

Whenever you divide a number into 1, you get the *reciprocal* of that number. In this example, when you divide 1,000,000 into 1, you get 0.000001, which is equal to power-of-10 notation with a negative exponent of 10^{-6}. The multiplier 10^{-6} indicates that the decimal point must be moved back (to the left) by six places, and therefore

$$32 \times \frac{1}{1,000,000} = 32 \times 0.000001 = 32 \times 10^{-6} = 0.0000032.$$

Now that you know exactly what a multiplier is, you have only to remember these simple rules:

1. A *negative exponent* tells you how many places *to the left* to move the decimal point.
2. A *positive exponent* tells you how many places *to the right* to move the decimal point.

Remember one important point: a negative exponent does not indicate a negative number; it simply indicates a fraction. For example, 4×10^{-3} meter means that 1 meter has been broken up into 1000 parts and we have 4 of those pieces, or 4/1000.

4–2–1 *Converting to Powers of 10*

To reinforce our understanding, let us try converting a few numbers to powers-of-10 notation.

■ **EXAMPLE A**

$$230,000,000 = 23\,\overset{7\ 6\ 5\ 4\ 3\ 2\ 1}{\overleftarrow{0,0\,0\,0,0\,0\,0}} = 23 \times 10^7$$

■ **EXAMPLE B**

$$760,000 = 76\,\overset{4\ 3\ 2\ 1}{\overleftarrow{0,0\,0\,0}} = 76 \times 10^4$$

■ **EXAMPLE C**

$$0.0019 = 0_{\circ}\overset{1\ 2\ 3\ 4}{\overrightarrow{0\,0\,1\,9.}} = 19 \times 10^{-4}$$

■ **EXAMPLE D**

$$0.00085 = 0_{\circ}\overset{1\ 2\ 3\ 4\ 5}{\overrightarrow{0\,0\,0\,8\,5.}} = 85 \times 10^{-5}$$

Most calculators have a key specifically for entering powers of 10. It is called the *exponent key* and operates as follows.

CALCULATOR KEYS

Name: Exponent entry key $\boxed{\text{EXP}}$ or $\boxed{\text{EE}}$

Function: Prepares calculator to accept next digits entered as a power-of-10 exponent. The sign of the exponent can be changed by using the change-sign key $\boxed{+/-}$.

Example: $76\,\boxed{\text{EXP}}\,^4$

Press keys: $\boxed{7}\ \boxed{6}\ \boxed{\text{EXP}}\ \boxed{4}$

Display shows: $\boxed{76.04}$

Example: 85×10^{-5}

Press keys: $\boxed{8}\ \boxed{5}\ \boxed{\text{EXP}}\ \boxed{5}\ \boxed{+/-}$

Display shows: $\boxed{85.\ -05}$

INTEGRATED MATH APPLICATION: ELECTRICITY

The capacity or capacitance of the capacitor shown can be calculated with the following formula:

$$C = \frac{(8.85 \times 10^{-12}) \times k \times A}{d}$$

where

C = Capacitance in farads (F)
k = Dielectric constant
A = Plate area in square meters (m^2)
d = Distance between the plates in meters

Calculate the capacitance of the capacitor shown.

■ *Solution:*

$$C = \frac{(8.85 \times 10^{-12}) \times k \times A}{d}$$

$$= \frac{(8.85 \times 10^{-12}) \times 6 \times 0.3}{0.0003}$$

$$= 5.31 \times 10^{-8} \quad \text{or} \quad 0.0531\ \mu\text{F (microfarads)}$$

Connecting lead

Dielectric (ceramic has a constant of 6)

CAPACITOR

Connecting lead

Metal plates (have an area of 0.3 square meter. The distance between the plates is 0.0003 m)

4–2–2   *Powers of 10 in Combination*

A power-of-10 multiplier raises the base number to the power of 10 indicated by the exponent. To get you used to working with powers of 10, let us do a few examples and include the calculator sequences.

■ **EXAMPLE A**

$$(2 \times 10^2) + (3 \times 10^3)$$
$$= 200 + 3000 = 3200$$

Steps:

$2 \times 10^2 = 2 \times 10 \times 10 = 200$

$3 \times 10^3 = 3 \times 10 \times 10 \times 10 = 3000$

$200 + 3000 = 3200$

Calculator sequence:

Answer: 3200 or

$\boxed{32.02}$ (32×10^2 or 3.2×10^3)

■ **EXAMPLE B**

$$(3 \times 10^6) - (2 \times 10^4)$$
$$= 3{,}000{,}000 - 20{,}000$$
$$= 2{,}980{,}000$$

Steps:

$3 \times 10^6 = 3 \times 10 \times 10 \times 10 \times 10 \times 10 \times 10$
$\quad = 3{,}000{,}000$

$2 \times 10^4 = 2 \times 10 \times 10 \times 10 \times 10$
$\quad = 20{,}000$

$3{,}000{,}000 - 20{,}000 = 2{,}980{,}000$

Calculator sequence: $\boxed{3}\ \boxed{\text{EXP}}\ \boxed{6}\ \boxed{-}\ \boxed{2}\ \boxed{\text{EXP}}\ \boxed{4}\ \boxed{=}$

Answer: 2,980,000 or

$\boxed{298.04}$ (2.98×10^6)

■ **EXAMPLE C**

$$\frac{1.6 \times 10^4}{4 \times 10^2} \quad \text{or}$$

$$(1.6 \times 10^4) \div (4 \times 10^2)$$

$$= 16,000 \div 400$$

$$= 40$$

Steps:

$$1.6 \times 10^4 = 1.6 \times 10 \times 10$$
$$\times 10 \times 10 = 16,000$$
$$4 \times 10^2 = 4 \times 10 \times 10$$
$$= 400$$
$$16,000 \div 400 = 40$$

Calculator sequence: ⎣1⎦ ⎣.⎦ ⎣6⎦ ⎣EXP⎦ ⎣4⎦ ⎣÷⎦
⎣4⎦ ⎣EXP⎦ ⎣2⎦ ⎣=⎦

Answer: 40

■ **EXAMPLE D**

$$(7.5 \times 10^{-6}) \times (1.86 \times 10^{-3})$$

$$= 0.0000075 \times 0.00186$$

$$= 0.00000001395 \text{ or}$$

$$1.395 \times 10^{-8}$$

Steps:

$$7.5 \times 10^{-6} = 0.0000007.5$$

$$1.86 \times 10^{-3} = 0.001.86$$

$$0.0000075 \times 0.00186 =$$
$$0.00000001395$$

Calculator sequence: ⎣7⎦ ⎣.⎦ ⎣5⎦ ⎣EXP⎦ ⎣6⎦
⎣+/−⎦ ⎣×⎦ ⎣1⎦ ⎣.⎦ ⎣8⎦
⎣6⎦ ⎣EXP⎦ ⎣3⎦ ⎣+/−⎦ ⎣=⎦

Answer: 1.395×10^{-8}

4–2–3 *Scientific and Engineering Notation*

As mentioned previously, powers of 10 are used in science and technology as a shorthand owing to the large number of zeros in many values. There are basically two systems or notations used, involving values that have exponents that are a power of 10. They are called **scientific notation** and **engineering notation.**

A number in *scientific notation* is expressed as a base number between 1 and 10 multiplied by a power of 10. In the following examples, the values on the left have been converted to scientific notation.

■ **EXAMPLE A**

$$32,000 = 3.2000_\circ = 3.2 \times 10^4$$

Scientific notation

Decimal point is moved to a position that results in a base number between 1 and 10. If decimal point is moved left, exponent is positive. If decimal point is moved right, exponent is negative.

Scientific Notation

A widely used floating-point system in which numbers are expressed as products consisting of a number between 1 and 10 multiplied by an appropriate power of 10.

Engineering Notation

A widely used floating-point system in which numbers are expressed as products consisting of a number that is greater than 1 multiplied by a power of 10 that is some multiple of 3.

EXAMPLE B

$$0.0019 = 0_{°}\widetilde{001}.9 = 1.9 \times 10^{-3}$$

Scientific notation

EXAMPLE C

$$114,300,000 = 1.\widetilde{14300000}_{°} = 1.143 \times 10^{8}$$

Scientific notation

EXAMPLE D

$$0.26 = 0_{°}\widetilde{2}.6 = 2.6 \times 10^{-1}$$

Scientific notation

As you can see from the preceding examples, the decimal point floats backward and forward, which explains why scientific notation is called a **floating-point number system.** Although scientific notation is used in science and technology, the engineering notation system, discussed next, is used more frequently.

In *engineering notation* a number is represented as a base number that is greater than 1 multiplied by a power of 10 that is some multiple of 3. In the following examples, the values on the left have been converted to engineering notation.

EXAMPLE A

$$32,000 = 32.\widetilde{000}_{°} = 32 \times 10^{3}$$

Decimal point is moved to a position that results in a base number that is greater than 1, and a power-of-10 exponent that is some multiple of 3.

EXAMPLE B

$$0.0019 = 0_{°}\widetilde{001}.9 = 1.9 \times 10^{-3}$$

EXAMPLE C

$$114,300,000 = 114.\widetilde{300000}_{°} = 1.143 \times 10^{6}$$

EXAMPLE D

$$0.26 = 0_{°}\widetilde{260}. = 260 \times 10^{-3}$$

■ INTEGRATED MATH APPLICATION: ELECTRICAL

Convert the following measurements to their engineering notation equivalents:

    a. 98,000 watts   **c.** 166,000,000 volts

    b. 0.18 ampere   **d.** 3300 ohms

■ *Solution:*

    a. 98,000 watts = 98 kilovolts   **c.** 166,000,000 volts = 166 megavolts

    b. 0.18 ampere = 180 milliamperes   **d.** 3300 ohms = 3.3 kilohms

---

CALCULATOR KEYS

Name: Normal, scientific, engineering modes

Function: Most calculators have different notation modes that affect the way an answer is displayed on the calculator's screen. Numeric answers can be displayed with up to 10 digits and a two-digit exponent. You can enter a number in any format.

Example: Normal notation mode is the usual way we express numbers, with digits to the left and right of the decimal, as in 12345.67.

    Sci (scientific) notation mode expresses numbers in two parts. The significant digits display with one digit to the left of the decimal. The appropriate power of 10 displays to the right of E, as in 1.234567E4.

    Eng (engineering) notation mode is similar to scientific notation; however, the number can have one, two, or three digits before the decimal, and the power-of-10 exponent is a multiple of three, as in 12.34567E3.

SELF-TEST EVALUATION POINT FOR SECTION 4–2

Now that you have completed this section, you should be able to:

■ **Objective 5.** *Explain the powers-of-10 method and how to convert to powers of 10.*

■ **Objective 6.** *Calculate the result of problems with powers of 10 in combination.*

■ **Objective 7.** *Describe the two floating-point number systems: scientific notation and engineering notation.*

Use the following questions to test your understanding of Section 4–2.

  1. Convert the following to powers of 10.

    a. 100 = ?   **d.** 1,000,000 = ?

    b. 1 = ?   **e.** $\dfrac{1}{1,000}$ = ?

    c. 10 = ?   **f.** $\dfrac{1}{1,000,000}$ = ?

  2. Convert the following to common numbers without exponents.

    a. 6.3×10^3 = ?

    b. $114,000 \times 10^{-3}$ = ?

    c. $7,114,632 \times 10^{-6}$ = ?

    d. 6624×10^6 = ?

  3. Perform the indicated operation on the following.

    a. $\sqrt{3 \times 10^6}$ = ?

    b. $(2.6 \times 10^{-6}) - (9.7 \times 10^{-9})$ = ?

    c. $\dfrac{(4.7 \times 10^3)^2}{3.6 \times 10^6}$ = ?

  4. Convert the following common numbers to engineering notation.

    a. 47,000 = ?

    b. 0.00000025 = ?

    c. 250,000,000 = ?

    d. 0.0042 = ?

There are two systems of weights and measures used in the United States. The older **customary system of weights and measures** uses units such as the mile, square foot, and ounce, whereas the newer **metric system of weights and measures** uses units such as the meter, liter, and gram. To begin let us examine the customary system in detail.

Customary System of Weights and Measures
A non-decimal system of weights and measures based on the foot and pound system of units.

Metric System of Weights and Measures
A decimal system of weights and measures based on the meter and the gram.

4–3–1   *Customary System of Weights and Measures*

Table 4–1 lists all the most frequently used customary units for length, area, volume, and weight. Many of these are probably already familiar to you. Study the conversions and test your understanding with the following example.

TABLE 4–1   U.S. Customary System

Linear Measure (Length)

12 inches	= 1 foot
3 feet	= 1 yard
$5\frac{1}{2}$ yards	= 1 rod
40 rods	= 1 furlong
8 furlongs (5280 feet)	= 1 statute mile

Therefore:
1 mile = 1760 yards = 5280 feet = 63,360 inches = 320 rods = 8 furlongs

Square Measure (Area)

144 square inches	= 1 square foot
9 square feet	= 1 square yard
$30\frac{1}{4}$ square yards	= 1 square rod
160 square rods	= 1 acre
640 acres	= 1 square mile

Dry Measure (Volume)

1728 cubic inches	= 1 cubic foot
27 cubic feet	= 1 cubic yard
2 pints	= 1 quart
8 quarts	= 1 peck
4 pecks	= 1 bushel

Liquid Measure (Volume)

60 minims	= 1 fluid dram
8 fluid drams	= 1 fluid ounce
16 fluid ounces	= 1 pint
4 gills	= 1 pint
2 pints	= 1 quart
4 quarts	= 1 gallon
$31\frac{1}{2}$ gallons	= 1 barrel
2 barrels	= 1 hogshead

Weight Measure

$27\frac{11}{32}$ grains	= 1 dram
16 drams	= 1 ounce
16 ounces	= 1 pound
100 pounds	= 1 short hundredweight
112 pounds	= 1 long hundredweight
20 short hundredweight	= 1 short ton = 2000 pounds
20 long hundredweight	= 1 long ton = 2240 pounds

Mariner's Measure

6 feet	= 1 fathom
1000 fathoms (approx.)	= 1 nautical mile
3 nautical miles	= 1 league

Surveyor's Measure

7.92 inches	= 1 link
100 links	= 1 chain

Wood Measure

16 cubic feet	= 1 cord foot
8 cord feet	= 1 cord
1 board foot (bf)	= 1 square foot of lumber, 1 inch thick

Time Measure

60 seconds	= 1 minute
60 minutes	= 1 hour
24 hours	= 1 day
7 days	= 1 week
4 weeks (28 to 31 days)	= 1 month
12 months (365–366 days)	= 1 year
100 years	= 1 century

Angular and Circular Measure

60 seconds	= 1 minute
60 minutes	= 1 degree
90 degrees	= 1 right angle
180 degrees	= 1 straight angle
360 degrees	= 1 circle

■ **EXAMPLE:**

Convert the following:

    a. How many feet are in a yard?
    b. How many square inches are in a square foot?
    c. How many pints are in a gallon?
    d. How many pounds are in a ton (short)?

■ *Solution:*

    a. 1 yard = 3 feet
    b. 1 square foot = 144 square inches
    c. 1 gallon = 8 pints
    d. 1 ton (short) = 2000 pounds

To demonstrate how much easier it is to work with the metric system rather than with the customary system, let us use a simple example.

■ **EXAMPLE:**

What is 20% of:

    a. 1 mile in feet (1 mile = 5280 feet)
    b. 1 kilometer in meters (1 kilometer = 1000 meters)

■ *Solution:*

    a. 20% of 5280 = 1056 feet
    b. 20% of 1000 = 200 meters

You more than likely needed the calculator for part (a) of the example but did not need it for part (b), proving that the metric system is easier to use, to represent, to count, and to perform mathematical operations on values; however, both systems are in use, and so a good understanding of each and a knowledge of how to convert between systems is important.

4–3–2   *Metric System of Weights and Measures*

The **metric system of weights and measures** listed in Table 4–2 was developed to make working with values easier by having only a few units (meter, liter, and gram) and prefixes that are multiples of 10 (milli = 1/1000, kilo = 1000).

    To help explain what we mean by units and prefixes, let us begin by examining how we measure length using the metric system. The standard **unit** of length in the metric system is the *meter* (abbreviated m). You have probably not seen the unit "meter" used a lot on its own. More

Unit
A determinate quantity adopted as a standard of measurement.

TABLE 4–2   **Metric System**

Linear Measure (Length)		Square Measure (Area)	
10 millimeters	= 1 centimeter	100 sq. millimeters	= 1 sq. centimeter
10 centimeters	= 1 decimeter	100 sq. centimeters	= 1 sq. decimeter
10 decimeters	= 1 meter	100 sq. decimeters	= 1 sq. meter
10 meters	= 1 decameter	100 sq. meters	= 1 sq. decameter
10 decameters	= 1 hectometer	100 sq. decameters	= 1 sq. hectometer
10 hectometers	= 1 kilometer	100 sq. hectometers	= 1 sq. kilometer
General		**General**	
1 centimeter = 10 millimeters		1 hectare (ha)	= 100 m × 100 m
1 meter = 100 cm = 1000 millimeters			= 10,000 sq. meters
1 kilometer = 1000 meters			

TABLE 4–2 *(continued)*

Cubic Measure (Volume)		General
1000 cu. millimeters	= 1 cu. centimeter	1 cubic centimeter (cm³, cc, or cu. cm) = 1 milli-
1000 cu. centimeters	= 1 cu. decimeter	liter, 1000 cu. centimeters = 1 liter
1000 cu. decimeters	= 1 cu. meter	

Weight

Liquid Measure (Volume)		Weight	
		10 milligrams	= 1 centigram
10 milliliters	= 1 centiliter	10 centigrams	= 1 decigram
10 centiliters	= 1 deciliter	10 decigrams	= 1 gram
10 deciliters	= 1 liter	10 grams	= 1 decagram
10 liters	= 1 decaliter	10 decagrams	= 1 hectogram
10 decaliters	= 1 hectoliter	10 hectograms	= 1 kilogram
10 hectoliters	= 1 kiloliter	100 kilograms	= 1 quintal
		10 quintals	= 1 ton

frequently you have heard and seen the terms *centimeter, millimeter,* and *kilometer.* All these words have two parts: a *prefix name* and *unit.* For example, with the name *centimeter, centi* is the prefix, and *meter* is the unit. Similarly, with the names *millimeter* and *kilometer, milli* and *kilo* are the prefixes, and *meter* is the unit. The next question, therefore, is: What are these prefixes? A **prefix** *is simply a power of 10 or multiplier that precedes the unit.* Figure 4–1 shows the names, symbols, and values of the most frequently used metric prefixes.

What does *centimeter* mean? If you look up the prefix *centi* in Figure 4–1, you can see that it is a prefix indicating a fraction (less than one). Its value or power of 10 is 10^{-2}, or one hundredth $\left(\frac{1}{100}\right)$. This means that one meter has been divided up into 100 pieces, and each of these pieces is a centimeter or a hundredth of a meter. Millimeters have been used in photography for many years because the width of film negatives is measured in millimeters (for example, 35 mm film). Looking up the prefix *milli* in Figure 4–1, you see that its power of 10 is 10^{-3}, or one thousandth $\left(\frac{1}{1000}\right)$. This prefix is used to indicate smaller fractions because *milli* measures length in thousandths of a meter, whereas *centi* measures length in hundredths of a meter. When the length of some object is less than 1 meter, therefore, the unit *meter* will have a prefix indicating the fractional multiplier, and this power of 10 will have a negative exponent value. Remember that these negative exponents do not indicate negative numbers; they simply indicate a fractional multiplier.

Looking at the number scale in Figure 4–1, you can see that when values are between 1 and 99, we do not need to use a prefix; for example, 1 meter, 62 meters, 84 meters, and so on. A value like 4500 meters, however, would be shortened to include the $1000(10^3)$ prefix *kilo,* as follows:

Prefix

An affix attached to the beginning of a word, base, or phrase.

Name	Pronunciation	Symbol	Value
tera	"terra"	T	10^{12} = 1,000,000,000,000
giga	"gigga"	G	10^{9} = 1,000,000,000
mega	"megga"	M	10^{6} = 1,000,000
kilo	"killoh"	k	10^{3} = 1,000
hecto	"heck-toe"	h	10^{2} = 100

No need for multiplier when a value is between 1 and 99.

centi	"sentee"	c	10^{-2} = 1/100
milli	"millie"	m	10^{-3} = 1/1,000
micro	"mike-row"	μ	10^{-6} = 1/1,000,000
nano	"nannoh"	n	10^{-9} = 1/1,000,000,000
pico	"peekoh"	p	10^{-12} = 1/1,000,000,000,000

No value

Number scale

+1,000,000 — Millions
+1,000 — Thousands — Values are greater than 100, so unit will have prefix.
+100 — Hundreds
+1 — Ones
+1/100 — Hundredths
+1/1,000 — Thousandths — Values are less than one or a fraction, so unit will have a fraction prefix.
+1/1,000,000 — Millionths
0 — Zero
− Negative numbers

FIGURE 4–1 **Metric Prefixes.**

$$4500 \text{ m} = 4.5 \times 1000 \text{ m} \quad \leftarrow (4.\overset{\frown}{5\,0\,0}.)$$

The value 4500 $\nearrow$ is shortened to 4.5 with the prefix kilo.

$$= 4.5 \times 10^3 \text{ m} \quad \leftarrow \text{(Because the 1000}$$
$$= 4.5 \text{ km} \qquad \text{or } 10^3 \text{ multiplier is } \textit{kilo}, \text{ we can substitute}$$
$$\textit{kilo}, \text{ or k, for } 10^3.)$$

Similarly, a length of 4,600,000 meters is shortened to

$$4{,}600{,}000 \text{ m} = 4.6 \times 1{,}000{,}000 \text{ m} \quad \leftarrow (4.\overset{\frown}{6\,0\,0\,0\,0\,0\,0}_\circ)$$
$$= 4.6 \times 10^6 \text{ m} \qquad \leftarrow (10^6 = \text{ mega, or M})$$
$$= 4.6 \text{ Mm}$$

Now that you have an understanding of metric prefixes, let us discuss some of the metric units and compare them with our known U.S. customary units.

1. Metric Units of Length

As already mentioned, the standard unit of length in the metric system is the meter. Table 4–3 shows how to convert a metric unit of length (millimeters, centimeters, meters, kilometers, and so on) to a U.S. customary unit of length (inches, feet, yards, and miles).

TABLE 4–3 Converting Metric Units to U.S. Customary Units (Metric → U.S.)

WHEN YOU KNOW	MULTIPLY BY	TO FIND
Length		
millimeters (mm)	0.04	inches (in.)
centimeters (cm)	0.4	inches (in.)
meters (m)	3.3	feet (ft)
meters (m)	1.1	yards (yd)
kilometers (km)	0.6	miles (mi)
Area		
sq. centimeters (cm^2)	0.16	sq. inches ($in.^2$)
sq. meters (m^2)	1.2	sq. yards (yd^2)
sq. kilometers (km^2)	0.4	sq. miles (mi^2)
hectares (ha) (10,000 m^2)	2.5	acres
Weight		
grams (g)	0.035	ounces (oz)
kilograms (kg)	2.2	pounds (lb)
tonnes (1000 kg) (t)	1.1	short tons
Volume		
milliliters (mL) ⎫	0.03	fluid ounces (fl oz)
liters (L) ⎪ Liquid volume	2.1	pints (pt)
liters (L) ⎬	1.06	quarts (qt)
liters (L) ⎭	0.26	gallons (gal)
cubic centimeters (cm^3) ⎫	16.387	cubic inches ($in.^3$)
cubic meters (m^3) ⎬ Dry volume	35	cubic feet (ft^3)
cubic meters (m^3) ⎭	1.3	cubic yards (yd^3)
Temperature		
Celsius (°C)	$\left(\dfrac{9}{5} \times °C\right) + 32$	Fahrenheit (°F)

■ INTEGRATED MATH APPLICATION: CARPENTRY

Use Table 4–3 to find how many feet are in 1 meter of lumber.

$$1 \text{ m} = ? \text{ feet}$$

■ *Solution:*

Referring to Table 4–3, you can see that "When You Know" meters, "Multiply by" 3.3 "To Find" feet (ft). Therefore,

$$1 \text{ m} \times 3.3 = 3.3 \text{ ft}$$

Other frequently used conversions are

$$1 \text{ cm} = 0.4 \text{ in.}$$
$$1 \text{ km} = 0.6 \text{ mi}$$

■ INTEGRATED MATH APPLICATION: CARPENTRY

The construction industry measures the volume of lumber in the unit *board feet*. As shown, *one board foot* (1 bf) is a piece of lumber that measures one foot by one foot by one inch. To find the amount of board feet in a load of lumber, therefore, you can use the following formula:

$$N = l \times w \times h$$

where
N = number of board feet
l = length in feet
w = width in feet
h = height in inches

1 board foot (1 bf)
1 inch
1 foot
1 foot

How many board feet are in an 8-foot 2-by-4, that measures:

$$1\frac{1}{2} \text{ inches by } 3\frac{1}{2} \text{ inches by 8 feet}$$

■ *Solution:*

Length = 8 feet	(Should be in feet, so okay)
Width = $3\frac{1}{2}$ inches	(Should be in feet, so must be converted)
Height = $1\frac{1}{2}$ inches	(Should be in inches, so okay)

$$3\frac{1}{2} \text{ inches of 12 inches}$$

$$\frac{3\frac{1}{2}}{12} = 3\frac{1}{2} \div 12$$

$$= \frac{7}{2} \div \frac{12}{1}$$

$$= \frac{7}{2} \times \frac{1}{12} = \frac{7}{24}$$

$$N = l \times w \times h$$

$$= 8 \times \frac{7}{24} \times 1\frac{1}{2} = 3\frac{1}{2} \text{ board feet}$$

TABLE 4–4   Converting U.S. Customary Units to Metric Units (U.S. → Metric)

WHEN YOU KNOW	MULTIPLY BY	TO FIND
Length		
inches (in.)	2.54	centimeters (cm)
feet (ft)	30	centimeters (cm)
yards (yd)	0.9	meters (m)
miles (mi)	1.6	kilometers (km)
Area		
sq. inches (in.^{2})	6.5	sq. centimeters (cm^2)
sq. feet (ft^2)	0.09	sq. meters (m^2)
sq. yards (yd^2)	0.8	sq. meters (m^2)
sq. miles (mi^2)	2.6	sq. kilometers (km^2)
acres	0.4	hectares (ha)
Weight		
ounces (oz)	28	grams (g)
pounds (lb)	0.45	kilograms (kg)
short tons (2000 lb)	0.9	tonnes (t)
Volume		
teaspoons (tsp)	5	milliliters (mL)
tablespoons (tbsp)	15	milliliters (mL)
fluid ounces (fl oz)	30	milliliters (mL)
cups (c) ⎫ Liquid volume	0.24	liters (L)
pints (pt)	0.47	liters (L)
quarts (qt)	0.95	liters (L)
gallons (gal) ⎭	3.8	liters (L)
cubic feet (ft^3) ⎫ Dry volume	0.03	cubic meters (m^3)
cubic yards (yd^3) ⎭	0.76	cubic meters (m^3)
Temperature		
Fahrenheit (°F)	$\frac{5}{9} \times (°F - 32)$	Celsius (°C)

Table 4–4 reverses the process by converting U.S. customary units to metric units. For example,

$$1 \text{ in.} = 2.54 \text{ cm}$$
$$1 \text{ ft} = 30 \text{ cm}$$
$$1 \text{ mi} = 1.6 \text{ km}$$

2. Metric Units of Area

The metric unit of area is the *square meter* (abbreviated m^2). For example, an area measuring 4 meters by 3 meters contains 12 square meters (12 m^2), as shown.

4 m by 3 m = ?

4 m × 3 m = 12 square meters or 12 m^2

How many square yards are in 12 square meters of Kentucky long grass sod?

■   *Solution:*

If we wanted to find out how many square yards (yd^2) are in 12 square meters, we would use Table 4–3 (metric → U.S.).

$$12\,m^2 = ?\,yd^2$$
$$12\,m^2 \times 1.2 = 14.4\,yd^2$$

3.   Metric Units of Weight

In the metric system, objects are weighed in *grams*. Referring to Table 4–3, you can see that a gram is not very heavy, only 0.035 of an ounce (oz). This is why we usually measure weight in kilograms (kilo = 10^3; therefore, 1 kilogram = 1000 grams). Using Table 4–3, you can see that

$$1\ kilogram\ (kg) = 2.2\ pounds\ (lb)$$

■   INTEGRATED MATH APPLICATION

A container measuring 1 foot long by 1 foot wide by 1 foot tall is filled with water and weighs 62.4 pounds. What is its metric weight in kilograms?

■   *Solution:*

$$62.4\,lb \times 0.45 = 28.1\,kg$$

4.   Metric Units of Volume

Volume
The amount of space occupied by a three-dimensional object as measured in cubic units.

Area is a space measured in two dimensions, whereas **volume** is a space measured in three dimensions, as follows:

Width × height × depth =
$10\,cm \times 10\,cm \times 10\,cm = 1000\,cm^3$ or 1000 cubic centimeters

The cube shown contains 1000 smaller 1 centimeter × 1 centimeter × 1 centimeter cubes. In the metric system a volume of 1000 cubic centimeters (1000 cm^3) is equal to 1 *liter* (L), which is the metric unit of *liquid volume*. Referring to Table 4–3, you can see that 1 liter is approximately equal to 1 quart (qt). This means that the next time you are at the gas station and get 2 gallons of gasoline or 8 quarts (4 quarts in 1 gallon), you have actually filled your tank with approximately 8 liters. *Dry volume* is also measured in the metric unit *meters cubed* (m^3).

■   INTEGRATED MATH APPLICATION: AUTOMOTIVE

An engine's cylinder displacement is measured in either cubic inches or liters. Determine the cylinder displacement of a 600 cubic inch engine in liters (round to the nearest liter).

■ *Solution:*

$$600 \text{ cu. in.} \times 16.387 = 9832.2 \text{ cm}^3$$

Since $1 \text{ cm}^3 = 1 \text{ mL}$:

$$9832.2 \text{ cm}^3 = 9832.2 \text{ mL}$$
$$9832.2 \text{ mL} \div 1000 = 9.8 \text{ L}$$
$$(\text{rounded}) = 10 \text{ liter engine}$$

5. Metric Unit of Temperature

The metric unit of temperature is **degrees Celsius** (symbolized by °C). Most people are more familiar with the U.S. customary temperature unit, which is **degrees Fahrenheit** (symbolized by °F). These two temperature scales are compared in Figure 4–2.

As you can see by comparing the scales, the metric Celsius scale has two easy points to remember: 0 °C is the freezing or melting point of water, and 100 °C is the boiling point of water. These same points on the Fahrenheit scale are not as easy to remember (the melting or freezing point of water is 32 °F, and the boiling point of water is 212 °F).

Table 4–3 shows how to convert a °C temperature to °F, and Table 4–4 shows how to convert a °F reading to °C.

■ **INTEGRATED MATH APPLICATION: METEOROLOGY**

If a thermometer reads 29 °C, what would this temperature be in degrees Fahrenheit?

■ *Solution:*

From Table 4–3, you can see that to convert a metric Celsius temperature to a U.S. customary temperature, you multiply the Celsius reading by $\frac{9}{5}$ and then add 32.

$$°F = \left(\frac{9}{5} \times °C\right) + 32$$
$$= (1.8 \times 29) + 32$$
$$= 52.2 + 32 = 84.2 \, °F$$

Steps:

$9 \div 5 = 1.82$

$1.8 \times 29 = 52.2$

$52.2 + 32 = 84.2 \, °F$

■ **INTEGRATED MATH APPLICATION: METEOROLOGY**

Convert 63 °F to Celsius.

Degrees Celsius

The international thermometer scale on which the interval between the melting point of ice (0 °C) and the boiling point of water (+100 °C) is divided into 99.9 degrees.

Degrees Fahrenheit

A thermometer scale on which the boiling point of water is at 212 degrees, and the freezing point of ice is at 32 degrees.

FIGURE 4–2 **Comparison of Fahrenheit and Celsius Temperature Scales.**

Solution:

Table 4–4 indicates that to convert °F to °C, you must subtract 32 from the Fahrenheit reading and then multiply the remainder by $\frac{5}{9}$.

$$°C = \frac{5}{9} \times (°F - 32)$$
$$= 0.55 \times (63 - 32)$$
$$= 0.55 \times 31 = 17.05\,°C$$

Steps:

$5 \div 9 = 0.55$

$63 - 32 = 31\,°C$

6. Electrical Units, Prefixes, and Conversions

Table 4–5 lists many of the more commonly used metric electrical quantities with their units and symbols. For example, electrical current (symbolized by I) is measured in amperes or amps (symbolized by A). As another example, electrical resistance (R) is measured in ohms (symbolized by the Greek capital letter omega, Ω).

If electrical values are much larger than one or are a fraction of one, the unit of electrical quantity is preceded by a prefix. Most electrical and electronic applications use engineering notation, so these prefixes are power-of-10 exponents that are some multiple of three (10^3, 10^6, 10^9, 10^{12}, and so on). As an example, Table 4–6 shows how these prefixes precede ampere, which is the electrical current unit. To help our understanding of these prefixes, let us do a few conversion problems.

TABLE 4–5 Metric Electrical Quantities, Their Units, and Their Unit Symbols

QUANTITY	SYMBOL	UNIT	SYMBOL
Charge	Q	coulomb	C
Current	I	ampere	A
Voltage	V	volt	V
Resistance	R	ohm	Ω
Capacitance	C	farad	F
Inductance	L	henry	H
Energy, work	W	joule	J
Power	P	watt	W
Time	t	second	s
Frequency	f	hertz	Hz

TABLE 4–6 Electrical Current Prefixes

NAME	ABBREVIATION	VALUE
Teraampere	TA	$10^{12} = 1{,}000{,}000{,}000{,}000$
Gigaampere	GA	$10^9 = 1{,}000{,}000{,}000$
Megaampere	MA	$10^6 = 1{,}000{,}000$
Kiloampere	kA	$10^3 = 1000$
Ampere	A	$10^0 = 1$
Milliampere	mA	$10^{-3} = \dfrac{1}{1000}$
Microampere	μA	$10^{-6} = \dfrac{1}{1{,}000{,}000}$
Nanoampere	nA	$10^{-9} = \dfrac{1}{1{,}000{,}000{,}000}$
Picoampere	pA	$10^{-12} = \dfrac{1}{1{,}000{,}000{,}000{,}000}$

Convert the following.

    a. 0.003 A = _____ mA (milliamperes)
    b. 0.07 mA = _____ μA (microamperes)
    c. 7333 mA = _____ A (amperes)
    d. 1275 μA = _____ mA (milliamperes)

■   *Solution:*

    a. 0.003 A = _____ mA. In this example, 0.003 A has to be converted so that it is represented in milliamperes (10^{-3}, or $\frac{1}{1000}$ of an ampere). The basic algebraic rule to remember is that the expressions on either side of the equals sign must be the same.

Left		Right	
Base	Exponent	Base	Exponent
0.003×10^0		= _____	$\times 10^{-3}$

    The exponent on the right in this problem will be decreased 1000 times (10^0 to 10^{-3}), so for the statement to balance, the number on the right will have to be increased 1000 times; that is, the decimal point will have to be moved to the right three places (0.003 to 3). Therefore,

$$0.003 \times 10^0 = 3 \times 10^{-3}$$

$$\text{or} \quad 0.003\,\text{A} = 3 \times 10^{-3}\,\text{A} \quad \text{or} \quad 3\,\text{mA}$$

    b. 0.07 mA = _____ μA. In this example the exponent is going from milliamperes to microamperes (10^{-3} to 10^{-6}), or 1000 times smaller, so the number must be made 1000 times greater.

$$0.070. \quad \text{or} \quad 70.0$$

Therefore, 0.07 mA = 70 μA.

    c. 7333 mA = _____ A. The exponent is going from milliamperes to amperes, increasing 1000 times, so the number must decrease 1000 times.

$$7333. \quad \text{or} \quad 7.333$$

Therefore, 7333 mA = 7.333 A.

    d. 1275 μA = _____ mA. The exponent is changing from microamperes to milliamperes, an increase of 1000 times, so the number must decrease by the same factor.

$$1.2750 \quad \text{or} \quad 1.275$$

Therefore, 1275 μA = 1.275 mA.

4–3–3   *Using the Calculator to Convert Weights and Measures*

Some calculators have the ability to convert between the customary and metric systems, and to convert prefixes within a system, as shown in the following examples.

CALCULATOR KEYS

Name: Convert key ▶

Function: Some calculators have a function that converts an expression from one unit to another.

$$expression\_unit \;\blacktriangleright\; \_unit2 \Rightarrow expression\_unit2$$

Examples: How many feet are in 3 meters?

Press keys: 3_m ▶ _ft ENTER

Answer: 9.842_ft

How many kilometers are in 4 light-years?

Press keys: 4_ltyr ▶ _km

Answer: 3.78421E13_km

How many ohms are in 4 kiloohms?

Press keys: 4_kΩ ▶ _Ω

Answer: 4000_Ω

   To convert a temperature value, you must use tmpCnv() instead of the ▶ operator.

Example:

Press keys: tmpCnv(100_°C to _°F)

Answer: 212_°F

4–3–4   *Measuring Devices*

Many instruments are available to measure properties. In this section we discuss the micrometer, vernier calipers, and multimeter.

1.   Measuring Physical Properties

   Several instruments are available for measuring the length of an object. The most basic devices are the ruler and tape measure, shown in Figure 4–3. The scale may be in customary units (inches and feet) and/or metric units (centimeters and meters). Smaller customary scales are available ($\frac{1}{32}$ and $\frac{1}{64}$ inch) when precision measurements have to be made.

   The steel ruler is ideal for measuring lengths to an accuracy of $\frac{1}{32}$ or $\frac{1}{64}$ inch; however, for greater accuracy, the micrometer, shown in Figure 4–4, and the vernier calipers, shown in Figure 4–5, can be used to measure lengths to an accuracy of $\frac{1}{1000}$ inch. Follow the step-by-step procedure, and associated example, to see how these instruments are used.

MEASURING LENGTH WITH A MICROMETER	EXAMPLE (Figure 4.4)
Step 1: Read the largest number on the sleeve, which measure $\frac{1}{10}$ (0.10) inch.	$2 \times 0.100 = 0.200$ inch
Step 2: Read the additional smaller scale graduations visible on the sleeve, which measure $\frac{25}{1000}$ (0.025) inch.	$3 \times 0.025 = 0.075$ inch
Step 3: Determine the total number of graduations on the thimble scale up to the sleeve's horizontal line, which measure $\frac{1}{1000}$ (0.001) each.	$14 \times 0.001 = 0.014$ inch
Step 4: Add all of the products to determine the measurement.	$\begin{array}{r} 0.200 \\ 0.075 \\ +\,0.014 \\ \hline 0.289 \\ \frac{289}{1000}\ inch \end{array}$

Steel Ruler

Tape Measure

FIGURE 4–3 **Basic Length Measurement Devices.**

FIGURE 4–4 **The Micrometer.**

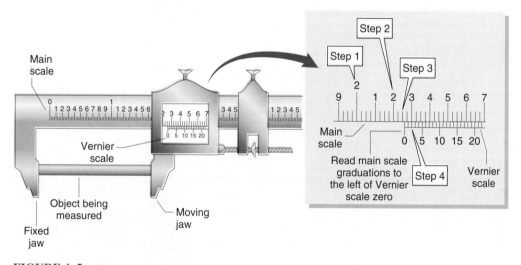

FIGURE 4–5 **The Vernier Caliper.**

■ **EXAMPLE**

Determine the length of the object being measured by the customary units micrometer shown.

■ *Solution:*

$$8 \times 0.100 = 0.800$$
$$0 \times 0.025 = 0.000$$
$$7 \times 0.001 = \underline{0.007}$$
$$0.807 \text{ inch}$$

EXAMPLE:

The metric micrometer shown is read in the same way as the customary units micrometer, except the graduations are:

Step 1: 1 mm (one millimeter)

Step 2: 0.5 mm ($\frac{5}{10}$ millimeter)

Step 3: 0.01 mm ($\frac{1}{100}$ millimeter)

Determine the length of the object being measured by the metric micrometers shown.

Step 1 : 7

Step 3 : 46

Step 2 : 1 half

■ *Solution:*

$$
\begin{array}{rl}
7 \times 1\,\text{mm} = & 7.00 \\
1 \times 0.5\,\text{mm} = & 0.50 \\
46 \times 0.01\,\text{mm} = & \underline{0.46} \\
& 7.96\ \text{millimeters}
\end{array}
$$

MEASURING LENGTH WITH VERNIER CALIPERS	EXAMPLE (Figure 4.5)
Step 1: Read inch graduations on the main scale to the left of vernier scale zero.	$2 \times 1.000 = 2.000$ inches
Step 2: Read $\frac{1}{10}$ inch graduations visible on the main scale to the left of vernier scale zero.	$2 \times 0.100 = 0.200$ inch
Step 3: Read $\frac{25}{1000}$ inch graduations on the main scale to the left of vernier scale zero.	$2 \times 0.025 = 0.050$ inch
Step 4: Determine which is the first graduation on the vernier scale to line up exactly with any line on the main scale. Vernier scale graduations measure $\frac{1}{1000}$ inch.	2 line on vernier scale aligns with $\frac{3}{10}$ line on main scale. $2 \times 0.001 = 0.002$ inch
Step 5: Add all the products to determine the measurement.	2.000 0.200 0.050 $\underline{+0.002}$ 2.252 inches

■ EXAMPLE

Determine the length of the object being measured by the vernier calipers shown.

■ *Solution:*

$$
\begin{array}{rl}
1 \times 1.000\,\text{in.} = & 1.000 \\
6 \times 0.100\,\text{in.} = & 0.600 \\
3 \times 0.025\,\text{in.} = & 0.075 \\
5 \times 0.001\,\text{in.} = & \underline{0.005} \\
& 1.680\ \text{inches}
\end{array}
$$

2. Measuring Electrical Properties

A multimeter is, as its name implies, a multipurpose meter. The step-by-step procedures listed in Figures 4–6, 4–7, and 4–8 show how this instrument can be used to measure electrical current, voltage, and resistance.

8. View the reading shown on the ammeter's display, being sure to note the unit of measurement.

7. Turn ON the power to the circuit by closing the power switch.

6. Connect the meter probes to the circuit so that the ammeter is placed in the path of circuit current flow.

5. Plug the black test probe into the COM (common) input jack.

1. Turn OFF the power to the circuit by opening the power switch.

2. Create an opening in the circuit where the ammeter probes can be inserted.

3. Select AMPS (A) on the digital multimeter's function switch.

4. Plug the red test probe into the high-current (10 amp) or low-current (300 milliamp) input jack depending on what value of current you expect. In electronic circuits current is generally a low value (300 mA jack is used), whereas in electrical circuits current is generally a high value (10 A jack is used).

NOTE: Connect the red test probe to the positive side of the circuit, and the black test probe to the negative side of the circuit. If you reverse the connections, the digital multimeter will display a negative sign.

FIGURE 4–6   Measuring Current with the Ammeter.

1. Select VOLTS (V) on the digital multimeter's function switch.

2. Plug the red test probe into the voltage (V) input jack.

3. Plug the black test probe into the common (COM) input jack.

4. Touch the probe tips across a power source or across a device.

5. View the reading shown on the voltmeter's display, being sure to note the unit of measurement.

NOTE: If test leads are reversed, a negative sign will show in the display.

FIGURE 4–7   Measuring Voltage with the Voltmeter.

1. Turn OFF power to the circuit ⚠ and remove the component to be tested.

2. Select resistance (Ω).

3. Plug the black test probe into the COM input jack. Plug the red test probe into the Ω input jack.

4. Connect the probe tips across the component or portion of the circuit for which you want to determine resistance.

5. View the reading, being sure to note the unit of measurement — ohms (Ω), kilohms (kΩ), or megohms (MΩ).

FIGURE 4–8   Measuring Resistance with the Ohmmeter.

■   **INTEGRATED MATH APPLICATION: ELECTRICAL**

Which multimeter is measuring the car battery voltage and which is monitoring the current being drawn by the music system? The product of voltage and current determines power:

$$P = V \times I$$

How much power is being consumed by the music system?

CHAPTER 4 / EXPONENTS AND THE METRIC SYSTEM

V = 13.6 V

12 V

12 Volt Car Battery

Car AM/FM Radio and CD Player (14 Ω)

1. First determine the current value being drawn by the device (follow "measuring current" procedure).

2. Then determine the voltage value being applied across device (follow "measuring voltage" procedure).

I = 970 mA

3. Multiply voltage value by current value to obtain power value ($P = V \times I$).

■ *Solution:*

A (ammeter) is monitoring current.

B (voltmeter) is measuring voltage.

$$P = V \times I$$
$$= 13.6\,V \times 970\,mA$$
$$= 13.2\,\text{watts}$$

SELF-TEST EVALUATION POINT FOR SECTION 4–3

Now that you have completed this section, you should be able to:

■ *Objective 8.* *List and describe how to convert between customary units of weight and measure.*

■ *Objective 9.* *List the metric prefixes and describe the value of each.*

■ *Objective 10.* *Describe the metric units for length, area, weight, volume, and temperature.*

■ *Objective 11.* *Convert U.S. customary units to metric units, and metric units to U.S. customary units.*

■ *Objective 12.* *List and describe many of the more frequently used electrical units and prefixes, and describe how to interconvert prefixes.*

■ *Objective 13.* *Show how the calculator can be used to convert between the customary and metric systems, and to convert prefixes.*

■ *Objective 14.* *Describe how the micrometer and vernier calipers are used to measure the size of an object.*

Use the following questions to test your understanding of Section 4–3.

1. Give the power-of-10 value for the following prefixes.
   a. kilo
   b. centi
   c. milli
   d. mega
   e. micro

2. Convert the following metric units to U.S. customary units.
   a. 15 cm to inches
   b. 23 kg to pounds
   c. 37 L to gallons
   d. 23 °C to degrees Fahrenheit

3. Convert the following U.S. customary units to metric.
   a. 55 miles per hour (mph) to kilometers per hour (km/h)
   b. 16 gal to liters
   c. 3 yd^2 to square meters
   d. 92 °F to degrees Celsius

4. List the names of the metric units for the following quantities.
   a. Length
   b. Weight
   c. Temperature
   d. Time
   e. Power
   f. Current
   g. Voltage
   h. Volume (liquid)
   i. Energy
   j. Resistance

5. Convert the following.
   a. 25,000 V = _____ kilovolts
   b. 0.014 W = _____ milliwatts
   c. 0.000016 μF = _____ nanofarads
   d. 3545 kHz = _____ megahertz

REVIEW QUESTIONS

Multiple Choice Questions

1. A base number's exponent indicates how many times the base number must be multiplied by:
   a. 2
   b. 3
   c. Itself
   d. Both (a) and (c)

2. What does 16^3 mean?
   a. 16×3
   b. $16 \times 16 \times 16$
   c. $16 \times (3 \times 3 \times 3)$
   d. $3 \times 3 \times 3 \times 3 \times 3 \times 3 \times 3 \times 3 \times 3 \times 3 \times 3 \times 3 \times 3 \times 3 \times 3 \times 3$

3. If $x^2 = 16$, then $\sqrt[2]{16} = x$. What is the value of x?
   a. 4
   b. 2
   c. 16
   d. 8

4. Calculate $\sqrt[5]{32}$.
   a. 2
   b. 1.3
   c. 4
   d. 6.4

5. Raise the following number to the power of 10 indicated by the exponent, and give the answer: 3.6×10^2.
   a. 3600
   b. 36
   c. 0.036
   d. 360

6. Convert the following number to powers of 10: 0.00029.
   a. 0.29×10^{-3}
   b. 2.9×10^{-4}
   c. 29×10^{-5}
   d. All the above

7. What is the power-of-10 value for the metric prefix *milli?*
   a. 10^{-3}
   b. 10^{-2}
   c. 10^3
   d. All the above

8. What is the metric unit for temperature?
   a. Fahrenheit
   b. Borg
   c. Celsius
   d. Rankine

9. Convert the following value of inductance to engineering notation: 0.016 henry (H).
   a. 1.6×10^2 H
   b. 16 mH
   c. 1.6 cH
   d. 16 kH

10. How many centimeters are in 8 inches?
    a. 2.03 m
    b. 3.2 cm
    c. 8 cm
    d. 20.32 cm

Communication Skill Questions

11. What is an exponent? (Introduction)
12. Define the following terms: (4–1)
    a. Square of a number
    b. Square root of a number
13. What is a power-of-10 exponent? (4–2)
14. Describe the scientific and engineering notation systems. (4–2)
15. What is the metric system? (4–3)
16. List the metric prefixes and their values. (4–3)
17. Explain how to convert the following: (4–3)
    a. Centimeters to inches
    b. Ounces to grams
    c. Yards to meters
    d. Miles to kilometers

18. What is the metric unit of: (4–3)
    a. Area
    b. Length
    c. Weight
    d. Volume
    e. Temperature
19. Give the units for the following electrical quantities: (4–3)
    a. Current
    b. Length
    c. Resistance
    d. Power
20. Arbitrarily choose values, and then describe the following conversions: (4–3)
    a. Amps to milliamps
    b. Volts to kilovolts

Practice Problems

21. Determine the square of the following values.
    a. 9^2
    b. 6^2
    c. 2^2
    d. 0^2
    e. 1^2
    f. 12^2
22. Determine the square root of the following values.
    a. $\sqrt{81}$
    b. $\sqrt{4}$
    c. $\sqrt{0}$
    d. $\sqrt{36}$
    e. $\sqrt{144}$
    f. $\sqrt{1}$
23. Raise the following base numbers to the power indicated by the exponent, and give the answer.
    a. 9^3
    b. 10^4
    c. 4^6
    d. 2.5^3
24. Determine the roots of the following values.
    a. $\sqrt[3]{343} = ?$
    b. $\sqrt{1296} = ?$
    c. $\sqrt[4]{760} = ?$
    d. $\sqrt[6]{46.656} = ?$
25. Convert the following values to powers of 10.
    a. $\dfrac{1}{100}$
    b. $1,000,000,000$
    c. $\dfrac{1}{1000}$
    d. 1000
26. Convert the following to proper fractions and decimal fractions.
    a. 10^{-4}
    b. 10^{-2}
    c. 10^{-6}
    d. 10^{-3}
27. Convert the following values to powers of 10 in both scientific and engineering notation.
    a. 475
    b. 8200
    c. 0.07
    d. 0.00045

28. Convert the following to whole-number values with a metric prefix.
    a. 0.005 A
    b. 8000 m
    c. 15,000,000 Ω
    d. 0.000016 s
29. Convert the following metric units to U.S. customary units.
    a. 100 km
    b. 29 m^2
    c. 67 kg
    d. 2 L
30. Convert the following U.S. customary units to metric.
    a. 4 in.
    b. 2 mi^2
    c. 4 oz
    d. 10 gal
31. Convert the following temperatures to °C.
    a. 32 °F
    b. 72 °F
32. Convert the following temperatures to °F.
    a. 6 °C
    b. 32 °C
33. State the metric unit of:
    a. Length
    b. Area
    c. Weight
    d. Dry volume
    e. Liquid volume
    f. Temperature
34. What is the metric unit of time?
35. Convert the following.
    a. 8000 ms = _____ microseconds
    b. 0.02 MV = _____ kilovolts
    c. 10 km = _____ meters
    d. 250 mm = _____ centimeters

Web Site Questions

Go to the Web site http://www.prenhall.com/cook, select the textbook *Introductory Mathematics,* select this chapter, and then follow the instructions when answering the multiple-choice practice problems.

Algebra, Equations, and Formulas

Back to the Future

Each part of a computer system is designed to perform a specific task. You would think that all these units were first thought of by some recent pioneer in the twentieth century, but in fact, two of these elements were first described in 1833.

Born in England in 1791, Charles Babbage became very well known for both his mathematical genius and eccentric personality. For many years Babbage occupied the Cambridge chair of mathematics, once held by Isaac Newton, and although he never delivered a single lecture, he wrote several papers on numerous subjects, ranging from politics to manufacturing techniques. He also helped develop several practical devices, including the tachometer and the railroad cowcatcher.

Babbage's ultimate pursuit, however, was that of mathematical accuracy. He delighted in spotting errors in everything from log tables (used by astronomers, mathematicians, and navigators) to poetry. In fact, he once wrote to poet Alfred Lord Tennyson, pointing out an inaccuracy in his line "Every moment dies a man—every moment one is born." Babbage explained to Tennyson that since the world population was actually increasing and not, as he indicated, remaining constant, the line should be rewritten to read "Every moment dies a man—every moment one and one-sixteenth is born."

In 1822, Babbage described in a paper and built a model of what he called "a difference engine," which could be used to calculate mathematical tables. The Royal Society of Scientists described his machine as "highly deserving of public encouragement," and a year later the government awarded Babbage £1500 for his project. Babbage originally estimated that the project should take 3 years; however, the design had its complications, and after 10 years of frustrating labor, in which the government grants increased to £17,000, Babbage was still no closer to completion. Finally, the money stopped and Babbage reluctantly decided to let his brainchild go.

In 1833, Babbage developed an idea for a much more practical machine, which he named "the analytical engine." It was to be a more general machine that could be used to solve a variety of problems, depending on instructions supplied by the operator. It would include two units, called a "mill" and a "store," both of which would be made of cogs and wheels. The store, which was equivalent to a modern-day computer memory, could hold up to 100 forty-digit numbers. The mill, which was equivalent to a modern computer's arithmetic and logic unit (ALU), could perform both arithmetic and logic operations on variables or numbers retrieved from the store, and the result could be stored in the store and then acted upon again or printed out. The program of instructions directing these operations would be fed into the analytical engine in the form of punched cards.

The analytical engine was never built. All that remains are the volumes of descriptions and drawings, and a section of the mill and printer built by Babbage's son, who also had to concede defeat. It was, unfortunately for Charles Babbage, a lifetime of frustration to have conceived the basic building blocks of the modern computer a century before the technology existed to build it.

Outline and Objectives

VIGNETTE: BACK TO THE FUTURE

INTRODUCTION

5–1   THE BASICS OF ALGEBRA

Objective 1: Define the following terms.
   a. Algebra
   b. Equation
   c. Formula
   d. Literal numbers

Objective 2: Describe the rules regarding the equality on both sides of the equals sign.

5–1–1 The Equality on Both Sides of the Equals Sign

Objective 3: Demonstrate how to remove a quantity by performing the same arithmetic operation on both sides of an equation or formula.

5–1–2 Treating Both Sides Equally

5–1–3 Directly Proportional and Inversely Proportional

5–2   TRANSPOSITION

5–2–1 Transposing Equations

Objective 4: Explain the steps involved in transposing an equation to determine the unknown.

Objective 5: Demonstrate how to develop an equation from a story problem.

Objective 6: Describe the process of factoring and how to remove parentheses within a formula or expression.

5–2–2 Transposing Formulas

Objective 7: Demonstrate the process of transposing a formula.

Objective 8: Describe how the terms *directly proportional* and *inversely proportional* are used to show the relationship between quantities in a formula.

Objective 9: Explain the proportionality among the three quantities in the Ohm's law formula and describe how to transpose the formula.

Objective 10: Describe the electric power formula and transpose the formula to develop alternative procedures.

5–3   SUBSTITUTION

Objective 11: Demonstrate how to use substitution to develop alternative formulas.

5–3–1 Circle Formula Example

5–3–2 Power Formula Example

5–4   EQUATIONS AND GRAPHING

Objective 12: Perform the following algebra skills:
   a. Create linear equations and express them in standard form
   b. Graph data and linear equations using the slope-intercept form and point-slope form
   c. Describe the properties of exponents
   d. Factor polynomials
   e. Solve quadratic equations using the quadratic formula

5–4–1 Linear Equations

5–4–2 Graphing Data

5–4–3 Graphing Linear Equations

5–4–4 Slope of a Line

5–4–5 Exponents

5–4–6 Factoring Patterns

5–4–7 Solving Quadratic Equations

5–5   RULES OF ALGEBRA—A SUMMARY

Objective 13: Summarize the terms, rules, properties and mathematical operations of algebra.

Introduction

I contemplated calling this chapter "using letters in mathematics" because the word *algebra* seems to make many people back away. If you look up *algebra* in the dictionary, it states that it is "a branch of mathematics in which letters representing numbers are combined according to the rules of mathematics." In this chapter you will discover how easy algebra is to understand and then see how we can put it to some practical use, such as rearranging formulas. As you study this chapter you will find that algebra is quite useful, and because it is used in conjunction with formulas, an understanding is essential for anyone entering a technical field.

5–1 THE BASICS OF ALGEBRA

Algebra

A generalization of arithmetic in which letters representing numbers are combined according to the rules of arithmetic.

Formula

A general fact, rule, or principle expressed usually in mathematical symbols.

As mentioned briefly in the introduction to this chapter, **algebra** by definition is a branch of mathematics in which letters representing numbers are combined according to the rules of mathematics. The purpose of using letters instead of values is to develop a general statement or **formula** that can be used for any values. For example, distance (d) equals velocity (v) multiplied by time (t), or

$$d = v \times t$$

where d = distance in miles
v = velocity or speed in miles per hour
t = time in hours

Using this formula, we can calculate how much distance was traveled if we know the speed or velocity at which we were traveling and the time for which we traveled at that speed. Therefore, if I were to travel at a speed of 20 miles per hour (20 mph) for 2 hours, how far, or how much distance, would I travel? Replacing the letters in the formula with values converts the problem from a formula to an **equation,** as shown:

Equation

A formal statement of the equality or equivalence of mathematical or logical expressions.

$$d = v \times t \qquad \leftarrow \text{(Formula)}$$
$$d = 20 \, \text{mph} \times 2 \, \text{h} \leftarrow \text{(Equation)}$$
$$d = 40 \, \text{mi} \qquad \leftarrow \text{(Answer)}$$

Literal Number

A number expressed as a letter.

Letters such as d, v, t and a, b, c are called **literal numbers** (letter numbers), and are used, as we have just seen, in general statements showing the relationship between quantities. They are also used in equations to signify an *unknown quantity*, as will be discussed in the following section.

5–1–1 *The Equality on Both Sides of the Equals Sign*

All equations or formulas can basically be divided into two sections that exist on either side of an equals sign, as shown:

Everything in the left half of the equation is equal to everything in the right half of the equation. This means that 8 times x (which is an unknown value) divided by 2 equals 16 divided by 1. Generally, it is not necessary to put 1 under the 16 in the right section because any number divided by 1 equals the same number $\left(\frac{16}{1} = 16 \div 1 = 16\right)$; however, it was included in this introduction to show that each section has both a top and a bottom. If the equation is written without the fraction bar or 1 in the denominator position, it appears as follows:

$$\frac{8 \times x}{2} = 16$$

Although the fraction bar is not visible, you should always assume that a number on its own on either side of the equals sign is above the fraction bar, as shown:

$$\frac{8 \times x}{2} = \frac{16}{}$$

Now that we understand the basics of an equation, let us see how we can manipulate it yet keep both sides equal to each other.

5–1–2   *Treating Both Sides Equally*

If you do exactly the same thing to both sides of an equation or formula, the two halves remain exactly equal, or in balance. This means that as long as you add, subtract, multiply, or divide both sides of the equation by the same number, the equality of the equation is preserved. For example, let us try adding, subtracting, multiplying, and dividing both sides of the following equation by 4 and see if both sides of the equation are still equal.

$$\boxed{2 \times 4 = 8} \; \leftarrow \text{(Original equation)}$$

1. Add 4 to both sides of the equation:

$$2 \times 4 = 8 \qquad \leftarrow \text{(Original equation)}$$
$$(2 \times 4) \;\fbox{$+\,4$}\; = 8 \;\fbox{$+\,4$}\; \leftarrow \text{(Add 4 to both sides.)}$$
$$8 + 4 = 8 + 4$$
$$12 = 12$$

Both sides of the equation remain equal.

2. Subtract 4 from both sides of the equation:

$$2 \times 4 = 8 \qquad \leftarrow \text{(Original equation)}$$
$$(2 \times 4) \;\fbox{$-\,4$}\; = 8 \;\fbox{$-\,4$}\; \leftarrow \text{(Subtract 4 from both sides.)}$$
$$8 - 4 = 8 - 4$$
$$4 = 4$$

Both sides of the equation remain equal.

3. Multiply both sides of the equation by 4:

$$2 \times 4 = 8 \qquad \leftarrow \text{(Original equation)}$$
$$(2 \times 4) \;\fbox{$\times\,4$}\; = 8 \;\fbox{$\times\,4$}\; \leftarrow \text{(Multiply both sides by 4.)}$$
$$8 \times 4 = 8 \times 4$$
$$32 = 32$$

Both sides of the equation remain equal.

4. Divide both sides of the equation by 4:

$$2 \times 4 = 8 \qquad \leftarrow \text{(Original equation)}$$

$$(2 \times 4) \oslash \widehat{4} = 8 \oslash \widehat{4} \leftarrow \text{(Divide both sides by 4.)}$$

$$\frac{2 \times 4}{4} = \frac{8}{4}$$

$$\frac{8}{4} = \frac{8}{4}$$

$$2 = 2$$

Both sides of the equation remain equal.

As you can see from the four preceding procedures, *if you add, subtract, multiply, or divide both halves of an equation by the same number, the equality of the equation is preserved.*

5–1–3   *Ratio and Proportion*

Proportional

The relation of one part to another or to the whole with respect to magnitude, quantity, or degree.

The word **proportional** is often used to describe a relationship between two quantities in a formula. For example, consider the following variation on the $d = v \times t$ formula:

$$\text{Time } (t) = \frac{\text{distance } (d)}{\text{velocity } (v)}$$

With this formula we can say that the time (t) it takes for a trip is directly proportional (symbolized by $\propto$) to the distance (d) that needs to be traveled. In symbols, the relationship appears as follows:

$$t \propto d \quad \text{(Time is directly proportional to distance.)}$$

Directly Proportional

The relation of one part to one or more other parts in which a change in one causes a similar change in the other.

The term **directly proportional** can therefore be used to describe the relationship between time and distance because if you increase the distance you have to travel it makes sense that it is going to take more time to travel that distance ($d \uparrow$ causes $t \uparrow$), and similarly, a decrease in distance will cause a corresponding decrease in time ($d \downarrow$ causes $t \downarrow$).

■   EXAMPLE

As an example, it will take 5 hours to travel 250 miles at 50 miles per hour:

$$5\,\text{h } (t) = \frac{250\,\text{mi } (d)}{50\,\text{mph } (v)}$$

Let us try doubling the distance and then halving the distance to see what effect it has on time.

■   *Solution:*

If time and distance are directly proportional to each other, they should change in proportion.

$$\text{Original example} \rightarrow \text{Time} = \frac{\text{distance}}{\text{velocity}} = \frac{250\,\text{mi}}{50\,\text{mph}} = 5\,\text{h}$$

$$t = \frac{d}{v} = \frac{\boxed{500\,\text{mi}}}{50\,\text{mph}} = \boxed{10\,\text{h}}$$

Doubling the distance from 250 mi to 500 mi should double the time it takes for the trip, from 5 h to 10 h.

Doubling the distance did double the time ($d \uparrow, t \uparrow$).

$$t = \frac{d}{v} = \frac{\boxed{125 \text{ mi}}}{50 \text{ mph}} = \boxed{2.5 \text{ h}}$$

Halving the distance from 250 mi to 125 mi should halve the time it takes for the trip, from 5 h to $2\frac{1}{2}$ h.

Halving the distance did halve the time $(d\downarrow, t\downarrow)$.

As you can see from this exercise, time is directly proportional to distance $(t \propto d)$. Therefore, *formula quantities are always directly proportional to one another when both quantities are above the fraction bars on opposite sides of the equals sign.*

On the other hand, **inversely proportional** means that the two quantities compared are opposite in effect. For example, consider again our time formula:

$$t = \frac{d}{v}$$

Observing the relationship between these quantities again, we can say that the time (t) it takes for a trip is *inversely proportional* (symbolized by $1/\propto$) to the speed or velocity (v) we travel. In symbols, the relationship appears as follows:

$$t \frac{1}{\propto} v \quad \text{(Time is inversely proportional to velocity.)}$$

The term *inversely proportional* can therefore be used to describe the relationship between time and velocity because if you were to increase your speed or velocity, you would obviously decrease the time it would take for the trip ($v\uparrow$ causes $t\downarrow$), and similarly, a decrease in velocity would cause a corresponding increase in time ($v\downarrow$ causes $t\uparrow$).

Inversely Proportional
The relation of one part to one or more other parts in which a change in one causes an opposite change in the other.

■ EXAMPLE

Using the previous example,

$$5 \text{ h} (t) = \frac{250 \text{ mi} (d)}{50 \text{ mph} (v)}$$

let us try doubling the speed we travel, or velocity, and then halving the velocity to see what effect it has on time. If time and velocity are inversely proportional to each other, a velocity change should have the opposite effect on time.

■ Solution:

$$\text{Original example} \rightarrow \text{Time} = \frac{\text{distance}}{\text{velocity}} = \frac{250 \text{ mi}}{50 \text{ mph}} = 5 \text{ h}$$

Doubling the speed from 50 mph to 100 mph should halve the time it takes to complete the trip, from 5 h to $2\frac{1}{2}$ h.

$$\rightarrow t = \frac{d}{v} = \frac{250 \text{ mi}}{\boxed{100 \text{ mph}}} = \boxed{2.5 \text{ h}}$$

Doubling the speed did halve the time $(v\downarrow, t\uparrow)$.

Halving the speed from 50 mph to 25 mph should double the time it takes to complete the trip, from 5 h to 10 h.

$$\rightarrow t = \frac{d}{v} = \frac{250 \text{ mi}}{\boxed{25 \text{ mph}}} = \boxed{10 \text{ h}}$$

Halving the speed did double the time $(v\downarrow, t\uparrow)$.

SECTION 5–1 / THE BASICS OF ALGEBRA

As you can see from this exercise, time is inversely proportional to velocity $\left(t\frac{1}{\propto}v\right)$. Therefore, *formula quantities are always inversely proportional to one another when one of the quantities is above the fraction bar and the other is below the fraction bar on opposite sides of the equals sign.*

■ **INTEGRATED MATH APPLICATION: ELECTRICAL**

The electrical resistance of a wire is directly proportional to its length, as shown in the example illustration. Wire 1 has a small length and therefore small resistance $(L_1 \downarrow, R_1 \downarrow)$. Similarly, wire 2 has a large length and therefore large resistance $(L_2 \uparrow, R_2 \uparrow)$. Referring to the equation below, you can see this proportional relationship. The length ratio has wire 1 above the fraction bar, and similarly, the resistance ratio also has wire 1 above the fraction bar—indicating the proportional relationship.

    a. Insert the values given in the illustration and determine whether both sides of the equation remain equal.

    b. If you double the length of wire 1 (36 feet) and so double its resistance (6 ohms), will the equation remain equal?

$$\frac{\text{Length of wire 1}}{\text{Length of wire 2}} = \frac{\text{Resistance of wire 1}}{\text{Resistance of wire 2}}$$

$$\boxed{\frac{L_1}{L_2} = \frac{R_1}{R_2}}$$

The ratio of wire 1's length divided by wire 2's length   is equal to   The ratio of wire 1's resistance divided by wire 2's resistance

Wire 1

Length = 18 feet
Resistance = 3 ohms

Wire 2

Length = 45 feet
Resistance = 7.5 ohms

■ *Solution:*

a. $\dfrac{L_1}{L_2} = \dfrac{R_1}{R_2}$

$\dfrac{18}{45} = \dfrac{3}{7.5}$

$0.4 = 0.4$

Yes

b. L_1 is doubled from 18 feet to 36 feet.

R_1 is doubled from 3 ohms to 6 ohms.

$\dfrac{36}{45} = \dfrac{6}{7.5}$

$0.8 = 0.8$

Yes

■ **INTEGRATED MATH APPLICATION: AUTOMOTIVE**

The pulley system shown details the power transfer from an engine to an air conditioning unit. With gears, the speed of a pulley is inversely proportional to its diameter (the larger the diameter, $d \uparrow$, the slower the speed, $s \downarrow$); therefore, the speed of pulley x is inversely proportional to its diameter $\left(s_x = \dfrac{1}{d_x}\right)$, and the speed of pulley y is inversely proportional to its diameter $\left(s_y = \dfrac{1}{d_y}\right)$.

The equation below describes this inverse relationship. The speed ratio has pulley x above the fraction bar; conversely, the diameter ratio has pulley x below the fraction bar—indicating the inversely proportional relationship.

$$\frac{\text{Speed of pulley } x}{\text{Speed of pulley } y} = \frac{\text{Diameter of pulley } y}{\text{Diameter of pulley } x}$$

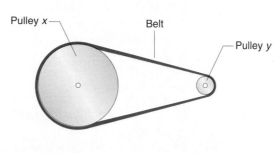

$$\frac{s_x}{s_y} = \frac{d_y}{d_x}$$

↑ The ratio of pulley x's speed divided by pulley y's speed

is equal to

↑ The ratio of pulley y's diameter divided by pulley x's diameter.

a. Insert the following values in the equation and determine whether both sides of the equation remain equal.

$$\text{Speed of pulley } x = 120 \text{ revolutions per minute (rpm)}$$
$$\text{Speed of pulley } y = 480 \text{ rpm}$$
$$\text{Diameter of pulley } y = 4 \text{ inches}$$
$$\text{Diameter of pulley } x = 16 \text{ inches}$$

b. If the diameter of pulley x was halved, its speed would be doubled. If these new values were inserted into the equation, would it remain equal?

■ *Solution:*

a.
$$\frac{s_x}{s_y} = \frac{d_y}{d_x}$$
$$\frac{120}{480} = \frac{4}{16}$$
$$0.25 = 0.25$$
Yes

b. d_x is halved, from 16 inches to 8 inches
s_x is doubled, from 120 rpm to 240 rpm.
$$\frac{240}{480} = \frac{4}{8}$$
$$0.5 = 0.5$$
Yes

SELF-TEST EVALUATION POINT FOR SECTION 5–1

Now that you have completed this section, you should be able to:

■ **Objective 1.** *Define the terms* algebra, equation, formula, *and* literal numbers.

■ **Objective 2.** *Describe the rules regarding the equality on both sides of the equals sign.*

■ **Objective 3.** *Demonstrate how to remove a quantity by performing the same arithmetic operation on both sides of an equation or formula.*

Use the following questions to test your understanding of Section 5–1.

1. Is the following equation true?
$$\frac{\frac{56}{14}}{2} = 2 \quad \text{or} \quad \frac{56 \div 14}{2} = 2$$

2. If we were to multiply the left side of the equation in Question 1 by 3 and the right side of the equation by 2, would the two sides be equal to each other?

3. Fill in the missing values.
$$\frac{144}{12} \times \boxed{} = \frac{36}{6} \times 2 \times \boxed{}$$
$$60 = 60$$

4. Determine the result and state whether the equation is equal.

$$\frac{(8-4)+26}{5} = \frac{81-75}{2}$$

5. Is the following equation balanced?

$$3.2\,\text{k}\Omega = 3200\,\Omega$$

5–2   TRANSPOSITION

Transposition

The transfer of any term of an equation from one side over to the other side with a corresponding change of the sign.

It is important to know how to *transpose*, or rearrange, equations and formulas so that you can determine the unknown quantity. This process of rearranging, called **transposition,** is discussed in this section.

5–2–1   *Transposing Equations*

As an example, let us return to the original problem introduced at the beginning of this chapter and try to determine the value of the unknown quantity *x*.

$$\frac{8 \times x}{2} = 16$$

To transpose the equation we must follow two steps:

Step 1.   Move the unknown quantity so that it is above the fraction bar on one side of the equals sign.

Step 2.   Isolate the unknown quantity so that it stands by itself on one side of the equals sign.

Looking at the first step, let us see if our unknown quantity is above the fraction bar on either side of the equals sign.

The unknown quantity
x is above the fraction bar. $$\frac{8 \times \widehat{x}}{2} = 16$$

Since step 1 is done, we can move on to step 2. Looking at the equation, you can see that *x* does not stand by itself on one side of the equals sign. To satisfy this step, we must somehow move the 8 above the fraction bar, and the 2 below the fraction bar away from the left side of the equation, so that *x* is on its own.

Let us begin by removing the 2. To remove a letter or number from one side of a formula or equation, simply remember this rule: *To move a quantity, simply apply to both sides the arithmetic opposite of that quantity.* Multiplication is the opposite of division, so to remove a "divide by 2" (÷ 2), simply multiply both sides by 2 (× 2), as follows:

$$\frac{8 \times x}{2} = 16 \qquad \leftarrow \text{(Original equation)}$$

$$\frac{8 \times x}{2} (\times\, 2) = 16 (\times\, 2) \quad \leftarrow \text{(Multiply both sides by 2.)}$$

$$\frac{8 \times x}{2} \times \frac{2}{1} = 16 \times 2 \quad \leftarrow \left(\text{Because } \tfrac{2}{2} = 1, \text{ the two 2s on the left side of the equation cancel.}\right)$$

$$(8 \times x) \times 1 = 16 \times 2 \qquad \left(\text{Because anything multiplied by 1 equals the same number, the 1 on}\right.$$

$$8 \times x = 16 \times 2 \qquad \text{the left side of the equation can be removed: } 8x \times 1 = 8x.)$$

Looking at the result so far, you can see that by multiplying both sides by 2, we effectively moved the 2 that was under the fraction bar on the left side to the right side of the equation above the fraction bar.

$$\frac{8 \times x}{2} = 16 \qquad \leftarrow \text{(Original equation)}$$

$$8 \times x = 16 \times 2 \leftarrow \text{(Result after both sides of the}$$
$$\text{equation were multiplied by 2)}$$

However, we still have not completed step 2, which was to isolate x on one side of the equals sign. To achieve this we need to remove the 8 from the left side of the equation. Once again we will do the opposite: Because the opposite of multiply is divide, to remove a "multiply by 8" we must divide both sides by 8 ($\div$ 8), as follows:

$$\frac{8 \times x}{2} = 16 \qquad \leftarrow \text{(Original equation)}$$

$$8 \times x = 16 \times 2 \qquad \leftarrow \text{(Equation after both}$$
$$\text{sides were multiplied by 2)}$$

$$\frac{8 \times x}{8} = \frac{16 \times 2}{8} \qquad \leftarrow \text{(Divide both sides by 8.)}$$

$$\frac{\cancel{8} \times x}{\cancel{8}} = \frac{16 \times 2}{8} \qquad \text{(Because } \frac{8}{8} = 1, \text{ the two 8s on the}$$
$$\text{left side of the equation cancel.)}$$

$$1 \times x = \frac{16 \times 2}{8} \qquad \text{(Anything multiplied by 1}$$
$$\text{equals the same number, so the}$$
$$\text{1 on the left side of the equation}$$
$$\text{can be removed: } 1 \times x = x.\text{)}$$

$$x = \frac{16 \times 2}{8}$$

Now that we have completed step 2, which was to isolate the unknown quantity so that it stands by itself on one side of the equation, we can calculate the value of the unknown x by performing the arithmetic operations indicated on the right side of the equation.

$$x = \frac{16 \times 2}{8} \qquad (16 \times 2 = 32)$$

$$x = \frac{32}{8} \qquad (32 \div 8 = 4)$$

$$x = 4$$

To double-check this answer, let us insert this value into the original equation to see if it works.

$$\frac{8 \times x}{2} = 16 \qquad \text{(Replace } x \text{ with 4, or substitute 4 for } x.\text{)}$$

$$\frac{8 \times 4}{2} = 16 \qquad (8 \times 4 = 32)$$

$$\frac{32}{2} = 16 \qquad (32 \div 2 = 16)$$

$$16 = 16 \qquad \text{(Answer checks out because } 16 = 16.\text{)}$$

1. Example with an Unknown Above the Fraction Bar

■ **EXAMPLE**

Determine the value of the unknown a in the following equation.

$$(2 \times a) + 5 = 23$$

■ *Solution:*

Step 1: Is the unknown quantity above the fraction bar? Yes.

Step 2: Is the unknown quantity isolated on one side of the equals sign? No.

Use the following steps to isolate the unknown, a.

	Steps:
$(2 \times a) + 5 = 23$	To remove the "+ 5," do the opposite: subtract 5 from both sides.
$(2 \times a) + \cancel{5} - \cancel{5} = 23 - 5$	$5 - 5 = 0$; any number plus 0 equals the same number: $(2 \times a) + 0 = 2 \times a$
$2 \times a = 23 - 5$	To remove the "2 $\times$," do the opposite: divide both sides by 2.
$\dfrac{\cancel{2} \times a}{\cancel{2}} = \dfrac{23 - 5}{2}$	$2 \div 2 = 1$; any number multiplied by 1 equals the same number: $1 \times a = a$
$1 \times a = \dfrac{23 - 5}{2}$	
$a = \dfrac{23 - 5}{2}$	Perform the arithmetic operations indicated to determine the value of a. $23 - 5 = 18 \qquad 18 \div 2 = 9$
$a = \dfrac{18}{2}$	
$a = 9$	

Double-check your answer by inserting it into the original equation.

$$(2 \times a) + 5 = 23 \qquad \text{(Replace } a \text{ with 9.)}$$
$$(2 \times 9) + 5 = 23$$
$$18 + 5 = 23$$
$$23 = 23 \qquad \text{(Answer checks out because } 23 = 23.\text{)}$$

2. Example with an Unknown Below the Fraction Bar

■ **EXAMPLE**

Determine the value of the unknown x in the following equation:

$$\frac{72}{x} = 12$$

■ *Solution:*

Step 1. Is the unknown quantity above the fraction bar? No.

Use the following steps to move x above the fraction bar.

	Steps:
$\dfrac{72}{x} = 12$	To move the x do the opposite: the opposite of "divide by x" is multiply by x.

$$\frac{72}{\cancel{x}} \left(\times \cancel{x} \right) = 12 \left(\times x \right)$$

Any number divided by itself $= 1$,
so $x \div x = 1$ and therefore the two x's
on the left side of the equation cancel.

$$72 \times 1 = 12 \times x$$
$$72 = 12 \times x$$

$$72 \times 1 = 72$$

Now that the unknown quantity x is above the fraction bar, we can proceed to step 2.

Step 2. Isolate the unknown quantity on one side of the equals sign.

$$72 = 12 \times x$$

To remove the "$12 \times$," simply do the opposite: divide both sides by 12.

$$\frac{72}{12} = \frac{\cancel{12} \times x}{\cancel{12}}$$

$12 \div 12 = 1, 1 \times x = x$; therefore, the 12s cancel.

$$\frac{72}{12} = 1 \times x$$

$$\frac{72}{12} = x$$

$$72 \div 12 = 6, x = 6$$

$$6 = x \quad \text{or} \quad x = 6$$

To double-check your answer, replace x with 6 in the original equation.

$$\frac{72}{x} = 12 \quad \text{(Original equation)}$$

$$\frac{72}{6} = 12$$

$$12 = 12$$

---

3. Example with an Unknown on Both Sides of the Equals Sign

In this section we will see how to deal with equations containing an unknown on both sides of the equals sign. In all equations like this, the same letter or literal number will have the same value no matter where it appears.

■ **EXAMPLE**

Determine the value of the unknown y in the following equation:

$$(6 \times y) + 7 = y + 27$$

■ *Solution:*

Step 1: Is the unknown above the fraction bar? Yes.

Step 2: Is the unknown isolated on one side of the equals sign? No.

To fulfill step 2 we need to isolate the unknown on one side of the equals sign. Since y exists on both sides, we must remove one. Studying the left side of the equation you can see that it has six y's $(6 \times y)$, whereas the right side of the equation has only one $y(y)$. The first step could therefore be to subtract $y(-y)$ from both sides.

$$(6 \times y) + 7 = y + 27 \qquad \leftarrow \text{(Original equation)}$$

Steps:

$$(6 \times y) + 7 \left(-y \right) = y + 27 \left(-y \right)$$

Subtract y from both sides:

$$(5 \times y) + 7 = 0 + 27$$

$(6 \times y) - y = 5 \times y$,

$$(5 \times y) + 7 = 27$$

$y - y = 0, 0 + 27 = 27$

$$(5 \times y) + 7 \left(-7 \right) = 27 \left(-7 \right)$$

To remove $+ 7$, subtract 7 from both sides.
$7 - 7 = 0, \quad 27 - 7 = 20$

$$5 \times y = 20$$

The opposite of multiply is divide. To remove "5 ×," divide both sides by 5.

$$\frac{5 \times y}{5} = \frac{20}{5}$$

$$5 \div 5 = 1$$

$$1 \times y = \frac{20}{5}$$

$$1 \times y = y$$

$$y = \frac{20}{5}$$

$20 \div 5 = 4$; therefore, $y = 4$.

$$y = 4$$

To double-check your answer, replace y with 4 in the original equation.

$$(6 \times y) + 7 = y + 27 \leftarrow \text{(Original equation)}$$
$$(6 \times 4) + 7 = 4 + 27 \leftarrow \text{(Replace } y \text{ with 4.)}$$
$$24 + 7 = 4 + 27$$
$$31 = 31$$

---

4. Story Problems

Story problems are probably the best practice because they are connected to our everyday life and help us to see more clearly how we can apply our understanding of mathematics. I have found the following steps helpful with story problems:

a. Determine what is unknown and assign it a literal number such as x.

b. Determine how many other elements are involved and how they relate to the unknown.

c. Construct an equation using the unknown and the other elements with their associated arithmetic operations.

d. Calculate the unknown using transposition if necessary.

■ **INTEGRATED MATH APPLICATION: PERSONAL FINANCE**

If you multiply the number of dollars in your wallet by 6 and then subtract $23, you will still end up with one fourth of $100. How many dollars do you have in your wallet?

■ *Solution:*

To start with step a, we must determine the unknown. The last sentence, "How many dollars do you have in your wallet?" states quite clearly that this is our unknown quantity. We will let x represent the number of dollars in our wallet.

Step b says to determine how many other elements are involved and how they relate. If we read again from the beginning, the example says:

"If you multiply the number of dollars in your wallet by 6 . . .

This part of the story indicates that our unknown x should be multiplied by 6; $x \times 6$.

. . . and then subtract $23, . . .

This part states that after our unknown has been multiplied by 6, it should have $23 subtracted; $(x \times 6) - 23$.

...you will still end up with...	This statement precedes a result and therefore an equals sign should follow our equation at this point; $(x \times 6) - 23 = $.
...one-fourth of \$100."	This is the result we end up with: $\frac{1}{4}$ of 100 or $\frac{1}{4} \times 100$.

Now, to complete step c, we must construct an equation using the unknown and the other elements with their associated arithmetic operations.

$$(x \times 6) - 23 = \frac{1}{4} \times 100$$

Now that we have an equation, we can move on to step d. Since x is not isolated on one side of the equals sign, we will have to transpose or rearrange the equation.

Step 1. Is the unknown above the fraction bar? Yes.

Step 2. Is the unknown isolated on one side of the equals sign? No.

$(x \times 6) - 23 = \frac{1}{4} \times 100$	*Steps:* Rather than multiply both sides by 4, let us reduce the fraction $\frac{1}{4}$ of 100. 100 is above the fraction bar; therefore, it can be written
$(x \times 6) - 23 = \frac{1 \times 100}{4}$	$1 \times 100 = 100$
$(x \times 6) - 23 = \frac{100}{4}$	$100 \div 4 = 25$; the fraction is now eliminated.
$(x \times 6) - 23 = 25$	
$(x \times 6) - 23 \boxed{+ 23} = 25 \boxed{+ 23}$	Add 23 to both sides to remove -23. $-23 + 23 = 0$, $(x \times 6) - 0 = x \times 6$, $25 + 23 = 48$
$x \times 6 = 48$	
$\dfrac{x \times 6}{\boxed{6}} = \dfrac{48}{\boxed{6}}$	Divide both sides by 6 to remove "multiply by 6."
$x = \dfrac{48}{6}$	$6 \div 6 = 1, 1 \times x = x$
$x = 8$	

You therefore started off with \$8 in your pocket. To double-check this answer, replace x with 8 in the original equation.

$$(x \times 6) - 23 = \frac{1}{4} \times 100 \quad \left(\frac{1}{4} \text{ of } 100 = 25 \right)$$
$$(8 \times 6) - 23 = 25 \quad (8 \times 6 = 48)$$
$$48 - 23 = 25 \quad (48 - 23 = 25)$$
$$25 = 25$$

Store owner A has four times as many nails as store owner B, who has 12,000. Combined, A and B have 60,000 nails. How many nails does store owner A have?

■   *Solution:*

A is the unknown, and as stated in the question, A is four times larger than B.

$$A = 4 \times B$$

If B has 12,000 nails, we can substitute 12,000 for B.

$$A = 4 \times 12,000$$

To calculate how many nails store owner A has, all we have to do is a simple multiplication.

$$A = 4 \times 12,000$$
$$A = 48,000$$

To prove this answer is true, we can return to the part of the question that states "combined, both A and B have 60,000 nails."

$$A + B = 60,000$$

We know A has 48,000 nails and B has 12,000 nails, so we can replace A and B with their equivalent values as follows:

$$A + B = 60,000$$
$$48,000 + 12,000 = 60,000$$
$$60,000 = 60,000$$

5. Factoring (Unmultiplying) and Distributing (Multiplying Through)

A factor is any number multiplied by another number that contributes to the result. For example, consider 3×8. In this multiplication example, 3 is a factor and 8 is a factor, and both will contribute to a result of 24.

Now let us reverse the process. What factors contribute to a result of 12? In this case we must determine which small numbers, when multiplied together, will produce 12. The answer could be

$$2 \times 6 = 12 \qquad \text{or} \qquad 3 \times 4 = 12$$

Are these, however, the smallest numbers that, when multiplied together, will produce 12? The answer is no because the 6 (in 2×6) can be broken into 2×3, and the 4 (in 3×4) can be broken into 2×2. Therefore, 12 has the factors $2 \times 2 \times 3$.

Now consider the equation $12 + 9$. What smaller number is a factor of both these numbers? The answer is 3 because you can get results of 12 and 9 by multiplying some other number by 3. Therefore, both 12 and 9 contain the factor 3. If we now remove the common factor 3 from both the numbers 12 and 9, we end up with the following equation:

If no symbol appears, it is assumed that it is a multiplication.

$12 + 9$ ← $12 \div 3 = ④, 9 \div 3 = ③$

$④ + ③$ ← We cannot let just $4 + 3$ represent $12 + 9$ because the factor 3 is not included.

$3 \times (4 + 3)$ or $3(4 + 3)$ ← This expression will be correct because $3 \times (4 + 3) = 3 \times 7 = 21$ and $12 + 9 = 21$.

Parentheses are used to group the addition $4 + 3$. The factor 3 outside the parentheses applies to everything inside the parentheses.

Factoring is therefore a reducing process that extracts the common factor from two or more larger numbers. As another example, remove the factor from the following equation:

$$27x - 9x$$

Steps:

9 is a factor common to both 27 and 9.

$$27x \div = 9 = 3x, 9x \div 9 = 1x$$

$$= 9(3x - 1x)$$

Place 9 outside the parentheses and the results of the division inside the parentheses.

We can check if $9(3x - 1x) = 27x - 9x$ simply by performing the arithmetic operations indicated:

$$27x - 9x = \boxed{18x}$$

$$9(3x - 1x) \quad \overset{\nearrow 1x = x}{}$$
$$= 9(3x - x) \leftarrow 3x - x = 2x$$
$$= 9 \times 2x$$
$$\boxed{= 18x}$$

Results are the same.

CALCULATOR KEYS

Name: Factor function

Function: Some calculators have a factor function that returns an expression factored with respect to all its variables.

factor (expression)

Examples: Extract the common factor from $(64x - 8y)$.
Press keys: Factor $(64x - 8y)$
Answer: $8(8x - y)$
Press keys: Factor $(16a - 4)$
Answer: $4(4a - 1)$

To reverse the factoring process and convert $9(3x - 1x)$ back to the original equation $27x - 9x$, simply reverse the arithmetic steps. Because we began by dividing both numbers by the common factor, simply do the opposite and multiply both values within the parentheses by the factor 9, as follows:

$$9(3x - 1x)$$
$$= 27x - 9x$$

Steps:

$$9 \times 3x = 27x,$$
$$9 \times 1x = 9x$$

To make any equation with parentheses easier to solve, always begin by removing the parentheses before doing any other operation.

This process of removing parentheses, or multiplying through, is called *distributing*, and **the distributive law of multiplication** is as follows.

$$\boxed{a \times (b + c) = (a \times b) + (a \times c)}$$

Distributive Law of Multiplication: The same result is obtained regardless of whether you multiply a whole or multiply each part of the whole.

■ EXAMPLE A

$$5(7x + 3x) \quad \longleftarrow 5 \times 7x = 35x, 5 \times 3x = 15x$$
$$= 35x + 15x$$

■ EXAMPLE B

$$6(4a - 3) \qquad 6 \times 4a = 24a, \quad 6 \times 3 = 18$$
$$= 24a - 18$$

To summarize, we now know about two new processes called *factoring* and *distributing*. To show how these two are the reverse of each other, let us do a few examples.

■ EXAMPLE A

Factoring:

$$12 + 18$$
$$= 6(2 + 3)$$

Common factor is ⑥
$$12 \div 6 = 2, 18 \div 6 = 3$$

Distributing:

$$6(2 + 3)$$
$$= 12 + 18$$

$$6 \times 2 = 12, 6 \times 3 = 18$$

■ EXAMPLE B

Factoring:

$$22y - 11$$
$$= 11(2y - 1)$$

Common factor is 11.
$$22y \div 11 = 2y, 11 \div 11 = 1$$

$$11(2y - 1)$$
$$= \ \ 22y - 11$$

Distributing:
$$11 \times 2y = 22y, 11 \times 1 = 11$$

Now that we know how to perform both of these operations, let us look at a practical application of them.

■ INTEGRATED MATH APPLICATION: AUTOMOTIVE

A motorbike uses a certain number of gallons of gasoline a week, a car uses twice that amount, and a truck uses four times as much as the motorbike and car combined. If all three use 100 gallons of gasoline a week, how much does each vehicle consume per week?

■ *Solution:*

Reading the problem carefully, you can see that each vehicle's consumption is related to the amount consumed by the motorbike. Therefore, if we can find the amount of gasoline consumed by the motorbike, we can calculate the amount consumed by the car and truck. We will represent the unknown quantity of gasoline that the bike consumes by x.

x ← Bike consumption

$2x$ ← Car consumption is twice that of the bike ($2 \times x$).

$4(x + 2x)$ ← Truck consumption is four times that of the
motorbike and car combined [$4 \times (x + 2x)$].
All three use 100 gallons per week.

$$x + 2x + 4(x + 2x) = 100 \text{ gal per week}$$

Now that we have an equation, the next step is to use transportation to determine the unknown value x.

	Steps:
$x + 2x + 4(x + 2x) = 100$	Distribute:
	$4 \times x = 4x, 4 \times 2x = 8x$
$x + 2x + 4x + 8x = 100$	Combine all the unknowns:
	$x + 2x + 4x + 8x = 15x$
$15x = 100$	Divide both sides by 15 to remove the "multiply by 15."
$\dfrac{\cancel{15}x}{\cancel{15}} = \dfrac{100}{15}$	$15 \div 15 = 1, 1 \times x = x$
$x = \dfrac{100}{15}$	$100 \div 15 = 6.67$
$x = 6.67$	

Therefore,

$$\text{Motorbike consumption} = x = 6.67 \text{ gal per week}$$

The car consumes twice that amount:

$$\text{Car consumption} = 2 \times x = 2 \times 6.67 = 13.34 \text{ gal per week}$$

The truck consumes four times as much as both combined:

$$\text{Truck consumption} = 4(x + 2x) = 4(6.67 + 2 \times 6.67)$$
$$= 4(6.67 + 13.34) \quad 4 \times 6.67 = 26.68$$
$$4 \times 13.34 = 53.36$$
$$= 26.68 + 53.36$$
$$= 80.0 \text{ gal per week}$$

To double-check our answer, let us add all the individual amounts to see if they equal the total:

$$\text{Motorbike} + \text{car} + \text{truck} = 100 \text{ gal per week}$$
$$6.67 + 13.34 + 80.0 = 100 \text{ gal per week}$$

Referring to the preceding example, you may have noticed that *you used factoring to develop the equation:*

Truck consumption is four times that of bike and car $= 4(x + 2x)$

and *you distributed to transpose the equation*:

$$4(x + 2x) = 4x + 8x$$

6. Square and Square Root

How do we remove the square or square root on one side of an equation? The answer once again is: Perform the opposite function. For example, if we square a number and then find its square root, we end up with the number with which we started:

$$\textcircled{4}\text{squared} = 4^2 = 4 \times 4 = 16$$
$$\text{Square root of } 16 = \sqrt{16} = \textcircled{4}$$

■ **EXAMPLE**

Determine the value of the unknown *I* in the following equation:

$$I^2 = 81$$

■ *Solution:*

To remove the square from the unknown quantity *I*, we simply take the square root of both sides.

$$I^2 = 81$$
$$\sqrt{I^2} = \sqrt{81} \quad \leftarrow \text{(Take square root of both sides.)}$$
$$I = \sqrt{81} \quad \leftarrow \text{(Square and square root cancel.)}$$
$$I = 9 \quad \leftarrow (\sqrt{81} = 9, \text{ since } 9 \times 9 = 81)$$

■ **EXAMPLE**

Determine the value of the unknown *x* in the following equation.

$$\sqrt{x} = 2 + 6$$

■ *Solution:*

$$\sqrt{x} = 2 + 6 \quad \leftarrow \text{(To remove the square root, square both sides.)}$$
$$\sqrt{x}^2 = (2 + 6)^2 \quad \leftarrow \text{(Square and square root cancel.)}$$
$$x = (2 + 6)^2$$
$$x = 8^2 \quad (8 \times 8 = 64)$$
$$x = 64$$

5–2–2   *Transposing Formulas*

As mentioned previously, *a formula is a general statement using letters and sometimes numbers that enables us to calculate the value of an unknown quantity.* Some of the simplest relationships are formulas. For example, to calculate the distance traveled, you can use the following formula, which was discussed earlier.

$$d = v \times t$$

where d = distance in miles
         v = velocity in miles per hour
         t = time in hours

Using this formula we can calculate distance if we know the speed and time. If we need only to calculate distance, this formula is fine. But what if we want to know what speed to go ($v = ?$) to travel a certain distance in a certain amount of time, or what if we want to calculate how long it will take for a trip ($t = ?$) of a certain distance at a certain speed? The answer is: Transpose or rearrange the formula so that we can solve for any of the quantities in the formula.

■ **EXAMPLE:**

As an example, calculate how long it will take to travel 250 miles when traveling at a speed of 50 miles per hour (mph). To solve this problem, let us place the values in their appropriate positions in the formula.

$$d = v \times t$$

$$250 \text{ mi} = 50 \text{ mph} \times t$$

The next step is to transpose just as we did before with equations to determine the value of the unknown quantity t.

Steps:

$250 = 50 \times t$ To isolate t, divide both sides by 50.

$$\frac{250}{50} = \frac{\cancel{50} \times t}{\cancel{50}} \qquad 50 \div 50 = 1, 1 \times t = t$$

$$\frac{250}{50} = t \qquad 250 \div 50 = 5$$

$$t = 5$$

It will therefore take 5 hours to travel 250 miles at 50 miles per hour ($250 \text{ mi} = 50 \text{ mph} \times 5 \text{ h}$).

All we have to do, therefore, is transpose the original formula so that we can obtain formulas for calculating any of the three quantities (distance, speed, or time) if two values are known. This time, however, we will transpose the formula instead of the inserted values.

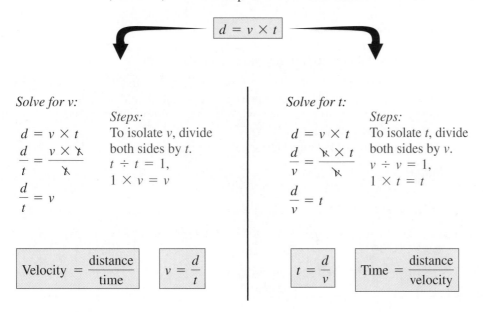

$$d = v \times t$$

Solve for v:

Steps:

$d = v \times t$ To isolate v, divide

$\dfrac{d}{t} = \dfrac{v \times \cancel{t}}{\cancel{t}}$ both sides by t.
   $t \div t = 1$,
   $1 \times v = v$

$\dfrac{d}{t} = v$

$$\boxed{\text{Velocity} = \frac{\text{distance}}{\text{time}}} \qquad \boxed{v = \frac{d}{t}}$$

Solve for t:

Steps:

$d = v \times t$ To isolate t, divide

$\dfrac{d}{v} = \dfrac{\cancel{v} \times t}{\cancel{v}}$ both sides by v.
   $v \div v = 1$,
   $1 \times t = t$

$\dfrac{d}{v} = t$

$$\boxed{t = \frac{d}{v}} \qquad \boxed{\text{Time} = \frac{\text{distance}}{\text{velocity}}}$$

$$\boxed{d = v \times t}$$ $$\boxed{v = \frac{d}{t}}$$ $$\boxed{t = \frac{d}{v}}$$

$$250 \text{ mi} = 50 \text{ mph} \times 5 \text{ h}$$ $$50 \text{ mph} = \frac{250 \text{ mi}}{5 \text{ h}}$$ $$5 \text{ h} = \frac{250 \text{ mi}}{50 \text{ mph}}$$

$$(250 = 250)$$ $$(50 = 50)$$ $$(5 = 5)$$

■ INTEGRATED MATH APPLICATION: AEROSPACE

The following formula is used to determine the thrust horsepower of a jet engine (H), when the engine is developing T pounds of thrust and is driving the plane at a velocity of V miles per hour.

$$H = \frac{T \times V}{375}$$

T = Pounds of thrust
V = Plane velocity in miles per hour
H = Thrust horsepower

   a. Calculate H when $T = 12,000$ lb and $V = 450$ mph.
   b. Transpose the formula to find V when H and T are known.
   c. Test the formula in (b) with the values given in (a).

■ *Solution:*

   a. $H = \dfrac{T \times V}{375} = \dfrac{12,000 \times 450}{375} = 14,400$ horsepower

   b. $V = \dfrac{375 \times H}{T}$

   c. $V = \dfrac{375 \times 14,400}{12,000} = 450$ miles per hour

---

■ INTEGRATED MATH APPLICATION: WATER MANAGEMENT

A water district adds a certain amount of chlorine (in pounds) to a reservoir that has a water flow rate through the reservoir (measured in millions of gallons per day). The formula is as follows:

$$C_A = 834 \times W \times C_C$$

C_A = Chlorine added in pounds
W = Water flow rate in millions of gallons per day
C_C = Chlorine concentration in parts per million (ppm)

   a. If a reservoir supplies 4.5 million gallons of water per day, and the chlorine concentration should be 24.3 ppm, how many pounds of chlorine should be added?
   b. Transpose the formula to determine chlorine concentration when the water flow rate and the amount of chlorine added is known.
   c. Test the formula in (b) with the values given in (a).

■ *Solution:*

   a. $C_A = 8.34 \times W \times C_C$
   $\quad\quad = 8.34 \times 4.5 \times 24.3 = 912$ pounds

   b. $C_C = \dfrac{C_A}{8.34 \times W}$

   c. $C_C = \dfrac{912}{8.34 \times 4.5} = 24.3$ parts per million

2. Ohm's Law Formula Example

In this section we will examine a law that was first discovered by Georg Ohm in 1826. It states that *the **electric current** in a circuit is directly proportional to the **voltage** source applied and inversely proportional to the **resistance** in the circuit.* Stated as a formula, it appears as follows:

Ohm's law formula
$$\text{Current} = \frac{\text{voltage}}{\text{resistance}}$$

Quantity	Symbol	Unit	Symbol
Current	I	Amperes	A
Voltage	V	Volts	V
Resistance	R	Ohms	Ω

To explain this formula, let us use the fluid analogy shown in Figure 5–1(a). An object will move only when a force is applied, and as you can see in Figure 5-1(a), the greater the pressure applied to the piston, the greater the amount of water flow through the pipe. Water flow is therefore directly proportional to pressure applied:

$$\text{Water flow} \propto \text{pressure applied}$$

In a similar way, the electric current (I) shown in Figure 5-1(b) is directly dependent on the voltage (V) force applied by the battery. For instance, a smaller voltage ($V \downarrow$) will cause a corresponding smaller circuit current ($I \downarrow$), and similarly, a larger voltage ($V \uparrow$) will result in a larger circuit current ($I \uparrow$). It can therefore be said that electric current,

(a) Fluid system

(b) Electric circuit

FIGURE 5–1   Current Is Proportional to Voltage (a) Fluid System (b) Electric Circuit.

which is the flow of electrons, is directly proportional to electric voltage, which is the force that moves electrons:

$$\text{Current } (I) \propto \text{voltage } (V)$$

Now, referring to the Ohm's law formula, you can see that current (I) and voltage (V) are directly proportional because these quantities are above the fraction bars on opposite sides of the equals sign.

$$\boxed{\text{Current } (I)} = \frac{\boxed{\text{voltage } (V)}}{\text{resistance } (R)}$$

The current (I) in an electric circuit is also inversely proportional to the opposition or resistance in the circuit. This can best be explained by the fluid analogy in Figure 5–2(a). If the valve is adjusted to offer a small opposition, a large amount of water will flow, whereas if the valve is adjusted to introduce a large amount of opposition, only a small amount of water will be permitted to flow. Water flow is therefore inversely proportional to the valve's opposition:

$$\text{Water flow } \frac{1}{\propto} \text{ valve opposition}$$

In a similar way, the electric current (I) shown in Figure 5-2(b) is inversely dependent on the opposition or resistance (R) offered by the resistor in the circuit. For instance, a small-value resistor will offer only a small opposition or resistance $(R\downarrow)$, which will permit a large circuit current $(I\uparrow)$, whereas a large-value resistor will offer a large opposition or resistance $(R\uparrow)$, which will allow only a small circuit current $(I\downarrow)$. It can be said, therefore, that electric current, which is the flow of electrons, is inversely proportional to electric resistance, which is the restriction to current flow:

$$\text{Current } (I) \frac{1}{\propto} \text{ resistance } (R)$$

(a) Fluid system

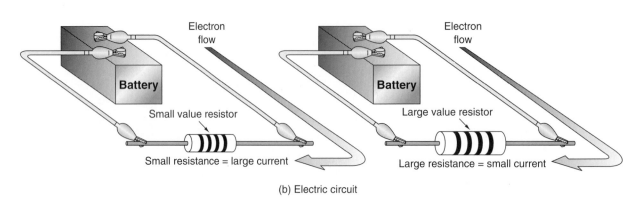

(b) Electric circuit

FIGURE 5–2   Current Is Inversely Proportional to Resistance (a) Fluid System (b) Electric Circuit.

CHAPTER 5 / ALGEBRA, EQUATIONS, AND FORMULAS

Once again, referring to the Ohm's law formula, you can see that current (I) and resistance (R) are inversely proportional to each other because one quantity is above the fraction bar and the other is below the fraction bar on opposite sides of the equals sign:

$$\boxed{\text{Current } (I)} = \frac{\text{voltage } (V)}{\boxed{\text{resistance } (R)}}$$

■ EXAMPLE

Calculate what voltage would cause 2 amperes (A) of electric current in a circuit having an opposition or resistance of 3 ohms (Ω).

■ *Solution:*

To solve this problem, begin by placing the values in their appropriate position within the formula. In this example,

$$\text{Current } (I) = \frac{\text{voltage } (V)}{\text{resistance } (R)}$$

Therefore,

$$2A = \frac{V?}{3\,\Omega}$$

or

$$2 = \frac{V}{3}$$

As you can see, we will first need to transpose the current formula ($I = V/R$) so that we can find a formula for voltage (V), because V is the unknown in this example.

$$\text{Current } (I) = \frac{\text{voltage } (V)}{\text{resistance } (R)}$$

$$I = \frac{V}{R}$$

Solve for V	*Solve for R*
$I = \dfrac{V}{R}$	$I = \dfrac{V}{R}$
$\dfrac{R \times I}{R} = \dfrac{V \times R}{R}$ (Multiply both sides by R.)	$\dfrac{R \times I}{R} = \dfrac{V \times R}{R}$ (Multiply both sides by R to bring R above the line.)
$\dfrac{R \times I}{\cancel{R}} = \dfrac{V \times \cancel{R}}{\cancel{R}}$ ($R \div R$ cancels.)	$\dfrac{R \times I}{\cancel{R}} = \dfrac{V \times \cancel{R}}{\cancel{R}}$ ($R \div R$ cancels.)
$\dfrac{R \times I}{} = \dfrac{V}{}$	$\dfrac{R \times I}{I} = \dfrac{V}{I}$ (Divide both sides by I to isolate R.)
$\boxed{V = I \times R}$	$\dfrac{R \times \cancel{I}}{\cancel{I}} = \dfrac{V}{I}$ ($I \div I$ cancels.)
	$\boxed{R = \dfrac{V}{I}}$

Now that we have a formula for voltage ($V = I \times R$), we can complete this problem:

$$V = I \times R \qquad \text{(Voltage} = \text{current} \times \text{resistance)}$$
$$= 2\,\text{A} \times 3\,\Omega$$
$$= 6\,\text{V}$$

This equation means that a voltage or electrical pressure of 6 volts would cause 2 amperes of electric current in a circuit having 3 ohms of resistance.

To test if these three formulas are correct, we can now plug into each formula the values of the preceding example, as follows:

$$I = \frac{V}{R} \qquad\qquad V = I \times R \qquad\qquad R = \frac{V}{I}$$

$$2\,\text{A} = \frac{6\,\text{V}}{3\,\Omega} \qquad 6\,\text{V} = 2\,\text{A} \times 3\,\Omega \qquad 3\,\Omega = \frac{6\,\text{V}}{2\,\text{A}}$$

As you can see from these formulas, the transposition has been successful.

■ INTEGRATED MATH APPLICATION: ELECTRICAL

Figure 5–3 shows a basic circuit for a home electrical system. If each household light and appliance has 120 volts connected across it, how much current will pass through lamp 1, lamp 2, the hair dryer, and the space heater?

■ *Solution:*

$$\text{Current through lamp 1 } (I_{L1}) = \frac{V_{L1}}{R_{L1}} = \frac{120\,\text{V}}{125\,\Omega} = 960\,\text{mA}$$

$$\text{Current through lamp 2 } (I_{L2}) = \frac{V_{L2}}{R_{L2}} = \frac{120\,\text{V}}{250\,\Omega} = 480\,\text{mA}$$

$$\text{Current through hair dryer } (I_{HD}) = \frac{V_{HD}}{R_{HD}} = \frac{120\,\text{V}}{40\,\Omega} = 3\,\text{A}$$

$$\text{Current through space heater } (I_{SH}) = \frac{V_{SH}}{R_{SH}} = \frac{120\,\text{V}}{12\,\Omega} = 10\,\text{A}$$

FIGURE 5–3 **Home Electrical System.**

3. Power Formula Example

Another formula frequently used in electricity and electronics is the power formula. We will also study this formula so that we can practice formula transposition again.

Electric power is the rate at which electric energy is converted into some other form of energy. Power, which is symbolized by P, is measured in *watts* (symbolized by W) in honor of James Watt. We hear watt used frequently in connection with the brightness of a lightbulb: for instance, a 60 watt bulb, a 100 watt bulb, and so on. This designation is a power rating and describes how much electric energy is converted every second. Because a lightbulb generates both heat and light energy, we would describe a 100 watt lightbulb as a device that converts 100 watts of electric energy into heat and light energy every second. Referring to Figure 5-1(b), you can see a lightbulb connected in an electric circuit. The amount of electric power supplied to that lightbulb is dependent on the voltage and the current. It is probably no surprise, therefore, that power is equal to the product of voltage and current. To state the formula:

Quantity	Symbol	Unit	Symbol
Power	P	Watts	W
Voltage	V	Volts	V
Current	I	Amperes	A

$$\boxed{\begin{array}{c}\text{Power} = \text{voltage} \times \text{current} \\ P = V \times I\end{array}}$$

This formula states that the amount of power delivered to a device is dependent on the electrical pressure or voltage applied across the device and the electric current flowing through the device.

■ EXAMPLE

What will be the electric current (I) in a circuit if a 60 watt lightbulb is connected across a 120 volt supply voltage?

■ *Solution:*

Let us begin by inserting the known values in their appropriate places in the power formula.

$$\text{Power}\,(P) = \text{voltage}\,(V) \times \text{current}\,(I)$$
$$60\,\text{W} = 120\,\text{V} \times ?\,\text{amperes}$$

As you can see, we first need to transpose the power formula ($P = V \times I$) so that we can find a formula for current (I), because I is the unknown in this example.

$$P = V \times I$$

Solve for V:

$$P = V \times I$$

$$\frac{P}{I} = \frac{V \times I}{I} \quad \text{(Divide both sides by } I.)$$

$$\frac{P}{I} = \frac{V \times \cancel{I}}{\cancel{I}} \quad \begin{array}{l}(I \div I = 1, \\ \text{and } V \times 1 = V.)\end{array}$$

$$\frac{P}{I} = V$$

$$\boxed{V = \frac{P}{I}}$$

This formula can be used to calculate V when P and I are known.

Solve for I:

$$P = V \times I$$

$$\frac{P}{V} = \frac{V \times I}{V} \quad \text{(Divide both sides by } V.)$$

$$\frac{P}{V} = \frac{\cancel{V} \times I}{\cancel{V}} \quad \begin{array}{l}(V \div V = 1, \\ \text{and } I \times 1 = I.)\end{array}$$

$$\frac{P}{V} = I$$

$$\boxed{I = \frac{P}{V}}$$

This formula can be used to calculate I when P and V are known.

Now that we have a formula for current ($I = P/V$), we can complete this problem:

$$I = \frac{P}{V} \quad \left(\text{Current} = \frac{\text{power}}{\text{voltage}}\right)$$

$$= \frac{60\,\text{W}}{120\,\text{V}}$$

$$= 0.5\,\text{A} \quad \text{or} \quad 500 \times 10^{-3}\,\text{A} \leftarrow (0.500.0)$$

$$= 500\ \text{mA}$$

This result means that a current of 500 milliamperes will exist in an electric circuit if a 120 volt battery is connected across a 60 watt light bulb.

CALCULATOR KEYS

Name: Solve function key

Function: Some calculators have a function that solves an expression for a specified variable.

Solve (*equation, variable*)

Example: Solve for R, when $I = V/R$
Press keys: Solve $(I = V/R, R)$
Answer: $R = V/I$

Solve for t, when $250 = 50 \times t$.
Press keys: Solve $(250 = 50 \times t, t)$
Answer: $t = 5$

Solve for y, when $(6 \times y) + 27 = y + 27$.
Press keys: Solve $((6 \times y) + 7 = y + 27, y)$
Answer: $y = 4$

■ INTEGRATED MATH APPLICATION: HEALTH CARE

The following rule, known as Young's formula, can be employed to determine a child's medicinal dosage.

$$C_D = \frac{C_A \times A_D}{C_A + 12}$$

C_D = Child's dose
C_A = Child's age
A_D = Adult's dose

a. Calculate an 8-year-old child's dosage if the adult's dosage is 125 milligrams.
b. Transpose the formula to determine the adult dosage if the child's age and dosage are known.
c. Test the formula in (b) with the values given in (a).

■ *Solution:*

a. $C_D = \dfrac{C_A \times A_D}{C_A + 12} = \dfrac{8 \times 125}{8 + 12} = 50 \text{ milligrams}$

b. $A_D = \dfrac{C_D(C_A + 12)}{C_A}$

c. $A_D = \dfrac{50(8 + 12)}{8} = 125 \text{ milligrams}$

◼ INTEGRATED MATH APPLICATION: SHEET METAL

The following formula determines how long a section of sheet metal should be to accommodate a right-angle inside bend. When lengths 1 and 2 and the thickness of the sheet metal are known:

$$L_T = L_1 + L_2 + \left(\frac{1}{2} \times T\right)$$

L_T = total length

a. Calculate L_T when $L_1 = 10\,\text{cm}$, $L_2 = 5\,\text{cm}$, and $I = \frac{1}{2}\,\text{cm}$.

b. How much of the total length in (a) is used for the right-angle inside bend?

c. Solve for L_1 and L_2.

◼ *Solution:*

a. $L_T = L_1 + L_2 + \left(\frac{1}{2} \times T\right)$

$$= 10\,\text{cm} + 5\,\text{cm} + \left(\frac{1}{2} \times 0.5\,\text{cm}\right)$$

$$= 15.25 \text{ centimeters}$$

b. 0.25 centimeter

c. $L_1 = L_T - L_2 - \left(\frac{1}{2} \times T\right)$

$$L_2 = L_T - L_1 - \left(\frac{1}{2} \times T\right)$$

SELF-TEST EVALUATION POINT FOR SECTION 5–2

Now that you have completed this section, you should be able to:

◼ *Objective 4.* Explain the steps involved in transposing an equation to determine the unknown.

◼ *Objective 5.* Demonstrate how to develop an equation from a story problem.

◼ *Objective 6.* Describe the process of factoring and how to remove parentheses within a formula or expression.

◼ *Objective 7.* Demonstrate the process of transposing a formula.

◼ *Objective 8.* Describe how the terms directly proportional *and* inversely proportional *are used to show the relationship between quantities in a formula.*

◼ *Objective 9.* Explain the proportionality among the three quantities in the Ohm's law formula, and describe how to transpose the formula.

◼ *Objective 10.* Describe the electric power formula, and transpose the formula to develop alternative procedures.

Use the following questions to test your understanding of Section 5–2.

1. Calculate the result of the following arithmetic operations involving literal numbers.
   a. $x + x =$
   b. $x \times x =$
   c. $7a + 4a =$
   d. $2x - x =$
   e. $\frac{x}{x} =$
   f. $x - x =$

2. Transpose the following equations to determine the unknown quantity.
   a. $x + 14 = 30$
   b. $8 \times x = \frac{80 - 40}{10} \times 12$
   c. $y - 4 = 8$
   d. $(x \times 3) - 2 = \frac{26}{2}$
   e. $x^2 + 5 = 14$
   f. $2(3 + 4x) = (2x + 13)$

3. Transpose the following formulas.
   a. $x + y = z, y = ?$
   b. $Q = C \times V, C = ?$
   c. $x_L = 2 \times \pi \times f \times L, L = ?$
   d. $V = I \times R, R = ?$

4. Determine the unknown in the following equations by using transposition.
   a. $I^2 = 9$
   b. $\sqrt{z} = 8$

Substitution
The replacement of one mathematical entity by another of equal value.

Substitution is a mathematical process used to develop alternative formulas by replacing or substituting one mathematical term with an equivalent mathematical term. To explain this process, let us try a couple of examples.

5–3–1 *Circle Formula Example*

A **circle** is a perfectly round figure in which every point is at an equal distance from the center.

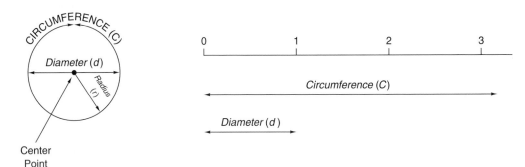

Circle
A closed plane curve, every point of which is equidistant from a fixed point within the curve.

Circumference
The perimeter of a circle.

Radius
A line segment extending from the center of a circle or sphere to the circumference.

Diameter
The length of a straight line passing through the center of an object.

Some of the elements of a circle are as follows.

a. The **circumference** (C) of a circle is the distance around the perimeter of the circle.

b. The **radius** (r) of a circle is a straight line drawn from the center of the circle to any point of the circumference.

c. The **diameter** (d) of a circle is a straight line drawn through the center of the circle to opposite sides of the circle. The diameter of a circle is therefore equal to twice the circle's radius ($d = 2 \times r$).

To review, *pi* (symbolized by π) is the name given to the ratio, or comparison, of a circle's circumference with its diameter. This constant will always be equal to approximately 3.14 because the circumference of any circle will always be about 3 times larger than the same circle's diameter. Stated mathematically:

$$\pi = \frac{\text{circumference } (C)}{\text{diameter } (d)} = 3.14$$

■ INTEGRATED MATH APPLICATION: AUTOMOTIVE

Determine the circumference of a tire that is 26 inches in diameter.

■ *Solution:*

To calculate the wheel's circumference, we will have to transpose the following formula to isolate C.

$$\pi = \frac{C}{d} \qquad \text{Multiply both sides by } d.$$

$$\pi \times d = \frac{C}{d} \times d \qquad d \div d = 1, \text{ and } C \times 1 = C.$$

$$\pi \times d = C$$

The circumference of the bicycle wheel is therefore:

$$C = \pi \times d$$

$$C = 3.14 \times 26 \text{ in.} = 81.64 \text{ inches}$$

In some examples, only the circle's radius will be known. In situations like this, we can substitute the expression $\pi = C/d$ to arrive at an equivalent expression comparing a circle's circumference with its radius.

$$\pi = \frac{C}{d} \qquad \text{Because } d = 2 \times r, \text{ we can substitute } 2 \times r \text{ for } d.$$

$$\pi = \frac{C}{d} \qquad \text{Therefore, this is equivalent to } \pi = \frac{C}{2 \times r}$$

■ INTEGRATED MATH APPLICATION: LANDSCAPING

How much decorative edging will you need to encircle a garden that measures 6 feet from the center to the edge?

■ *Solution:*

In this example r is known, so to calculate the garden's circumference, we will have to transpose the following formula to isolate C:

$$\pi = \frac{C}{2 \times r} \qquad\qquad \text{Multiply both sides by } 2 \times r.$$

$$\pi \times (2 \times r) = \frac{C}{2 \times r} \times (2 \times r) \quad (2 \times r) \div (2 \times r) = 1, \text{ and } C \times 1 = C.$$

$$\pi \times (2 \times r) = C$$

The edging needed for the circumference of the garden will therefore be:

$$C = \pi \times (2 \times r)$$
$$C = 3.14 \times (2 \times 6\,\text{ft}) = 37.68\,\text{feet}$$

To continue our example of the circle, how could we obtain a formula for calculating the area within a circle? The answer is best explained with the diagram shown in Figure 5–4. In Figure 5-4(a) a circle has been divided into eight triangles, and then these triangles have been separated as shown in Figure 5-4(b). Figure 5-4(c) shows how each of these triangles can be thought of as having an A and a B section that if separated and re-arranged will form a square. The area of a square is easy to calculate because it is simply the product of the square's width and height (area $= w \times h$). The area of the triangles shown in Figure 5-4(c), therefore, will be

$$\text{Area} = \frac{1}{2} \times (w \times h)$$

Referring to Figure 5-4(d), you can see that the width of all eight triangles within a circle is equal to the circle's circumference ($w = C$), and the height of the triangles is equal to the circle's radius ($h = r$). The area of the circle, therefore, can be calculated with the following formula:

$$\text{Area of a circle} = \frac{1}{2} \times (C \times r)$$

Using both substitution and transposition, let us now try to reduce this formula with $C = \pi \times (2 \times r)$.

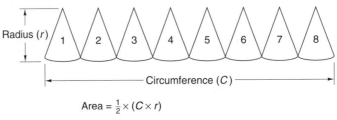

FIGURE 5–4   Calculating the Area of a Circle.

$$\text{Area of a circle} = \frac{1}{2} \times (C \times r) \qquad \leftarrow \text{Because } C = \pi \times (2 \times r), \text{ we can replace } C \text{ with } \pi \times (2 \times r).$$

$$= \frac{1}{2} \times \left[\pi \times (2 \times r) \times r \right] \qquad \leftarrow \text{We can now use transposition to reduce.}$$

$$= \frac{1 \times \pi \times 2 \times r \times r}{2} \qquad \leftarrow 2 \div 2 = 1$$

$$= 1 \times \pi \times r \times r \qquad \leftarrow r \times r = r^2$$

$$= 1 \times \pi \times r^2 \qquad \leftarrow 1 \times (\pi \times r^2) = \pi \times r^2$$

$$\text{Area of a circle} = \pi \times r^2$$

■ EXAMPLE

Calculate the area of a circle that has a radius of 14 inches.

■ *Solution:*

$$\text{Area of a circle} = \pi \times r^2$$
$$= 3.14 \times (14 \text{ in.})^2$$
$$= 3.14 \times 196 \text{ in.}^2$$
$$= 615.4 \text{ square inches}$$

■ EXAMPLE

Calculate the radius of a circle that has an area of 624 square centimeters (cm^2).

■ *Solution:*

To determine the radius of a circle when its area is known, we have to transpose the formula area of a circle $= \pi \times r^2$ to isolate r.

$$\text{Area of a circle} = \pi \times r^2$$
$$A = \pi \times r^2 \qquad \leftarrow \text{Divide both sides by } \pi.$$
$$\frac{A}{\pi} = \frac{\pi \times r^2}{\pi} \qquad \leftarrow \pi \div \pi = 1, 1 \times r^2 = r^2$$
$$\frac{A}{\pi} = r^2 \qquad \leftarrow \text{Take the square root of both sides.}$$
$$\sqrt{\frac{A}{\pi}} = \sqrt{r^2} \qquad \leftarrow \sqrt{r^2} = r$$
$$\sqrt{\frac{A}{\pi}} = r$$

Now we can calculate the radius of a circle that has an area of 624 cm^2.

$$r = \sqrt{\frac{A}{\pi}} = \sqrt{\frac{624}{3.14}} = \sqrt{198.7} = 14.1 \text{ centimeters}$$

■ INTEGRATED MATH APPLICATION: MACHINING

Pomeroy's formula can be used to approximate the power needed by a metal punch machine:

$$P \approx \frac{T^2 \times d \times N}{3.78}$$

d = hole diameter

T = metal thickness

All four will be punched simultaneously ($N = 4$)

where
P = power required (in horsepower)
T = metal thickness
d = hole diameter
N = number of holes to be punched simultaneously
$\approx$: approximately equal to

a. If the hole's radius is known instead of its diameter, what would be the new formula?

b. Transpose Pomeroy's formula to determine how many holes a certain punch machine can manage (solve for N).

c. Determine P when:

$$I = 3 \text{ cm}$$
$$d = 0.5 \text{ cm}$$
$$N = 4$$

■ *Solution:*

a. $P \approx \dfrac{T^2 \times d \times N}{3.78}$ Substitute $2 \times r$ for d (diameter = twice the radius)

$P \approx \dfrac{T^2 \times (2 \times r) \times N}{3.78}$

b. $N \approx \dfrac{3.78 \times P}{d \times T^2}$

c. $P \approx \dfrac{T^2 \times d \times N}{3.78} = \dfrac{3^2 \times 0.5 \times 4}{3.78} = 4.76 \text{ horsepower}$

5–3–2   *Power Formula Example*

In Section 5–2 you were introduced to two electrical formulas: the Ohm's law formula, which states

$$\text{Current } (I) = \frac{\text{voltage } (V)}{\text{resistance } (R)}$$

and the power formula, which states

$$\text{Power } (P) = \text{voltage } (V) \times \text{current } (I)$$

Using transposition we were able to develop the following alternative formulas that allowed us to calculate any quantity if two quantities were known.

	OHM'S LAW FORMULA			POWER FORMULA	
FORMULA	CALCULATES	WHEN THESE ARE KNOWN:	FORMULA	CALCULATES	WHEN THESE ARE KNOWN:
$I = \dfrac{V}{R}$	Current (I) in amperes (A)	Voltage (V) and resistance (R)	$P = V \times I$	Power (P) in watts (W)	Voltage (V) and current (I)
$V = I \times R$	Voltage (V) in volts (V)	Current (I) and resistance (R)	$V = \dfrac{P}{I}$	Voltage (V) in volts (V)	Power (P) and current (I)
$R = \dfrac{V}{I}$	Resistance (R) in ohms (Ω)	Voltage (V) and current (I)	$I = \dfrac{P}{V}$	Current (I) in amperes (A)	Power (P) and voltage (V)

Although it seems from this table that we have covered all possible alternatives, there are a few missing. For example, consider the power formula $P = V \times I$. When trying to calculate power, we may not always have voltage (V) and current (I). We may, for example, have values only for voltage (V) and resistance (R), or current (I) and resistance (R). In situations such as this the mathematical process called *substitution* can be used to develop alternative power formulas by replacing or substituting one mathematical term for an equivalent mathematical term. The following process shows how we can substitute terms in the $P = V \times I$ formula to arrive at alternative power formulas for wattage calculations.

Original power formula:
Calculates power (P) when
voltage (V) and current (I)
are known.

P = power in watts (W)
V = voltage in volts (V)
I = current in amperes (A)

$$P = V \times I$$

$P = V \times \boxed{I}$ ← (Because $I = V/R$,
                    substitute V/R
                    for I.)

$P = \widehat{V} \times I$ ← (Because $V = I \times R$,
                   substitute V for $I \times R$.)

$P = V \times \left(\dfrac{V}{R}\right)$ ← ($V \times V = V^2$.)

$P = \widehat{I \times R} \times I$ ← ($I \times I = I^2$.)

$$\boxed{P = \dfrac{V^2}{R}}$$

$$\boxed{P = I^2 \times R}$$

This formula calculates power (P) when
voltage (V) and resistance (R) are known.

This formula calculates power (P) when
current (I) and resistance (R) are known.

Let us now look at two applications to see how we can use these formulas to calculate unknown quantities.

■ INTEGRATED MATH APPLICATION: ELECTRICAL

The household heater shown here has a resistance of 7 ohms and is connected to the 120 volt source at the wall outlet. Calculate the amount of power dissipated by the heater.

Electric heater

120 V

■ *Solution:*

In this example, both voltage (120 V) and resistance (7 Ω) are known, and the unknown to be calculated is power (P). We must therefore look at all the power formulas to select one that includes voltage (V) and resistance (R).

$$P = V \times I$$

$\left(P = \dfrac{V^2}{R}\right)$ ← This power formula is
             the one to use because
             it includes V and R.

$$P = I^2 \times R$$

$$\text{Power } (P) = \frac{\text{voltage}^2 \, (V^2)}{\text{resistance } (R)}$$

Replace the literal
numbers V and
R with values.

$$= \frac{(120 \, V)^2}{7 \, \Omega} \qquad 120^2 = 14{,}400$$

$$= \frac{14{,}400}{7 \, \Omega}$$

$$= 2057 \, \text{W} \qquad 2.\overset{\frown}{057}$$

or approximately 2 kW

■ INTEGRATED MATH APPLICATION: ELECTRICAL

The lightbulb shown here is rated at 6 volts, 60 milliamps.

    a. Using Ohm's law, calculate the resistance of the electric lamp or lightbulb.
    b. Using the power formula, calculate the lamp's wattage at its rated current and resistance.

■ *Solution:*

Part (a) asks us to use Ohm's law to calculate the resistance of the electric lamp.

Ohm's law

$$I = \frac{V}{R}$$

$$V = I \times R$$

$$\boxed{R = \frac{V}{I}}$$

$$\text{Resistance } (R) = \frac{\text{voltage } (V)}{\text{current } (I)}$$

$$= \frac{6\,\text{V}}{60\,\text{mA}}$$

$$= \frac{6\,\text{V}}{60 \times 10^{-3}\,\text{A}}$$

$$= 100\,\Omega$$

Lightbulb

Part (b) directs us to use the power formula (P) to calculate the lamp's wattage at its rated current (I) and resistance (R). I and R are known, so we will use the following formula:

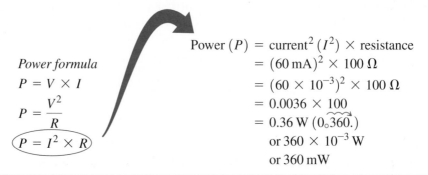

Power formula

$$P = V \times I$$

$$P = \frac{V^2}{R}$$

$$\boxed{P = I^2 \times R}$$

$$\text{Power } (P) = \text{current}^2\,(I^2) \times \text{resistance}$$

$$= (60\,\text{mA})^2 \times 100\,\Omega$$

$$= (60 \times 10^{-3})^2 \times 100\,\Omega$$

$$= 0.0036 \times 100$$

$$= 0.36\,\text{W}\ (0.0360.)$$

$$\text{or } 360 \times 10^{-3}\,\text{W}$$

$$\text{or } 360\,\text{mW}$$

As a review of the Ohm's law and power formula combinations, Figure 5–5 shows all the possible formulas for the four electrical properties.

SELF-TEST EVALUATION POINT FOR SECTION 5–3

Now that you have completed this section, you should be able to:

■ ***Objective 11.*** *Demonstrate how to use substitution to develop alternative formulas.*

Use the following questions to test your understanding of Section 5–3.

  1. Use substitution to calculate the value of the unknown.

$$x = y \times z \quad \text{and} \quad a = x \times y$$

Calculate a if $y = 14$ and $z = 5$.

  2. If $I = V \times R$ and $P = V \times I$, show the transposition and substitution steps that were followed to develop the formula circle shown in Figure 5–5. (The development follows parts (a) to (d) in the figure.)

  3. What percentage of 12 gives 2.5?

$$(x\% \text{ of } 12 = 2.5)$$

  4. 60 is 22% of what number?

$$(22\% \text{ of } x = 60)$$

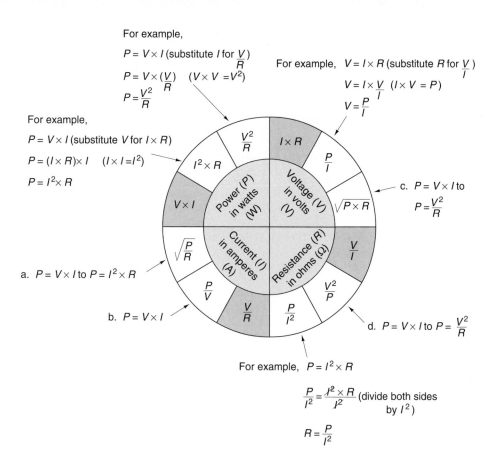

For example,

$P = V \times I$ (substitute I for $\frac{V}{R}$)

$P = V \times (\frac{V}{R})$ $\quad (V \times V = V^2)$

$P = \frac{V^2}{R}$

For example, $\quad V = I \times R$ (substitute R for $\frac{V}{I}$)

$V = I \times \frac{V}{I}$ $\quad (I \times V = P)$

$V = \frac{P}{I}$

For example,

$P = V \times I$ (substitute V for $I \times R$)

$P = (I \times R) \times I$ $\quad (I \times I = I^2)$

$P = I^2 \times R$

c. $P = V \times I$ to $P = \frac{V^2}{R}$

a. $P = V \times I$ to $P = I^2 \times R$

b. $P = V \times I$

d. $P = V \times I$ to $P = \frac{V^2}{R}$

For example, $\quad P = I^2 \times R$

$\frac{P}{I^2} = \frac{I^2 \times R}{I^2}$ (divide both sides by I^2)

$R = \frac{P}{I^2}$

FIGURE 5–5 **The Ohm's Law and Power Formula Circle.**

5–4 EQUATIONS AND GRAPHING

As stated previously, when algebraic terms or expressions are separated by an equals sign, the resulting algebraic statement is called an equation.

5–4–1 *Linear Equations*

In the simplest equations, the **literal number** (letters used to represent quantities) are raised to the first power, and are referred to as **linear equations.** For example:

$d = v \times t \quad \leftarrow$ Each of these literal numbers is raised to the first power:
$$d^1 = v^1 \times t^1$$
This first power is not normally included since $d^1 = d$.

A linear equation can appear to be different, but in fact be exactly the same. For instance, $x + y = 2$ is the same as $x = 2 - y$. To avoid confusion a **standard form** (STF) was adopted so that an alternate correct answer was not mistaken for an incorrect one.

Standard form (STF): $\quad Ax + By = C$

Variable terms are on the left of the equals sign.

Number is always on the right of the equals sign.

Coefficients (A, B, C) must be integers.

A is generally always positive.

Literal Number

A letter used to represent quantities or variables. If a salad has tomatoes, lettuce, and cucumbers, $S = T + L + C$.

Linear Equation

If the literal numbers in an equation are raised to the first power, they are called linear equations. The first power is not normally included, since $b^1 = b$.

Standard Form

A linear equation can appear to be very different from another equation when in fact they are equal. To avoid confusion, a standard (STF) was adopted: $ax + by = c$.

Coefficient

The numerical factors in a term.

Integer

A number without a fractional part.

The numerical **coefficient** of an algebraic term is the real number part of the term. For example, with the term $6x$, the number 6 is the numerical coefficient. An **integer** is a whole number without a fractional part, including zero. For example, $3, 0, -6$, and $+12$ are all examples of integers, whereas 18.3 and $-6\frac{1}{2}$ are not integers because they have fractional parts.

■ **EXAMPLE:**

Convert the following linear equation into standard form.

$$3x - 4y - 3 = 6x + 5y - 8$$

■ *Solution:*

$$
\begin{array}{ll}
3x - 4y - 3 = 6x + 5y - 8 & (-3x) \\
-4y - 3 = 3x + 5y - 8 & (+4y) \\
-3 = 3x + 9y - 8 & (+8) \\
+5 = 3x + 9y & \\
\text{STF:} \quad 3x + 9y = 5 &
\end{array}
$$

5–4–2 *Graphing Data*

In many areas of science and technology, you will often see data shown visually in the form of a graph. To obtain this data, a *practical analysis* is performed on a device, in which the input is physically changed in increments and the output results are recorded in a table.

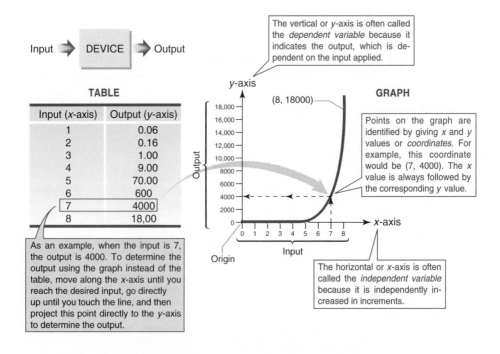

Cartesian Coordinate System

Two perpendicular number lines used to place points in a plane.

The graph shown would accommodate any data in which the input and output data are positive, but what sort of graph could be used if positive and negative values had to be plotted? The answer is a **Cartesian coordinate system**. With this system, the previous positive vertical x-axis and horizontal y-axis are still used, but are extended beyond the zero point in the opposite direction to accommodate negative values.

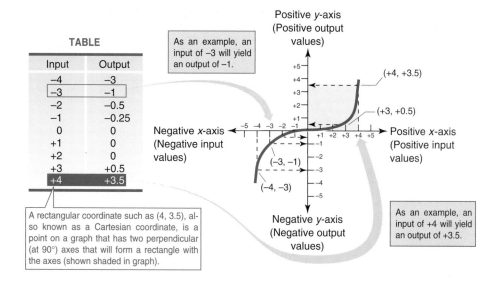

TABLE

Input	Output
−4	−3
−3	−1
−2	−0.5
−1	−0.25
0	0
+1	0
+2	0
+3	+0.5
+4	+3.5

As an example, an input of −3 will yield an output of −1.

Positive y-axis (Positive output values)

Negative x-axis (Negative input values)

Positive x-axis (Positive input values)

Negative y-axis (Negative output values)

As an example, an input of +4 will yield an output of +3.5.

(+4, +3.5)
(+3, +0.5)
(−3, −1)
(−4, −3)

A rectangular coordinate such as (4, 3.5), also known as a Cartesian coordinate, is a point on a graph that has two perpendicular (at 90°) axes that will form a rectangle with the axes (shown shaded in graph).

5–4–3 *Graphing Linear Equations*

To combine our understanding of linear equations and graphs, let us step through the process of converting an equation to a graph.

Consider the linear equation:

$$y = 2x + 2$$

This equation is, in fact, the formula you would use to determine the output (y) from a device for any given input (x). For example, if you applied 3 to the input $(x = 3)$, the output would be:

$$x \rightarrow \boxed{\begin{array}{c} \text{DEVICE} \\ y = 2x + 2 \end{array}} \rightarrow y$$

$$y(\text{output}) = 2(3) + 2 \qquad x(\text{input}) = 3$$
$$= 6 + 2$$
$$= 8$$

So how can we graph the response of this device? The answer is to pick some example input values, calculate the outputs for these inputs, and plot the points on a graph.

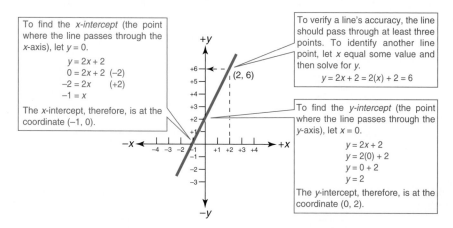

To find the *x-intercept* (the point where the line passes through the x-axis), let $y = 0$.

$$y = 2x + 2$$
$$0 = 2x + 2 \quad (-2)$$
$$-2 = 2x \quad (+2)$$
$$-1 = x$$

The x-intercept, therefore, is at the coordinate (−1, 0).

To verify a line's accuracy, the line should pass through at least three points. To identify another line point, let x equal some value and then solve for y.

$$y = 2x + 2 = 2(x) + 2 = 6$$

(2, 6)

To find the *y-intercept* (the point where the line passes through the y-axis), let $x = 0$.

$$y = 2x + 2$$
$$y = 2(0) + 2$$
$$y = 0 + 2$$
$$y = 2$$

The y-intercept, therefore, is at the coordinate (0, 2).

5–4–4 *Slope of a Line*

The steepness of the line in a graph is an indication of how much or little the output is changing for a given input. For example, in the distance versus time graph that follows, we can see how fast we were going by seeing how steep the line is.

Distance (miles)

Line a: 40 miles in 2 hours
Line b: 30 miles in 2 hours
Line c: 20 miles in 2 hours
Line d: 10 miles in 2 hours

A steeper line means you have a larger output (*y*) for a given input (*x*).

Time (hours)

Slope of a Line

The ratio of the vertical change to the horizontal change between two points plotted on a line.

The **slope of a line** is defined as the change in *y* that results from an increase in *x*.

$$\text{Slope} = \frac{\Delta y}{\Delta x}$$

Δy = change in *y*, called *rise*

Δx = change in *x*, called *run*

Change is generally symbolized by the Greek letter delta, Δ.

The change in input (Δx) value will always be positive, but the change in output (Δy) value can be either positive or negative based on whether *y* increased or decreased. To explain this point, take a look at the following two examples.

When a slope rises from left to right, the slope is said to be positive, and so *y* should be a positive value.

When a slope falls from left to right, the slope is said to be negative, and so *y* should be a negative value.

Slope-Intercept Form

A form used to express the slope of a line in which *m* = slope and *b* indicates the *y*-intercept, SIF: $y = mx + b$.

Any linear equation containing two variables can be written in the **slope-intercept form (SIF).**

$$\text{Slope-Intercept Form (SIF):} \quad y = mx + b$$

m = slope

b = *y*-intercept

In this appropriately named form, *m* represents the slope $(\Delta y/\Delta x)$ and *b* the *y*-intercept. The main characteristic of an equation in slope-intercept form is that it is solved for *y*, which means that *y* should be by itself on the left side of the equation.

■ **EXAMPLE**

Referring to the graph, determine the equation for the line and give it in slope-intercept form.

■ *Solution:*

Slope rises from left to right and so y is positive.

$$\text{Slope} = \frac{\Delta y}{\Delta x} = \frac{2}{1} = 2$$

$$y\text{-intercept} = +2$$

The equation of the line in slope-intercept form is:

$$\text{(SIF)} \quad y = 2x + 2$$

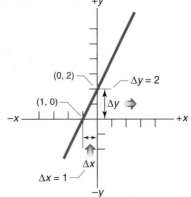

■ **EXAMPLE**

Convert the following equation to standard form (STF) and slope-intercept form (SIF), and then graph the line.

$$3y - 2x = 9$$

■ *Solution:*

First, convert the equation to standard form (STF), which is $Ax + By = C$.

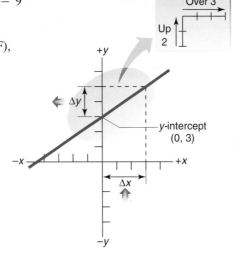

$$
\begin{array}{ll}
3y - 2x = 9 & (+2x) \\
3y = 9 + 2x & (-9) \\
3y - 9 = 2x & (-3y) \\
-9 = 2x - 3y & \\
\text{STF:} \quad 2x - 3y = -9 &
\end{array}
$$

Now, convert to slope-intercept form (SIF), which is: $y = mx + b$.

$$
\begin{array}{ll}
3y - 2x = 9 & (+2x) \\
3y = 9 + 2x & (\div 3) \\
y = \dfrac{2x + 9}{3} & \\
y = \dfrac{2}{3}x + \dfrac{9}{3} &
\end{array}
$$

$$\text{SIF:} \quad y = \frac{2}{3}x + 3$$

Using the SIF, it is easy to graph the line, as shown. The y-intercept is $+3$, the slope is positive $\left(\frac{2}{3}\right)$, and so the line rises from left to right, up 2 from the y-intercept and over 3.

Linear equations can also be represented in **point-slope form (PSF),** in which the point given may not be the y-intercept.

Point-Slope Form (PSF): $\quad y - y_1 = m(x - x_1) \qquad m = \text{slope}$

Subscripts (y_1 and x_1) are used to denote the point you are given.

Point-Slope Form

A form used to express the slope of a line in which x_1, y_1 indicates the point and m indicates slope. PSF: $y - y_1 = m(x - x_1)$.

■ EXAMPLE

The line shown in the graph crosses through point $(-5, 2)$ and has a slope of $-\frac{1}{5}$. What is the equation for this line in point-slope form?

■ *Solution:*

Point $(-5, 2)$ is not the y-intercept, $x_1 = -5$, $y_1 = 2$, and $m = -\frac{1}{5}$. Therefore:

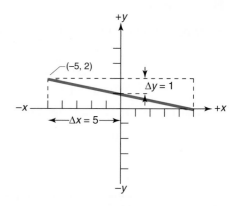

$$y - y_1 = m(x - x_1)$$

$$y - 2 = -\frac{1}{5}(x - (-5))$$

$$y - 2 = -\frac{1}{5}(x + 5)$$

$$\text{PSF:} \quad y - 2 = -\frac{1}{5}(x + 5)$$

■ EXAMPLE

Convert the point-slope form (PSF) equation in the previous example into standard form (STF).

■ *Solution:*

$$\text{PSF:} \quad y - 2 = -\frac{1}{5}(x + 5) \qquad \text{STF:} \quad Ax + Bx = c$$

To convert to standard form (STF), the coefficients must be integers (no fractions).

$$y - 2 = -\frac{1}{5}(x + 5) \qquad\qquad (\times 5)$$

$5y - 10 = -1(x + 5)$ (distribute the negative, $-1 \cdot (x + 5)$)

$5y - 10 = -x - 5$ $(+10)$

$\qquad 5y = -x + 5$ $(+x)$

$x + 5y = 5$

$\text{STF:} \quad x + 5y = 5$

5–4–5 *Exponents*

Many literal numbers are raised to a power, and so it is important to understand how to deal with these exponents if you are to transpose equations and formulas. In this section, we will review some of the rules discussed previously, and examine some new equalities:

1. Multiplication of Exponents

$$\boxed{a^2 \cdot a^3 = a^{2+3} = a^5}$$

Explanation:

$$a^2 \cdot a^3 = (a \cdot a) \cdot (a \cdot a \cdot a)$$
$$= a \cdot a \cdot a \cdot a \cdot a = a^5 \qquad (\text{``}a\text{'' as a factor 5 times})$$

2. Division of Exponents

$$\frac{a^3}{a^2} = a^{3-2} = a^1 = a$$

Explanation:

$$\frac{a^3}{a^2} = \frac{a \cdot \cancel{a} \cdot \cancel{a}}{\cancel{a} \cdot \cancel{a}} = a \cdot 1 \cdot 1 = a \qquad (a \div a = 1)$$

3. Negative Exponents

$$a^{-2} = \frac{1}{a^2}$$

Explanation:

$$2^2 = 2 \cdot 2 = 4$$
$$2^{-2} = \frac{1}{2} \cdot \frac{1}{2} = \frac{1}{4} = \frac{1}{2^2}$$

4. Exponents Raised to a Power

$$(a^3)^2 = a^{3 \times 2} = a^6$$

Explanation:

$$(a^3)^2 = (a \cdot a \cdot a)^2 = (a \cdot a \cdot a) \cdot (a \cdot a \cdot a) = a \cdot a \cdot a \cdot a \cdot a \cdot a = a^6$$

5. Root Power

$$\sqrt[3]{a^2} = \left(\sqrt[3]{a}\right)^2$$

Example:

$$\sqrt[3]{64^2} = \left(\sqrt[3]{64}\right)^2$$
$$\sqrt[3]{4096} = (4)^2$$
$$16 = 16$$

Whether you square "a" and then find the cube root of the result, or find the cube root of "a" and then square the result, the final answer is the same.

6. Fractional Power

$$a^{\frac{2}{3}} = \sqrt[3]{a^2}$$

Explanation:

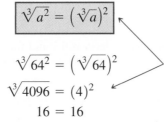

■ **EXAMPLE**

The reciprocal of a number will perform the opposite or inverse action on a value. For instance:

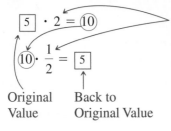

$\frac{1}{2}$ is the reciprocal of 2. $\frac{1}{2}$ will perform the opposite function of 2, since 2 will double a value whereas $\frac{1}{2}$ will half a value.

Original Value   Back to Original Value

Therefore, since $\frac{1}{3}$ is the reciprocal of 3:

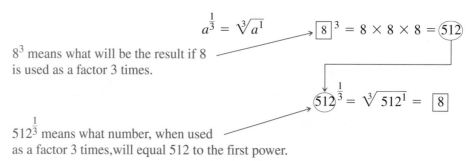

$$a^{\frac{1}{3}} = \sqrt[3]{a^1}$$

$8^3 = 8 \times 8 \times 8 = \boxed{512}$

8^3 means what will be the result if 8 is used as a factor 3 times.

$$512^{\frac{1}{3}} = \sqrt[3]{512^1} = \boxed{8}$$

$512^{\frac{1}{3}}$ means what number, when used as a factor 3 times, will equal 512 to the first power.

5–4–6   *Factoring Patterns*

In this section, we will take our understanding of factoring a step further.

1. Basic Patterns

A three term (trinomial) expression can be factored if you can find two numbers whose sum is the coefficient of the second term, and whose product is the third term.

$$x^2 + bx + c = (x + m) \cdot (x + n)$$

■ **EXAMPLE:**

$$x^3 + 13x + 40 = (x + ?) \cdot (x + ?)?$$

■ *Solution:*

In this expression, the coefficient of the second term is 13 and the coefficient of the third term is 40. Using trial and error, test to see if you can find two numbers whose sum $(+)$ is 13, and whose product $(\times)$ is 40.

$$12 + 1 = 13 \,(\text{Yes}), \quad 12 \times 1 = 12 \,(\text{No})$$
$$11 + 2 = 13 \,(\text{Yes}), \quad 11 \times 2 = 22 \,(\text{No})$$
$$10 + 3 = 13 \,(\text{Yes}), \quad 10 \times 3 = 30 \,(\text{No})$$
$$9 + 4 = 13 \,(\text{Yes}), \quad \;\;9 \times 4 = 36 \,(\text{No})$$
$$8 + 5 = 13 \,(\text{Yes}), \quad \;\;8 \times 5 = 40 \,(\text{Yes})$$

The two common factors for 13 and 40, therefore, are 8 and 5. These can be placed in the factoring pattern as follows:

$$x^2 + 13x + 40 = (x + 8) \cdot (x + 5)$$

To test if the factoring pattern in the previous example is correct, let us multiply through to see if we obtain our original equation.

Steps:

$(x + 8) \cdot (x + 5)$	$x^2 + ? + ? + ?$	$x \cdot x = x^2$
$(x + 8) \cdot (x + 5)$	$x^2 + 5x + ? + ?$	$x \cdot 5 = 5x$
$(x + 8) \cdot (x + 5)$	$x^2 + 5x + 8x + ?$	$8 \cdot x = 8x$
$(x + 8) \cdot (x + 5)$	$x^2 + 5x + 8x + 40$	$8 \cdot 5 = 40$

$$(x + 8) \cdot (x + 5) = x^2 + (5x + 8x) + 40 = x^2 + 13x + 40$$

The order in which 8 and 5 are inserted into the factoring pattern is immaterial as proved here:

$$(x + 8)(x + 5) \qquad \text{or} \qquad (x + 5)(x + 8)$$
$$= x^2 + 5x + 8x + 40 \qquad\qquad = x^2 + 8x + 5x + 40$$
$$= x^2 + 13x + 40 \qquad\qquad = x^2 + 13x + 40$$

In summary, here are the factoring pattern combinations, all of which can be obtained by using the "Factor" function on your calculator, as previously described.

Factoring Patterns (Trinomials)

$$\square + \square + \square \qquad = (x + m)(x + n)$$

$$\square + \square - \square \qquad = (x + m)(x - n) \text{ or } (x - m)(x + n)$$

$$\square - \square + \square \text{ or } \square - \square - \square = (x - m)(x - n)$$

2. Difference of Perfect Squares

A perfect square is a number such as 16 or 25, since it is a result of a number multiplied by itself. For example:

$$4 \times 4 = 16$$
$$5 \times 5 = 25$$
$$\longrightarrow \text{Perfect squares}$$

If you have one perfect square subtracted (difference) from another, you can factor it using the following pattern.

$$\boxed{x^2 - y^2 = (x - y)(x + y)}$$

Example:

$$x^2 - 25$$
$$= x^2 - 5^2 \qquad (x^2 \text{ and } 5^2 \text{ are perfect squares})$$
$$= (x - 5)(x + 5)$$

You can factor the difference of perfect squares $[x^2 - 4 = (x - 2)(x + 2)]$, but you cannot factor the sum of perfect squares $(x^2 + 4 = x^2 + 4)$.

3. Sum of Perfect Cubes

$$\boxed{x^3 + y^3 = (x + y)(x^2 - xy + y^2)}$$

Example:

$$x^3 + 125$$
$$= x^3 + 5^3$$
$$= (x + 5)(x^3 - 5x + 5^2)$$

4. Difference of Perfect Cubes

$$\boxed{x^3 - y^3 = (x - y)(x^2 + xy + y^2)}$$

Example:

$$x^3 - 64$$
$$= x^3 - 4^3$$
$$= (x - 4)(x^2 + 4x + 4^2)$$

As stated previously, if the literal numbers in an equation are raised to the first power, they are called linear equations. If the literal numbers are raised to the second power, they are called **second-degree equations** or **quadratic equations.** In standard form (STF) a quadratic equation looks like this:

$$\text{STF:} \quad ax^2 + bx + c = 0 \quad \text{(Quadratic Equation)}$$

a is the coefficient of x^2

b is the coefficient of x

c is a constant

Whenever you solve for the unknown (x) in a quadratic equation, there will always be one of two possible answers. To prove this, let us look at an example.

■ **EXAMPLE**

Solve the following equation:

$$3x^2 + 4x = -1$$

■ *Solution:*

The first step is to set the equation so that it is equal to zero:

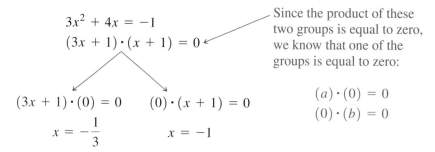

$$3x^2 + 4x = -1$$
$$(3x + 1) \cdot (x + 1) = 0$$

Since the product of these two groups is equal to zero, we know that one of the groups is equal to zero:

$$(3x + 1) \cdot (0) = 0 \qquad (0) \cdot (x + 1) = 0$$
$$x = -\frac{1}{3} \qquad\qquad x = -1$$

$$(a) \cdot (0) = 0$$
$$(0) \cdot (b) = 0$$

This equation, like all quadratic equations, has two possible solutions $\left(-\frac{1}{3} \text{ and } -1 \right)$.
To see if this is true, insert each into the original equation $(3x^2 + 4x = -1)$.

When your quadratic equation is set to zero $(ax^2 + bx + c = 0)$, you can solve for x by inserting your a, b, and c coefficients into the following **quadratic formula.**

$$x = \frac{-b \pm \sqrt{b^2 - 4ac}}{2a}$$

■ **EXAMPLE**

Solve the equation $2x^2 + 12x - 18 = 0$

■ *Solution:*

$$a = 2, b = 12, c = -18$$
$$x = \frac{-b \pm \sqrt{b^2 - 4ac}}{2a}$$

$$= \frac{-(12) \pm \sqrt{(12)^2 - 4(2)(-18)}}{2(2)}$$

$$= \frac{-12 \pm \sqrt{144 - (-144)}}{4}$$

$$= \frac{-12 \pm \sqrt{288}}{4}$$

$$= \frac{-12 \pm 12\sqrt{2}}{4}$$

$$= \frac{(-12) + \left(12\sqrt{2}\right)}{4} \qquad = \frac{(-12) - \left(12\sqrt{2}\right)}{4}$$

$$= \frac{-12}{4} + \frac{12\sqrt{2}}{4} \qquad = \frac{-12}{4} - \frac{12\sqrt{2}}{4}$$

$$= (-3) + \left(3\sqrt{2}\right) \qquad = (-3) - \left(3\sqrt{2}\right)$$

The following memory aid may help you to remember the equation formula.

The bad boy $(-b)$	$\rightarrow x = -b$
could not decide yes or no $(\pm)$	$\rightarrow x = -b \pm$
on whether to go to the radical party $(\sqrt{})$	$\rightarrow x = -b \pm \sqrt{}$
where he could square dance	$\rightarrow x = -b \pm \sqrt{b^2}$
to the bad music of the Four Aces	$\rightarrow x = -b \pm \sqrt{b^2 - 4ac}$
until it was all over at 2 a.m.	$\rightarrow x = \dfrac{-b \pm \sqrt{b^2 - 4ac}}{4}$

SELF-TEST EVALUATION POINT FOR SECTION 5–4

Now that you have completed this section, you should be able to:

■ **Objective 12.** *Perform the following algebra skills:*
a. Create linear equations and express them in standard form
b. Graph data and linear equations using the slope-intercept form and point-slope form
c. Describe the properties of exponents
d. Factor polynomials
e. Solve quadratic equations using the quadratic formula

Use the following questions to test your understanding of Section 5–4.

1. Write the equation of a line in slope-intercept form that has a slope of -2 and a y-intercept of 3.

2. Factor the expression: $x^2 - 8x + 15$.

3. Solve the following quadratic equation using the quadratic formula:
$$4x^2 + 12x = 0$$

4. Simplify the following expression using exponential rules:
$$\left(4x^4y^2\right)^2$$

5–5   RULES OF ALGEBRA—A SUMMARY

As a review of the lessons learned in this chapter, Table 5–1 summarizes the rules of algebra with many examples. Study every problem, since these will provide you with excellent practice for the applications that follow.

TABLE 5–1   Rules of Algebra—A Summary

Monomial (One Term): $2a$

Binomial (Two Terms): $2a + 3b$

Trinomial (Three Terms): $2a + 3b + c$

Polynomial (Any Number of Terms).

Order of Operations: Parentheses, Exponents, Multiplication, Division, Addition, Subtraction.

   (Memory Aid: Please, Excuse, My, Dear, Aunt, Sally)

Basic Rules:

$a + a = 2a$	$a - a = 0$	$a \cdot a = a^2$	$a \div a = 1$	$\sqrt{a^2} = a$
$a + 1 = a + 1$	$a - 1 = a - 1$	$a \cdot 1 = a$	$a \div 1 = a$	
$a + 0 = a$	$a - 0 = a$	$a \cdot 0 = 0$	$a \div 0 = $ Impossible	$0 \div a = 0$

Parentheses First: $(3 + 6) \cdot 2 = 9 \cdot 2 = 18$ $9 - (2 \cdot 3) = 9 - 6 = 3$ $6 + (9 - 7) = 6 + 2 = 8$

No Parentheses—Step 1:   Multiplications and Divisions, left to right. $3 + 6 \cdot 4 = 3 + 24 = 27$

              Step 2:   Additions and Subtractions, left to right. $4 \cdot 2 + 3 - 4 \div 2 = 8 + 3 - 2 = 9$

Combine Only Like Terms: $2x + x + y = 3x + y$ $3a + 4b - a = 2a + 4b$

Commutative Property of Addition—order of terms makes no difference: $a + b = b + a$

Associative Property of Addition—grouping of terms makes no difference: $(a + b) + c = a + (b + c)$

Coefficients: $x + x + x = 3x$ $x + y + x + y = 2x + 2y$

Exponents: $x \cdot x \cdot x = x^3$ $x \cdot y \cdot x \cdot y = x^2 \cdot y^2 = x^2 y^2$

$$a^{-2} \cdot a^3 = a^{2+3} = a^5 \qquad \frac{a^3}{a^2} = a^{3-2} = a^1 = a$$

$$a^{-2} = \frac{1}{a^2} \qquad (a^3)^2 = a^{3 \times 2} = a^6$$

$$\sqrt[3]{a^2} = \left(\sqrt[3]{a}\right)^2 \qquad a^{\frac{2}{3}} = \sqrt[3]{a^2}$$

Commutative Property of Multiplication—order of factors makes no difference: $a \cdot b = b \cdot a$

Associative Property of Multiplication—grouping of factors makes no difference: $(ab)c = a(bc)$

Distributive Property of Multiplication—$a(b + c) = ab + ac$ $2(a - 3b) = 2 \cdot a - 2 \cdot 3b = 2a - 6b$

Transposition—Step 1: Ensure unknown is above fraction bar. $x \cdot 8 = 16$ (divide both sides by $8, \div 8$)

         Step 2: Isolate unknown. $(x \cdot 8) \div 8 = (16) \div 8, \qquad x = 2$

Unknown Above Fraction Bar	Unknown Below Fraction Bar	Unknown on Both Sides	Combining Unknowns
$(2 \cdot a) + 5 = 23 \quad (-5)$	$\dfrac{72}{x} = 12 \qquad (\cdot x)$	$(6 \cdot y) + 7 = y + 27 \quad (-7)$	$x + 2x + 4(x + 2x) = 100$
$2 \cdot a = 18 \quad (\div 2)$	$72 = 12 \cdot x \quad (\div 12)$	$6y = y + 20 \quad (-y)$	$x + 2x + 4x + 8x = 100$
$a = 9$	$6 = x$	$5y = 20 \quad (\div 5)$	$15x = 100$
		$y = 4$	$x = 6.67$

Positive and Negative Number Rules

Addition $\oplus$ $(+) + (+) = (+), (-) + (-) = (-)$

           $(+) + (-)$ or $(-) + (+) = $ **Difference**

Subtraction $\ominus$ Change sign of second number, and then add.

Multiplication $\otimes$ $(+) \cdot (+) = (+), (-) \cdot (-) = (+)$

           $(+) \cdot (-)$ or $(-) \cdot (+) = (-)$

Division $\ominus$ $(+) \div (+) = (+), (-) \div (-) = (+)$

           $(+) \div (-)$ or $(-) \div (+) = (-)$

Positive and Negative Numbers

Addition $\oplus$ $2x + 5x = 7x \qquad -6a + (+3a) = -3a$

Subtraction $\ominus$ $-3y^2 - (-2y^2) = -3y^2 + (+2y^2) = -y^2$

Multiplication $\otimes$ $-3(6a) = -3 \cdot (+6a) = -18a$

Division $\ominus$ $14a \div (-7) = -2a$

Adding Polynomials: $(4a^2 + 6b^2 + x - 2) + (7a^2 - b^2 + 3)$

$$
\begin{array}{r}
4a^2 + 6b^2 + x - 2 \\
+ \quad 7a^2 - b^2 \quad + 3 \\
\hline
11a^2 + 5b^2 + x + 1
\end{array}
$$

Subtracting Polynomials: $(6x^2 + y^2 - 4) - (-4x^2 + 2y^2 + 6)$

$$
\begin{array}{rl}
6x^2 + y^2 - 4 & \text{Rule: Add} \\
- \quad -4x^2 + 2y^2 + 6 & \text{the Opposite} \rightarrow
\end{array}
\qquad
\begin{array}{r}
6x^2 + y^2 - 4 \\
+ \quad 4x^2 - 2y^2 - 6 \\
\hline
10x^2 - y^2 - 10
\end{array}
$$

Multiplying Monomials, Binomials

$x^2 \cdot x^3 \cdot x = x \cdot x \cdot x \cdot x \cdot x \cdot x = x^{2+3+1} = x^6$ $4x^2 \cdot x^3 = 4 \cdot x \cdot x \cdot x \cdot x \cdot x = 4x^{2+3} = 4x^5$

$(x^3)^2 = (x \cdot x \cdot x) \cdot (x \cdot x \cdot x) = x^{3 \cdot 2} = x^6$ $(-6y)^2 \cdot y^3 = (-6)^2 \cdot y^2 \cdot y^3 = 36y^5$

$2(a + 2b) = 2a + 4b$ $-2(2a - 3b) = (-2 \cdot 2a) - (-2 \cdot 3b) = -4a - (-6b) = p - 4a + 6b$

Multiplying Polynomials

$$
\begin{array}{r}
x^2 + 2x - 1 \\
\times \qquad x - 3 \\
\hline
-3x^2 - 6x + 3 \\
x^3 + 2x^2 - x \\
\hline
x^3 - x^2 - 7x + 3
\end{array}
\quad
\begin{array}{l}
(x^2 + 2x - 1) \cdot (x - 3) = ? \\
\leftarrow -3(x^2 + 2x - 1) \\
\leftarrow x(x^2 - 2x - 1)
\end{array}
$$

$$
\begin{array}{r}
a - 3 \\
\times \quad a - 3 \\
\hline
-3a + 9 \\
a^2 - 3a \\
\hline
a^2 - 6a + 9
\end{array}
\quad
\begin{array}{l}
(a - 3)^2 = ? \\
\leftarrow -3(a - 3) \\
\leftarrow a(a - 3)
\end{array}
$$

164   CHAPTER 5 / ALGEBRA, EQUATIONS, AND FORMULAS

<u>Dividing Polynomials</u>

$$\frac{a^4}{a} = \frac{{}^1\cancel{a} \cdot a \cdot a \cdot a}{{}_1\cancel{a}} = a^3 \qquad \frac{-6ab^3}{2ab} = -\frac{{}^{-3}\cancel{6} \cdot \cancel{a} \cdot \cancel{b} \cdot b \cdot b}{{}_1\cancel{2} \cdot \cancel{a} \cdot \cancel{b}} = -3b^2$$

$$\frac{2ab - 4a^2b^2}{2a} = \frac{2ab}{2a} - \frac{4a^2b^2}{2a} = b - 2ab^2 \qquad \frac{2a^2 + 6a + 8ab}{2a} = \frac{2a^2}{2a} + \frac{6a}{2a} + \frac{8ab}{2a} = a + 3 + 4b$$

<u>Linear Equations</u> (literals raised to first power):

Standard Form (STF): $Ax + By = C$ Slope $= \dfrac{\Delta y}{\Delta x}$ ("╱" slope $= +$, "╲" slope $= -$)

Slope-Intercept Form (SIF): $y = mx + b$

Point-Slope Form (PSF): $y - y_1 = m(x - x_1)$

<u>Factoring Patterns</u> $(x^2 + bx + c = (x + m)(x + n))$:

$$\square + \square + \square = (x + m)(x + n)$$

$$\square + \square - \square = (x + m)(x - n) \text{ or } (x - m)(x + n)$$

$$\square - \square + \square \text{ or } \square - \square - \square = (x - m)(x - n)$$

Difference of Perfect Squares: $x^2 - y^2 = (x - y)(x + y)$

Sum of Perfect Cubes: $x^3 + y^3 = (x + y)(x^2 - xy + y^2)$

Difference of Perfect Cubes: $x^3 - y^3 = (x - y)(x^2 + xy + y^2)$

<u>Quadratic Equations</u> (literals raised to second power):

Standard Form (STF): $ax^2 + bx + c = 0$

Quadratic Formula: $x = \dfrac{-b \pm \sqrt{b^2 - 4ac}}{2a}$

■ INTEGRATED MATH APPLICATION: ELECTRONICS

The **period** of a waveform is the time it takes to complete one full cycle. A cycle is made up of a positive and negative alternation of voltage or current. The **frequency** of a waveform indicates the number of repetitions in one second. As an example, the electricity available at the wall outlet in your home is an alternating current (ac) sine wave that has a frequency of 60 hertz (Hz). This means that 60 cycles arrive every second, and so each cycle's period would be $\frac{1}{60}$th of a second. Expressed mathematically:

Frequency = 60 hertz

Period $= \dfrac{1}{\text{frequency}} = \dfrac{1}{60} = 16.67$ milliseconds

Period = 16.67 milliseconds

Frequency $= \dfrac{1}{\text{period}} = \dfrac{1}{16.67 \text{ ms}} = 60$ Hz

Period

Time taken to complete one complete cycle of a periodic or repeating waveform.

Frequency

Rate or recurrences of a periodic wave normally within a unit of one second, measured in hertz (cycles/second).

$$\text{Frequency (in hertz)} = \frac{1}{\text{period (in seconds)}} \quad \text{and} \quad \text{Period} = \frac{1}{\text{frequency}}$$

a. Determine the period of the waveforms shown.

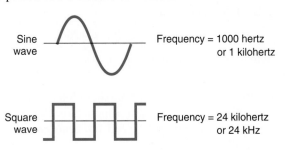

Sine wave — Frequency = 1000 hertz or 1 kilohertz

Square wave — Frequency = 24 kilohertz or 24 kHz

b. Determine the frequency of the waveforms shown.

- ■ *Solution:*

a. Period $= \dfrac{1}{\text{frequency}} = \dfrac{1}{1 \times 10^3} = 1 \times 10^{-3}$ seconds

or 0.001 second ($\frac{1}{1000}$ th of a second)

Period $= \dfrac{1}{\text{frequency}} = \dfrac{1}{24 \times 10^3} = 4.2 \times 10^{-5}$ seconds or 0.000042 second

b. $F = \dfrac{1}{P} = \dfrac{1}{1 \times 10^{-3}} = 1 \times 10^3$ hertz or 1000 Hz (1000 cycles every second)

$F = \dfrac{1}{P} = \dfrac{1}{560 \times 10^{-6}} = 1785.7$ hertz or 1.8 kilohertz

---

■ INTEGRATED MATH APPLICATION: SHEET METAL

The formula for calculating the water pressure in pounds per square foot is:

$$P = 62.4 \times D$$

where P = pressure in pounds per square foot.

        D = depth of water in feet.

        62.4: weight of 1 cubic foot of water.

Water tank

a. Calculate the pressure applied to the bottom of a tank that holds 6 feet of water.

b. If a tank is constructed to withstand 1200 pounds per square foot of pressure, what is the maximum safe height of the tank?

- ■ *Solution:*

a. $P = 62.4 \times D = 62.4 \times 6 = 374.4$ pounds per square foot

b. $D = \dfrac{P}{62.4} = \dfrac{1200}{62.4} = 19.2$ feet

■ INTEGRATED MATH APPLICATION: HEALTH CARE

The following formula is used to determine the maximum heart rate of a person during exercise, which depends on the person's age.

$$H_R = (-0.8 \times A) + 176$$

where H_R = heart rate in beats per minute (bpm)

        A = age

a. A 40-year-old patient when exercising on a treadmill has a heart rate of 140 beats per minute. Is this value below the person's maximum target heart rate?

b. At what age should your heart rate not exceed 160 beats per minute?

Solution:

a. $H_R(\text{max}) = (-0.8 \times A) + 176 = (-0.8 \times 40) + 176 = 144$ beats per minute

$\qquad\qquad$ 140 is below the maximum target heart rate of 144.

b. $A = 220 - 1.25 \times H_R(\text{max})$

$\qquad = 220 - 1.25 \times 160$

$\qquad = 220 - 200$

$\qquad = 20$ years old

SELF-TEST EVALUATION POINT FOR SECTION 5–5

Now that you have completed this section, you should be able to:

■ **Objective 12.** *Summarize the terms, rules, properties, and mathematical operations of algebra.*

Use the following questions to test your understanding of Section 5–5.

1. What is the order of operations?

2. Give an example of a:
   a. Monomial   **c.** Trinomial
   b. Binomial   **d.** Polynomial

3. Determine the following:
   a. $a \div 1 = ?$
   b. $a + a = ?$
   c. $a - a = ?$
   d. $a \times a = ?$
   e. $3x + b - 2x = ?$

4. Determine the unknown:
   a. $16 + a = 5 \times a, a = ?$
   b. $\dfrac{42}{b} = 3 \times 7$

REVIEW QUESTIONS

Multiple Choice Questions

1. If you perform exactly the same mathematical operation on both sides of an equation, the equality is preserved.
   a. True   **b.** False

2. Are the two sides of the following equation, $(3 - 2) \times 16 = 32 \div 2$, equal?
   a. Yes
   b. No
   c. Only if the right half is doubled
   d. Only if the left half is doubled

3. $x + x = $ _____, $x \times x = $ _____, $x \div x = $ _____, and $x - x = $ _____.
   a. $x^2, 2x, 1, 0$ **c.** $1x, 2x, 1, 0$
   b. $2x, \sqrt{x}, 0, 1$ **d.** $2x, x^2, 1, 0$

4. What is the value of the unknown in the following equation: $3a + 9 = 2a + 10$?
   a. 2   **c.** 1
   b. 10   **d.** 3

5. Calculate the value x for the following equation: $x + y^2 = y^2 + 9$.
   a. 3   **c.** 9
   b. 6   **d.** 12

6. What is the value of the unknown, s, in the following equation: $\sqrt{s^2} + 14 = 26$?
   a. 12   **c.** 4
   b. 26   **d.** 5

7. What is the correct transposition for the following formula: $a = \dfrac{b - c}{x}, x = ?$
   a. $x = \dfrac{b - a}{c}$ **c.** $x = \dfrac{b - c}{a}$
   b. $x = (b - a) \times c$ **d.** $x = x(b - c)$

8. If $V = I \times R$ and $P = V \times I$, develop a formula for P when I and R are known.
   a. $P = \dfrac{I}{R}$ **c.** $P = I^2 \times R$
   b. $P = I^2 \times R^2$ **d.** $P = \sqrt{\dfrac{I}{R}}$

9. If $E = M \times C^2, M = ?$ and $C = ?$
   a. $\sqrt{\dfrac{E}{C}}, \dfrac{E^2}{M}$ **c.** $E \times C^2, \sqrt{C^2 \times E}$
   b. $\dfrac{E}{C^2}, \sqrt{\dfrac{E}{M}}$ **d.** $\dfrac{C^2}{E}, \sqrt{\dfrac{M}{E}}$

10. If $P = \dfrac{V^2}{R}$, transpose to obtain a formula to solve for V.

    a. $V = \sqrt{P \times R}$ **c.** $V = P \times R^2$

    b. $V = \dfrac{P^2}{R}$ **d.** $V = \sqrt{P} \times R$

Communication Skill Questions

11. Define the following terms: (5–1)

    a. Algebra   **c.** Equation

    b. Formula   **d.** Literal number

12. Why is it important to treat both sides of an equation equally? (5–1)

13. Describe how the following three factors can be combined in a formula: distance (d), velocity (v), and time (t). (5–2)

14. What two steps must be followed to transpose an equation? (5–2)

15. Describe the factoring process. (5–2)

16. Define the terms *proportional* and *inversely proportional*. (5–2)

17. Explain how current (I), voltage (V), and resistance (R) can be combined in a formula. (5–2)

18. Transpose the formula from question 22 to solve for I, V, and R. (5–2)

19. If $P = V \times I$, describe how you would solve for V and I. (5–2)

20. Define substitution. (5–3)

21. If $\pi = C/d$, and area $= \frac{1}{2} \times (C \times r)$, describe how substitution and transposition can be employed to obtain a formula for area when only the radius of a circle is known. (5–3)

22. Explain how the parentheses can be removed from the following examples: (5–2)

    a. $4(3b - 1b)$ **b.** $2(3x \times 2x)$

23. If $a = b \times c$, explain how you would solve for b and c. (5–2)

24. Describe the relationship between square and square root. (5–2)

25. Why is formula manipulation important? (5–1)

Practice Problems

26. Calculate the result of the following arithmetic operations involving literal numbers.

    a. $x + x + x = ?$ **g.** $4y \times 3y^2 = ?$

    b. $5x + 2x = ?$ **h.** $a \times a = ?$

    c. $y - y = ?$ **i.** $2y \div y = ?$

    d. $2x - x = ?$ **j.** $6b \div 3b = ?$

    e. $y \times y = ?$ **k.** $\sqrt{a^2} = ?$

    f. $2x \times 4x = ?$ **l.** $(x^3)^2 = ?$

27. Transpose the following equations to determine the unknown value.

    a. $4x = 11$

    b. $6a + 4a = 70$

    c. $5b - 4b = \dfrac{7.5}{1.25}$

    d. $\dfrac{2z \times 3z}{4.5} = 2z$

28. Apply transposition and substitution to Newton's second law of motion.

    a. $F = ma$ [Force (F) = mass (m) × acceleration (a)] If $F = m \times a$, $m = ?$ and $a = ?$

    b. If acceleration (a) equals change in speed (Δv) divided by time (t), how would the $F = ma$ formula appear if acceleration (a) was substituted for $\dfrac{\Delta v}{t}$?

29. **a.** Using the formula power (P) = voltage (V) × current (I), calculate the value of electric current through a hair dryer that is rated as follows: power = 1500 watts and voltage = 120 volts.

$$\boxed{P = V \times I}$$

    where P = power in watts

             V = voltage in volts

             I = current in amperes

    b. Using Ohm's formula voltage (V) = current (I) × resistance (R), calculate the resistance or opposition to current flow offered by the hairdryer in Question 14a.

30. The following formula is used to calculate the total resistance (R_T) in a parallel circuit containing two parallel resistances $(R_1$ and $R_2)$.

$$R_T = \dfrac{1}{\dfrac{1}{R_1} + \dfrac{1}{R_2}}$$

    a. Calculate R_T (total resistance) if the resistance of $R_1 = 6$ ohms and the resistance of $R_2 = 12$ ohms.

    b. Indicate the calculator sequence used for part (a).

    c. If R_2 and R_T were known and we needed to calculate the value of R_1, we would have to transpose the total parallel resistance formula to solve for R_1. The following steps show how to transpose the total resistance formula to calculate R_1 when R_T and R_2 are known. Describe what has occurred in each of the steps.

$$R_T = \cfrac{1}{\dfrac{1}{R_1} + \dfrac{1}{R_2}} \qquad \text{(Original formula)}$$

1. $R_T \times \left(\dfrac{1}{R_1} + \dfrac{1}{R_2} \right) = \cfrac{1}{\dfrac{1}{R_1} + \dfrac{1}{R_2}} \times \left(\dfrac{1}{\cancel{R_1}} + \dfrac{1}{\cancel{R_2}} \right)$

$R_T \times \left(\dfrac{1}{R_1} + \dfrac{1}{R_2} \right) = 1$

2. $\cfrac{\cancel{R_T} \times \left(\dfrac{1}{R_1} + \dfrac{1}{R_2} \right)}{\cancel{R_T}} = \dfrac{1}{R_T}$

$\dfrac{1}{R_1} + \dfrac{1}{R_2} = \dfrac{1}{R_T}$

or $\dfrac{1}{R_T} = \dfrac{1}{R_1} + \dfrac{1}{R_2}$

3. $\dfrac{1}{R_T} - \dfrac{1}{R_2} = \dfrac{1}{R_1} + \dfrac{1}{\cancel{R_2}} - \dfrac{1}{\cancel{R_2}}$

$\dfrac{1}{R_T} - \dfrac{1}{R_2} = \dfrac{1}{R_1}$

or $\dfrac{1}{R_1} = \dfrac{1}{R_T} - \dfrac{1}{R_2}$

4. $\dfrac{1}{R_1} = \boxed{\dfrac{1}{R_T} - \dfrac{1}{R_2}}$

$\dfrac{\dfrac{1}{R_T} - \dfrac{1}{R_2}}{} = \overline{\quad\textit{lowest common denominator (LCD)}\quad}$

$\dfrac{\dfrac{1}{R_T} - \dfrac{1}{R_2}}{} = \dfrac{}{R_T \times R_2}$

$\dfrac{\dfrac{1}{R_T} - \dfrac{1}{R_2}}{} = \dfrac{R_2 - R_T}{R_T \times R_2}$

$\dfrac{1}{R_T} - \dfrac{1}{R_2} = \boxed{\dfrac{R_2 - R_T}{R_T \times R_2}}$

$\dfrac{1}{R_1} = \dfrac{R_2 - R_T}{R_T \times R_2}$

(LCD) method 1: fractions
$R_T \times R_2 = R_T \times R_2$

Steps:
$(R_T \times R_2) \div R_T = R_2,$
$R_2 \times 1 = R_2$

$(R_T \times R_2) \div R_2 = R_T,$
$R_T \times 1 = R_T$

5. $\dfrac{1}{\cancel{R_1}} \times \cancel{R_1} = \dfrac{R_2 - R_T}{R_T \times R_2} \times R_1$

$1 = \dfrac{R_2 - R_T}{R_T \times R_2} \times R_1$

6. $1 \times R_T \times R_2 = \dfrac{R_2 - R_T}{\cancel{R_T} \times \cancel{R_2}} \times R_1 \times \cancel{R_T} \times \cancel{R_2}$

$R_T \times R_2 = (R_2 - R_T) \times R_1$

7. $\dfrac{R_T \times R_2}{R_2 - R_T} = \dfrac{(\cancel{R_2} - \cancel{R_T}) \times R_1}{\cancel{R_2} - \cancel{R_T}}$

$\dfrac{R_T \times R_2}{R_2 - R_T} = R_1 \quad \text{or} \quad R_1 = \dfrac{R_T \times R_2}{R_2 - R_T}$

d. Using the values from part (a), double-check the new formula.

e. If we wanted to obtain a formula for calculating R_2 when both R_T and R_1 are known, which of the steps would we change, and in what way?

f. Give the formula for calculating R_2 when both R_T and R_1 are known.

g. Using the values from part (a), double-check this new formula.

Web Site Questions

Go to the Web site http://www.prenhall.com/cook, select the textbook *Introductory Mathematics*, select this chapter, and then follow the instructions when answering the multiple-choice practice problems.

Geometry and Trigonometry

Finding the Question to the Answer!

More than 350 years ago in 1637, Pierre de Fermat, a French mathematician and physicist, wrote an apparently simple theorem in the margins of a book. He also added that he had discovered marvelous proof of it but lacked enough space to include it in the margin. Later that night he died and took with him a mystery that mathematicians have been trying to solve ever since. Many of the brightest minds in mathematics have struggled to find the proof, and many have concluded that Fermat must have been mistaken despite his considerable mathematical ability.

Fermat's last theorem, as it has been appropriately named, has to do with equations of the form x to the nth power, plus y to the nth power = z to the nth power.

$$x^n + y^n = z^n$$

A power of 2 (square) is familiar as the Pythagorean theorem, which states that the sum of the squares of the lengths of two sides of a right-angle triangle is equal to the square of the length of the other side:

$$c^2 = a^2 + b^2$$

For example,

$$5^2 = 3^2 + 4^2$$
$$25 = 9 + 16$$

Fermat's last theorem seemed not to work because, for example,

$$? = 3^3 + 4^3$$
$$91 = 27 + 64$$

and there is not a whole number that can be cubed to equal 91.

Now, after thousands of claims of success that proved untrue, mathematicians say the daunting challenge, perhaps the most famous of unsolved mathematical problems, has been surmounted. The conqueror is Andrew Wiles, 40, a British mathematician who works at Princeton University. Wiles announced the result at the last of three lectures at Cambridge University in England. Within a few minutes of the conclusion of his final lecture, computer mail messages were winging around the world, sending mathematicians into a frenzy. It had apparently taken 7 years for Wiles to solve the problem. He continued a chain of ideas that was begun in 1954 by several other mathematicians.

After being criticized by many in the eighteenth and nineteenth centuries and ridiculed in the twentieth century by mathematicians with supercomputers, Fermat was elevated finally in 1993 to the status of "a genius before his time."

Outline and Objectives

VIGNETTE: FINDING THE QUESTION TO THE ANSWER!

INTRODUCTION

6–1 BASIC GEOMETRY TERMS

Objective 1: Describe the purpose of geometry, and define many of the basic geometric terms.

Objective 2: Demonstrate how to use a protractor to measure and draw an angle.

6–2 PLANE FIGURES

Objective 3: Define the term *plane figure,* and identify the different types of polygons.

Objective 4: List the formulas for determining the perimeter and area of plane figures.

6–2–1 Quadrilaterals

Objective 5: Explain how the Pythagorean theorem relates to right-angle triangles.

Objective 6: Determine the length of one side of a right-angle triangle when the lengths of the other two sides are known.

Objective 7: Describe how vectors are used to represent the magnitude and direction of physical quantities and how they are arranged in a vector diagram.

Objective 8: Explain how the Pythagorean theorem can be applied to a vector diagram to calculate the magnitude of a resultant vector.

6–2–2 Right-Angle Triangles

6–2–3 Vectors and Vector Diagrams

6–2–4 Other Triangles

6–2–5 Circles

6–2–6 Other Polygons

6–3 TRIGONOMETRY

Objective 9: Define the trigonometric terms:
- a. Opposite
- b. Adjacent
- c. Hypotenuse
- d. Theta
- e. Sine
- f. Cosine
- g. Tangent

Objective 10: Demonstrate how the sine, cosine, and tangent trigonometric functions can be used to calculate:
- a. The length of an unknown side of a right-angle triangle if the length of another side and the angle theta are known
- b. The angle theta of a right-angle triangle if the lengths of two sides of the triangle are known

6–3–1 Sine of Theta

6–3–2 Cosine of Theta

6–3–3 Tangent of Theta

6–3–4 Summary

6–4 SOLID FIGURES

Objective 11: Define the term *solid figure*, and identify the different types.

Objective 12: List the formulas for determining the lateral surface area, total surface area, and volume of solid figures.

6–4–1 Prisms

6–4–2 Cylinders

6–4–3 Spheres

6–4–4 Pyramids and Cones

Introduction

Geometry is one of the oldest branches of mathematics and deals with the measurement, properties, and relationships of points, lines, angles, plane figures or surfaces, and solid figures. The practical attributes of geometry were used by the ancient Egyptians to build the pyramids, and today, it is widely used in almost every branch of engineering technology. Unlike algebra, which deals with unknowns that are normally represented by a letter, geometry deals with a physical point, line, surface, or figure.

The first section of this chapter deals with the basic elements of geometry, the second section covers plane figures, the third trigonometry, and the final section examines solid figures.

6–1   BASIC GEOMETRIC TERMS

Figure 6–1(a) shows some examples of line segments. In the first example you can see that a line segment exists between ending points A and B. An abbreviation is generally used to represent a line segment, in which a bar is placed over the two end-point letters, and so "line segment AB" could be written as $\overline{AB}$.

Figure 6–1(a) also shows an example of a vertical line that extends straight up and down, a horizontal line that extends from left to right, and two parallel lines that extend in the same direction while remaining at the same distance from each other.

In Figure 6–1(b) an angle is formed between two intersecting line segments or lines, which are also called **sides.** The common point at which they connect is called the **vertex.** The angle symbol $\angle$ is generally used as an abbreviation, and so "angle ABC" could be written as $\angle ABC$. The unit of measurement for the opening between the sides, or angle, is the degree (symbolized " ° ").

Figure 6–1(c) shows how a protractor can be used to measure angles. An image of a protractor has been printed in the back of this text and can, if copied, be used for measuring and drawing angles. In this example, we are using the protractor to measure $\angle ABC$, from Figure 6–1(b). The three-step procedure for measuring an angle using a protractor is as follows:

Sides

The lines that form an angle.

Vertex

A point where two lines meet.

MEASURING AN ANGLE WITH A PROTRACTOR

Step 1:   Place protractor's center mark on vertex.

Step 2:   Rotate protractor until zero-degree mark is on one line.

Step 3:   Read the measure in degrees where the line intersects the protractor's scale.

You may have noticed that the protractor has a reversing upper and lower scale, so that an angle can be read either clockwise or counterclockwise. Thus if you were to measure the angle in Figure 6–1(b), you would use the lower scale of the protractor as shown in Figure 6–1(c), moving counterclockwise from 0°. On the other hand, if the angle was in a reverse direction as shown with the 90° angle in Figure 6–1(d), you would use the upper scale of the protractor moving clockwise from 0°.

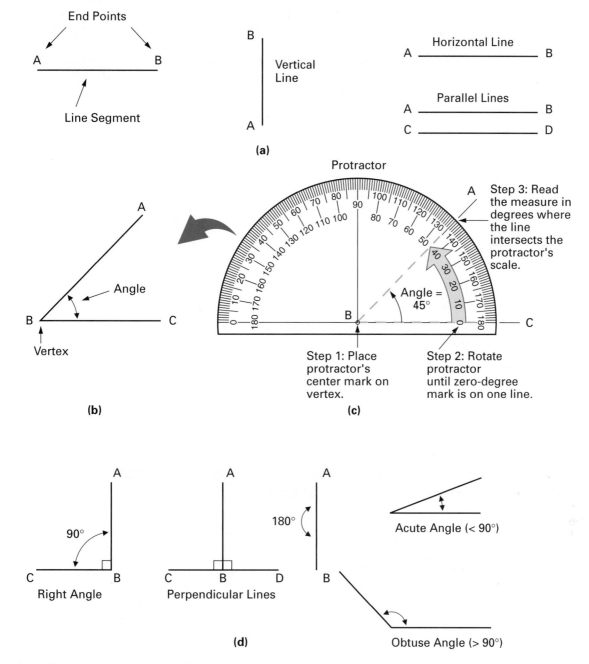

FIGURE 6–1 **Basic Geometric Terms.**

The protractor can also be used to draw an angle. The three-step procedure for drawing an angle using a protractor is as follows:

DRAWING AN ANGLE WITH A PROTRACTOR

Step 1: Draw a straight line on the paper representing one side of the angle.

Step 2: Place the protractor's center mark on the end of the line that is to serve as the angle's vertex.

Step 3: Draw a dot on the paper next to the protractor's scale equal to the measure of the desired angle, and then draw a line connecting this dot to the vertex.

Figure 6–1(d) shows a few more geometric basics. A right angle is one of 90° and can be symbolized with a small square in the corner. Two lines are perpendicular to each other when the point at which they intersect forms right angles. The angle measure of a straight angle is 180°. The terms **acute angle** and **obtuse angle** describe angles that are relative to a right angle (90°). An acute angle is one that is less than a right angle (< 90°), while an obtuse angle is one that is greater than a right angle (> 90°).

■ **INTEGRATED MATH APPLICATION: ELECTRICAL**

What are the angles of the elbows being used to connect conduit:

   a. A to B
   b. C to D

■ *Solution:*

   a. 90°
   b. 135°

■ **INTEGRATED MATH APPLICATION: MACHINING**

The technical drawing details a part to be fabricated. What is the angle of the flare?

■ *Solution:*

   30°

SELF-TEST EVALUATION POINT FOR SECTION 6–1

Now that you have completed this section, you should be able to:

■ *Objective 1.* *Describe the purpose of geometry, and define many of the basic geometric terms.*

■ *Objective 2.* *Demonstrate how to use a protractor to measure and draw an angle.*

Use the following questions to test your understanding of Section 6–1:

1. Using a protractor, measure:
   a. ∠AXB **d.** ∠AXE **g.** ∠AXH **j.** ∠IXG
   b. ∠AXC **e.** ∠AXF **h.** ∠AXI **k.** ∠IXF
   c. ∠AXD **f.** ∠AXG **i.** ∠IXH **l.** ∠IXE

2. Which of the angles in the previous question are right angles, obtuse angles, and acute angles?

3. Draw:
   a. A vertical line   **e.** An angle of 35°
   b. A horizontal line   **f.** An angle of 172°
   c. Three parallel lines **g.** An obtuse angle
   d. A right angle   **h.** An acute angle

6–2   PLANE FIGURES

A plane figure is a perfectly flat surface, such as a tabletop, floor, wall, or windowpane. In this section we will examine the details of several flat plane surfaces, which, as you will discover, have only two dimensions—length and width.

6–2–1 *Quadrilaterals*

A **polygon** is a plane geometric figure that is made up of three or more line segments. Quadrilaterals, such as the square, rectangle, rhombus, parallelogram, and trapezoid, are all examples of four-sided polygons.

The information given in Figure 6–2 summarizes the facts and formulas for the square, which has four right angles and four equal length sides, and is a **parallelogram** (which means it has two sets of parallel sides). The box in Figure 6–2 lists the formulas for calculating a square's perimeter, area, and side length.

Polygon

A plane geometry figure bounded by three or more line segments.

Parallelogram

A quadrilateral in which both pairs of opposite sides are parallel.

■ **INTEGRATED MATH APPLICATION: ELECTRONICS**

The capacitor shown has two same-size metal foil plates that are separated by a nonconductive insulator called a *dielectric*. Unwrapped, the foil strips measure 6.2 centimeters by 1.3 centimeters. What is the capacitor's plate area?

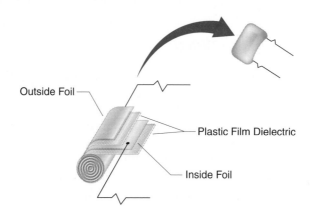

■ *Solution:*

$$\text{Area}(A) = \text{length}\ (l) \times \text{width}\ (w)$$
$$= 6.2 \text{ square centimeters}(\text{cm}^2)$$

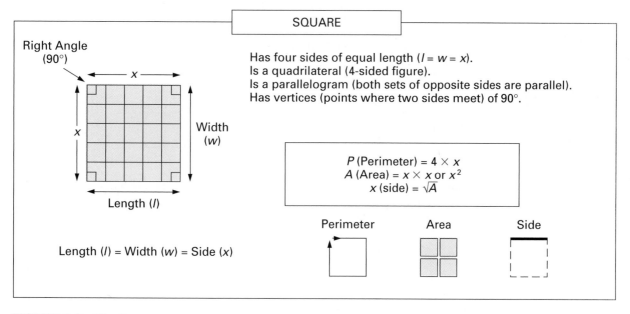

FIGURE 6–2 The Square.

How many packs of tongue-and-groove hardwood flooring would be needed to cover an 18-foot by 12-foot room? A pack can cover 16 square feet, and you should allow 25% extra for end and side matching.

■ *Solution:*

$$18 \text{ feet} \times 12 \text{ feet} = 216 \text{ square feet } (216 \text{ ft}^2)$$
$$+ \ 25\% \text{ of } 216 \quad = \quad \underline{54 \text{ ft}^2}$$
$$270 \text{ ft}^2$$

$$270 \text{ ft}^2 \div 16 \text{ ft}^2/\text{pack} = 16.9 \ (17 \text{ packs})$$

Figures 6–3, 6–4, 6–5, and 6–6 on pp. 178–180 summarize the facts and formulas for other quadrilaterals. In addition, these diagrams include, when necessary, a visual breakdown of the formulas so that it is easier to see how the formula was derived.

6–2–2   *Right-Angle Triangles*

Right-Angle Triangle

A triangle that contains one right or 90° angle.

The **right-angle triangle,** or right triangle, shown in Figure 6–7(a) (p. 182), has three sides and three corners. Its distinguishing feature, however, is that two of the sides of this triangle are at right angles (at 90°) to each other. The small square box within the triangle is placed in the corner to show that sides *A* and *B* are *square* or at *right angles* to each other.

If you study this right triangle you may notice two interesting facts about the relative lengths of sides *A*, *B*, and *C*. These observations are as follows.

1. Side *C* is always longer than side *A* or side *B*.
2. The total length of side *A* and side *B* is always longer than side *C*.

FIGURE 6–3   The Rectangle.

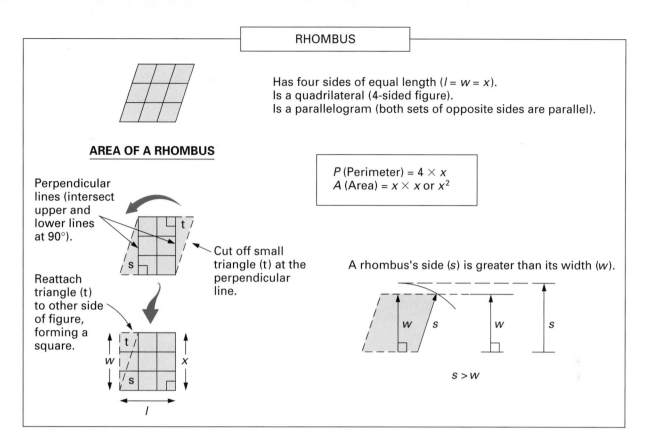

Has four sides of equal length ($l = w = x$).
Is a quadrilateral (4-sided figure).
Is a parallelogram (both sets of opposite sides are parallel).

AREA OF A RHOMBUS

P (Perimeter) = $4 \times x$
A (Area) = $x \times x$ or x^2

Perpendicular lines (intersect upper and lower lines at 90°).

Cut off small triangle (t) at the perpendicular line.

Reattach triangle (t) to other side of figure, forming a square.

A rhombus's side (s) is greater than its width (w).

$s > w$

FIGURE 6–4   The Rhombus.

PARALLELOGRAM

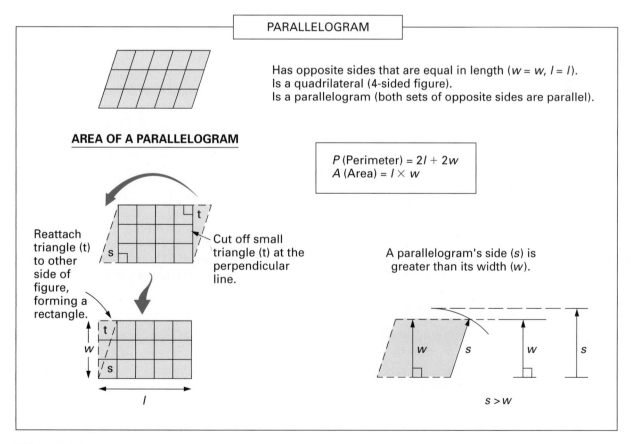

Has opposite sides that are equal in length ($w = w$, $l = l$).
Is a quadrilateral (4-sided figure).
Is a parallelogram (both sets of opposite sides are parallel).

AREA OF A PARALLELOGRAM

P (Perimeter) = $2l + 2w$
A (Area) = $l \times w$

Reattach triangle (t) to other side of figure, forming a rectangle.

Cut off small triangle (t) at the perpendicular line.

A parallelogram's side (s) is greater than its width (w).

$s > w$

FIGURE 6–5   The Parallelogram.

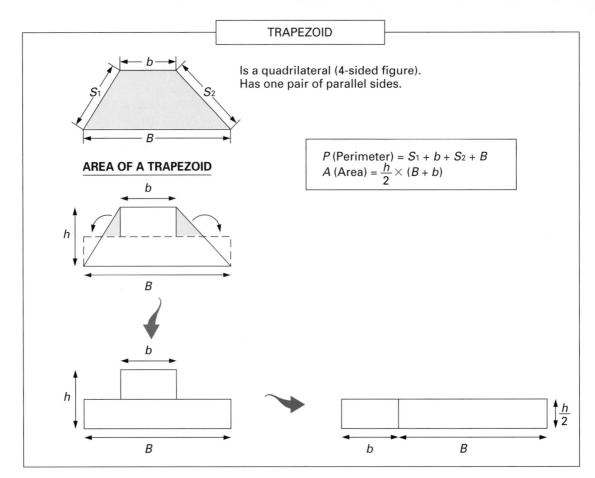

FIGURE 6–6　**The Trapezoid.**

The right triangle in Figure 6–7(b) has been drawn to scale to demonstrate another interesting fact about this triangle. If side A were to equal 3 cm and side B were to equal 4 cm, side C would equal 5 cm. This demonstrates a basic relationship among the three sides and accounts for why this right triangle is sometimes referred to as a *3, 4, 5 triangle*. As long as the relative lengths remain the same, it makes no difference whether the sides are 3 cm, 4 cm, and 5 cm or 30 km, 40 km, and 50 km. Unfortunately, in most applications our lengths of A, B, and C will not work out as easily as 3, 4, 5.

1. Pythagoras's Theorem

Pythagorean Theorem

A theorem in geometry: The square of the length of the hypotenuse of a right triangle equals the sum of the squares of the lengths of the other two sides.

When a relationship exists among three quantities, we can develop a formula to calculate an unknown when two of the quantities are known. It was Pythagoras who first developed this basic formula or equation (known as the **Pythagorean theorem**), which states that *the square of the length of the hypotenuse (side C) of a right triangle equals the sum of the squares of the lengths of the other two sides:*

$$C^2 = A^2 + B^2 \qquad \begin{pmatrix} 5^2 = 3^2 + 4^2 \\ 25 = 9 \ + 16 \end{pmatrix}$$

By using the rules of algebra, we can transpose this formula to derive formulas for sides C, B, and A.

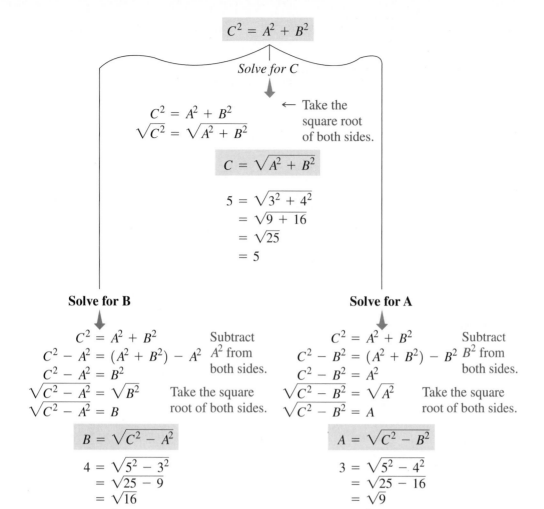

$$C^2 = A^2 + B^2$$

Solve for C

$$\begin{aligned} C^2 &= A^2 + B^2 \\ \sqrt{C^2} &= \sqrt{A^2 + B^2} \end{aligned}$$ ← Take the square root of both sides.

$$C = \sqrt{A^2 + B^2}$$

$$\begin{aligned} 5 &= \sqrt{3^2 + 4^2} \\ &= \sqrt{9 + 16} \\ &= \sqrt{25} \\ &= 5 \end{aligned}$$

Solve for B

$$\begin{aligned} C^2 &= A^2 + B^2 \\ C^2 - A^2 &= (A^2 + B^2) - A^2 \\ C^2 - A^2 &= B^2 \\ \sqrt{C^2 - A^2} &= \sqrt{B^2} \\ \sqrt{C^2 - A^2} &= B \end{aligned}$$

Subtract A^2 from both sides.

Take the square root of both sides.

$$B = \sqrt{C^2 - A^2}$$

$$\begin{aligned} 4 &= \sqrt{5^2 - 3^2} \\ &= \sqrt{25 - 9} \\ &= \sqrt{16} \\ &= 4 \end{aligned}$$

Solve for A

$$\begin{aligned} C^2 &= A^2 + B^2 \\ C^2 - B^2 &= (A^2 + B^2) - B^2 \\ C^2 - B^2 &= A^2 \\ \sqrt{C^2 - B^2} &= \sqrt{A^2} \\ \sqrt{C^2 - B^2} &= A \end{aligned}$$

Subtract B^2 from both sides.

Take the square root of both sides.

$$A = \sqrt{C^2 - B^2}$$

$$\begin{aligned} 3 &= \sqrt{5^2 - 4^2} \\ &= \sqrt{25 - 16} \\ &= \sqrt{9} \\ &= 3 \end{aligned}$$

Let us now test these formulas with a few examples.

■ INTEGRATED MATH APPLICATION: PAINTING

A 4-foot ladder has been placed in a position 2 feet from a wall. How far up the wall will the ladder reach?

■ *Solution:*

In this example, *A* is unknown, and therefore

$$\begin{aligned} A &= \sqrt{C^2 - B^2} \\ &= \sqrt{4^2 - 2^2} \\ &= \sqrt{16 - 4} \\ &= \sqrt{12} \\ &= 3.46 \quad \text{or} \quad \approx 3.5 \text{ feet} \end{aligned}$$

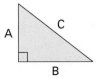

Is a triangle (3-sided figure).
Has one right angle (indicated by square).

P (Perimeter) = A + B + C

A (Area) = $\dfrac{A \times B}{2}$

Sides (A, B, C)
Pythagorean theorem: $C^2 = A^2 + B^2$

$C = \sqrt{A^2 + B^2}$ $B = \sqrt{C^2 - A^2}$ $A = \sqrt{C^2 - B^2}$

(a)   AREA OF A RIGHT TRIANGLE

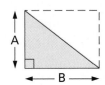

(b)   SIDES OF A RIGHT TRIANGLE

Pythagorean theorem: $A^2 + B^2 = C^2$
Square of side A + Square of side B = Square of side C.
Example: 9 + 16 = 25

A = 3
B = 4
C = 5

$3^2 = 9$ + $4^2 = 16$ = $5^2 = 25$

FIGURE 6–7   The Right Triangle.

■   **INTEGRATED MATH APPLICATION: CONSTRUCTION**

A 16-foot flagpole is casting a 12-foot shadow. What is the distance from the end of the shadow to the top of the flagpole?

■   *Solution:*

In this example, *C* is unknown, and therefore

$$C = \sqrt{A^2 + B^2}$$
$$= \sqrt{16^2 + 12^2}$$
$$= \sqrt{256 + 144}$$
$$= \sqrt{400}$$
$$= 20 \text{ feet}$$

■ INTEGRATED MATH APPLICATION: GEOLOGY

A 200-foot piece of string is stretched from the top of a 125-foot cliff to a point on the beach. What is the distance from this point to the cliff?

■ *Solution:*

In this example, B is unknown, and therefore

$$B = \sqrt{C^2 - A^2}$$
$$= \sqrt{200^2 - 125^2}$$
$$= \sqrt{40,000 - 15,625}$$
$$= \sqrt{24,375}$$
$$= 156 \text{ feet}$$

■ INTEGRATED MATH APPLICATION: CARPENTRY

Calculate the area of the patio deck shown. Also determine the length of side C for a railing.

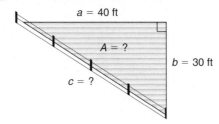

■ *Solution:*

$$A = \frac{a \times b}{2} = \frac{40 \times 30}{2} = 600 \text{ square feet}$$
$$c = \sqrt{a^2 + b^2} = \sqrt{40^2 + 30^2} = \sqrt{1600 + 900} = \sqrt{2500} = 50 \text{ feet}$$

6–2–3   *Vectors and Vector Diagrams*

A **vector or phasor** is an arrow used to represent the magnitude and direction of a quantity. Vectors are generally used to represent a physical quantity that has two properties. For example, Figure 6–8(a) shows a motorboat heading north at 12 miles per hour. Figure 6–8(b) shows how vector **A** could be used to represent the vessel's direction and speed. The size of the vector represents the speed of 12 miles per hour by being 12 centimeters long, and because we have made the top of the page north, vector **A** should point straight up so that it represents the vessel's direction. Referring to Figure 6–8(a) you can see that there is another factor that also needs to be considered, a 12 mile per hour easterly tide. This tide is represented in our vector diagram in Figure 6–8(b) by vector **B,** which is 12 centimeters long and pointing east. Since the motorboat is pushing north at 12 miles per hour and the tide is pushing east at 12 miles per hour, the resultant course will be northeast, as indicated by vector **C** in Figure 6–8(b). Vector **C** was determined by **vector addition** (as seen by the dashed lines) of vectors **A** and **B.** The result, vector **C,** is called the **resultant vector.** This resultant vector indicates the motorboat's course and speed. The direction or course, we can see, is northeast because vector **C** points in a direction midway between north and east. The speed of the motorboat, however, is indicated by the length or magnitude of vector **C.** This length, and therefore the motorboat's speed, can be calculated by using the Pythagorean theorem. Probably your next question is: How is the **vector diagram** in Figure 6–8(b) similar to a right-angle triangle? The answer is to redraw Figure 6–8(b), as shown in Figure 6–8(c). Because the dashed line x is equal in length to vector **A,** vector **A** can be put in the position of line x to form a right-angle triangle with vector **B** and vector **C.** Now that we have a right-angle

Vector or Phasor

A quantity that has magnitude and direction and that is commonly represented by a directed line segment whose length represents the magnitude and whose orientation in space represents the direction.

Vector Addition

Determination of the sum of two out-of-phase vectors using the Pythagorean theorem.

Resultant Vector

A vector derived from or resulting from two or more other vectors.

Vector Diagram

A graphic drawing using vectors that shows arrangement and relations.

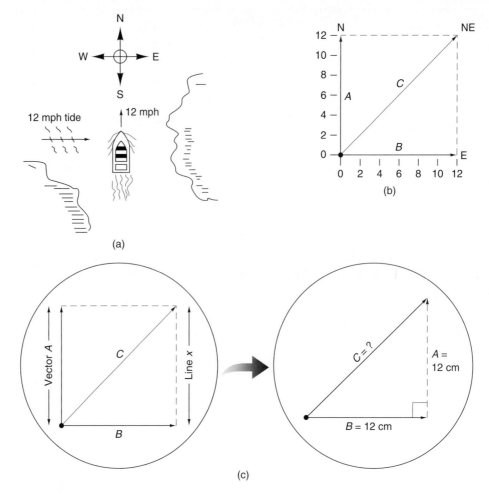

FIGURE 6–8 **Vectors and Vector Diagrams.**

triangle with two known lengths and one unknown length, we can calculate the unknown vector's length and therefore the motorboat's speed.

$$C = \sqrt{A^2 + B^2}$$
$$= \sqrt{12^2 + 12^2}$$
$$= \sqrt{144 + 144}$$
$$= \sqrt{288} = 16.97$$

The length of vector **C** is 16.97 cm, and because each 1 mile per hour of the motorboat was represented by 1 centimeter, the motorboat will travel at a speed of 16.97 miles per hour in a northeasterly direction.

■ INTEGRATED MATH APPLICATION: ELECTRONICS

An electrical circuit has two elements that oppose current flow; one form of opposition is called *resistance* (symbolized by R), and the other form of opposition is called *reactance* (symbolized by X). These two oppositions are out of sync, or not in phase, with each other, as seen by the vector diagram in Figure 6–9(a), and therefore cannot simply be added. Using the Pythagorean theorem, vectorially add the two forms of opposition (R and X) to determine the total opposition to current flow, which is called *impedance* (symbolized by Z).

FIGURE 6–9   An Electrical Application of the Pythagorean Theorem.

(a)   (b)

■   *Solution:*

First, redraw the vector diagram to produce a right-angle triangle, as shown in Figure 6–9(b). Then calculate Z:

$$Z = \sqrt{R^2 + X^2}$$
$$= \sqrt{30^2 + 16^2}$$
$$= \sqrt{900 + 256}$$
$$= \sqrt{1156}$$
$$= 34 \ \Omega$$

The total opposition or impedance (Z) to current flow is therefore 34 ohms.

6–2–4   *Other Triangles*

Figures 6–10, 6–11, and 6–12 summarize the facts and formulas for other triangle types.

SCALENE TRIANGLE

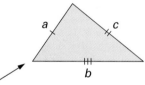

Is a triangle (3-sided figure).
Has no sides that are equal.

A different number of check marks on the sides of the triangle are used to indicate sides of different sizes.

P (Perimeter) $= a + b + c$
A (Area) $= \dfrac{b \times h}{2}$

If the length of the sides are known:

$$A = \sqrt{s \times (s - a) \times (s - b) \times (s - c)}$$

In which s (half perimeter) $= \dfrac{a + b + c}{2}$

AREA OF A SCALENE TRIANGLE

Scalene triangle occupies half the area of the rectangle below.

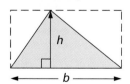

FIGURE 6–10   The Scalene Triangle.

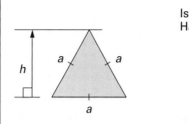

EQUILATERAL TRIANGLE

Is a triangle (3-sided figure).
Has three equal sides.

$$P \text{ (Perimeter)} = 3 \times a$$
$$A \text{ (Area)} = \frac{a \times h}{2} \quad \text{or} \quad A = \frac{a^2\sqrt{3}}{4}$$

FIGURE 6–11   The Equilateral Triangle.

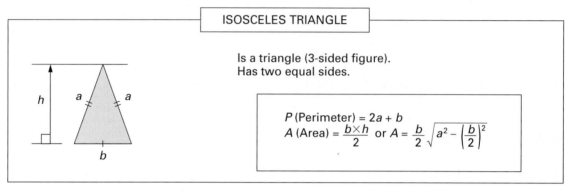

ISOSCELES TRIANGLE

Is a triangle (3-sided figure).
Has two equal sides.

$$P \text{ (Perimeter)} = 2a + b$$
$$A \text{ (Area)} = \frac{b \times h}{2} \quad \text{or} \quad A = \frac{b}{2}\sqrt{a^2 - \left(\frac{b}{2}\right)^2}$$

FIGURE 6–12   The Isosceles Triangle.

6–2–5   *Circles*

Details relating to the circle have been discussed in previous chapters. These facts and formulas are summarized in Figure 6–13.

6–2–6   *Other Polygons*

Figure 6–14(a) shows that if lines are drawn from the center of any polygon to each vertex, equivalent triangles are formed, which enables us to determine the polygon's area. In Figure 6–14(b), the formula given allows us to determine the number of degrees in the angles for each polygon type. From the examples shown we can see that each angle in an equilateral triangle is 60°; for a square each angle is 90°, as expected; for a pentagon each angle is 108°; a hexagon has six 120° angles; and an octagon has eight 135° vertices.

■   **INTEGRATED MATH APPLICATION: MACHINING**

Find the area of each of the following parts.

a. ← 0.5 cm

2 cm

b.

$r = \frac{1}{2}$ in.

$1\frac{1}{2}$ in.

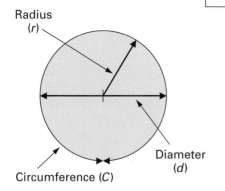

Radius (r)

Circumference (C)

Diameter (d)

Is a perfectly round figure.
Every point is at an equal distance from the center.

Diameter $(d) = 2 \times r$
Circumference $(C) = \pi \times d$ or $c = \pi \times (2r)$
Area $(A) = \pi r^2$

Every circle's circumference is 3.14 times larger than its diameter. This constant is called pi (symbolized π).

$$\pi = \frac{\text{Circumference } (C)}{\text{Diameter } (d)} = 3.14$$

$\pi = \dfrac{C}{d}$ $C = \pi \times d$ $d = \dfrac{C}{\pi}$

Since $d = 2r$, then:

$\pi = \dfrac{C}{2r}$ $C = \pi \times 2r$ $2r = \dfrac{C}{\pi}$

or $r = \dfrac{C}{2\pi}$

AREA OF A CIRCLE

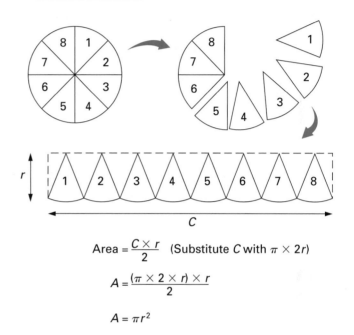

Area $= \dfrac{C \times r}{2}$ (Substitute C with $\pi \times 2r$)

$$A = \frac{(\pi \times 2 \times r) \times r}{2}$$

$$A = \pi r^2$$

FIGURE 6–13   The Circle.

c.

2.5 cm

5 cm

4 cm

d. 1.2 cm |← 1.4 cm

2.2 cm

5.4 cm

9.8 cm

■ *Solution:*

a. Large circle: $C = 2$ cm, $r = 1$ cm, $A = \pi r^2 = \pi \times 1^2 = 3.142$ cm^2
   Small circle: $C = 0.5$ cm, $r = 0.25$ cm, $A = \pi r^2 = \pi \times 0.25^2 = 0.19$ cm^2
               Total area $= 3.142 - 0.19 = 2.95$ cm^2

b. Half circle: $r = 0.5$ in., $A = \pi \times r^2 = \pi \times 0.5^2 = 0.79, 0.79 \div 2 = 0.4$ in.^{2}
   Rectangle: $A = l \times w = 1.5 \times 0.5 = 0.75$ in.^{2}
               Total area $= 0.75 + 0.4 = 1.15$ in.^{2}

(a) Polygon Area

TRIANGLE
(3 sides)

SQUARE
(4 sides)

PENTAGON
(5 sides)

HEXAGON
(6 sides)

OCTAGON
(8 sides)

Equivalent triangles are created by drawing lines from the center of the polygon to each vertex.

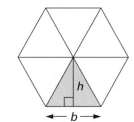

P (Perimeter) = # of Sides $\times b$
A (Area) = $\dfrac{b \times h}{2} \times$ # of Sides

(b) Number of Degrees in Each Vertex of a Polygon

x (Degrees in Vertex) = $\dfrac{180° \times (\text{# of Sides} - 2)}{\text{# of Sides}}$

TRIANGLE:
(Equilateral) $x = \dfrac{180° \times (3-2)}{3} = \dfrac{180°}{3} = 60°$ $x = 60°$
$3x = 180°$

SQUARE: $x = \dfrac{180° \times (4-2)}{4} = \dfrac{360°}{4} = 90°$ $x = 90°$
$4x = 360°$

PENTAGON: $x = \dfrac{180° \times (5-2)}{5} = \dfrac{540°}{5} = 108°$ $x = 108°$
$5x = 540°$

HEXAGON: $x = \dfrac{180° \times (6-2)}{6} = \dfrac{720°}{6} = 120°$ $x = 120°$
$6x = 720°$

OCTAGON: $x = \dfrac{180° \times (8-2)}{8} = \dfrac{1{,}080°}{8} = 135°$ $x = 135°$
$8x = 1{,}080°$

FIGURE 6–14   Other Polygons.

c. Circle area: $A = \pi r^2 = \pi \times 2.5^2 = 19.6\,\text{cm}^2$

Hexagon area: $A = \dfrac{b \times h}{2} \times 6 = \dfrac{4\,\text{cm} \times 5\,\text{cm}}{2} \times 6 = 60\,\text{cm}^2$

Total area $= 60 - 19.6 = 40.4\,\text{cm}^2$

d. Circle area: $A = \pi r^2 = \pi \times 2.2^2 = 15.2\,\text{cm}^2 \div 2 = 7.6\,\text{cm}^2$

Trapezoid area: (upper base) $= 1.2\,\text{cm} + 4.4\,\text{cm} + 1.4\,\text{cm} = 7\,\text{cm}$

(lower base) $= 9.8\,\text{cm}$

$$A = \frac{h}{2} \times (B + b)$$

$$= \frac{5.4\,\text{cm}}{2} \times (9.8\,\text{cm} + 7\,\text{cm})$$

$$= 2.7\,\text{cm} \times 16.8\,\text{cm} = 45.36\,\text{cm}^2$$

SELF-TEST EVALUATION POINT FOR SECTION 6–2

Now that you have completed this section, you should be able to:

■ **Objective 3.** *Define the term plane figure, and identify the different types of polygons.*

■ **Objective 4.** *List the formulas for determining the perimeter and area of plane figures.*

■ **Objective 5.** *Explain how the Pythagorean theorem relates to right-angle triangles.*

■ **Objective 6.** *Determine the length of one side of a right-angle triangle when the lengths of the other two sides are known.*

■ **Objective 7.** *Describe how vectors are used to represent the magnitude and direction of physical quantities and how they are arranged in a vector diagram.*

■ **Objective 8.** *Explain how the Pythagorean theorem can be applied to a vector diagram to calculate the magnitude of a resultant vector.*

Use the following questions to test your understanding of Section 6–2:

1. Define the terms:
   a. Plane figure
   b. Polygon

2. Sketch the following and list the formulas for perimeter and area:
   a. Square
   b. Rectangle
   c. Rhombus
   d. Parallelogram
   e. Trapezoid
   f. Right triangle
   g. Scalene triangle
   h. Equilateral triangle
   i. Isosceles triangle
   j. Circle

3. Calculate the lengths of the unknown sides in the following triangles.

4. Calculate the magnitude of the resultant vectors in the following vector diagrams.

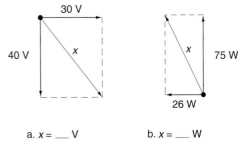

a. x = ___ V   b. x = ___ W

c.
x = ___ mm
y = ___ mm

6–3   TRIGONOMETRY

In the previous section you saw how Pythagoras's theorem could be used to find the length of one side of a right triangle when the lengths of the other two sides are known. **Trigonometry** is the study of the properties of triangles, and in this section we will examine the right-angle triangle in a little more detail, and see how angles and sides are related.

Trigonometry
The study of the properties of triangles.

FIGURE 6–15   Names Given to the
Sides of a Right Triangle.

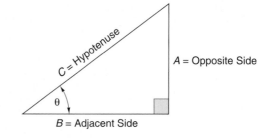

A = Opposite Side

B = Adjacent Side

Until this point we have called the three sides of our right triangle *A, B,* and *C.* Figure 6–15 shows the more common names given to the three sides of a right-angle triangle. Side *C,* called the **hypotenuse,** is always the longest of the three sides. Side *B,* called the **adjacent side,** always extends between the hypotenuse and the vertical side. An angle called **theta** (θ, a Greek letter) is formed between the hypotenuse and the adjacent side. The side that is always opposite the angle θ is called the **opposite side.**

Angles are always measured in degrees because all angles are part of a circle, like the one shown in Figure 6–16(a). A circle is divided into 360 small sections called **degrees.** In Figure 6–16(b) our right triangle has been placed within the circle. In this example the triangle occupies a 45° section of the circle and therefore theta equals 45 degrees ($\theta = 45°$). Moving from left to right in Figure 6–16(c) you will notice that angle θ increases from 5° to 85°. The length of the hypotenuse (H) in all these examples remains the same; however, the adjacent side's length decreases ($A\downarrow$) and the opposite side's length increases ($O\uparrow$) as angle θ is increased ($\theta\uparrow$). This relationship between the relative length of a triangle's sides and theta means that we do not have to know the length of two sides to calculate a third. If we have the value of just one side and the angle theta, we can calculate the length of the other two sides.

6–3–1   *Sine of Theta (Sin θ)*

In the preceding section we discovered that a relationship exists between the relative length of a triangle's sides and the angle theta. **Sine** is a comparison between the length of the opposite side and the length of the hypotenuse. Expressed mathematically,

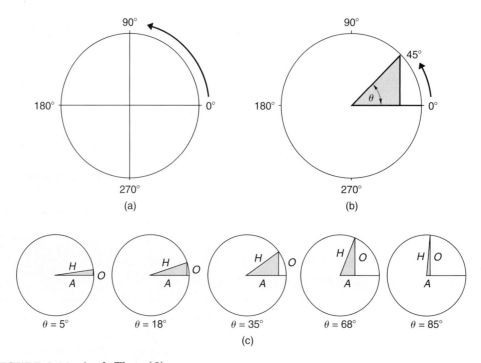

FIGURE 6–16   Angle Theta (θ).

Hypotenuse

The side of a right-angle triangle that is opposite the right angle.

Adjacent

The side of a right-angle triangle that has a common endpoint, in that it extends between the hypotenuse and the vertical side.

Theta

The eighth letter of the Greek alphabet, used to represent an angle.

Opposite

The side of a right-angle triangle that is opposite the angle theta.

Degree

A unit of measure for angles equal to an angle with its vertex at the center of a circle and its sides cutting off $\frac{1}{360}$ of the circumference.

Sine

The trigonometric function that for an acute angle is the ratio between the leg opposite the angle when it is considered part of a right triangle, and the hypotenuse.

$$\text{Sine of theta } (\sin \theta) = \frac{\text{opposite side } (O)}{\text{hypotenuse } (H)}$$

Because the hypotenuse is always larger than the opposite side, the result will always be less than 1 (a decimal fraction). Let us use this formula in a few examples to see how it works.

CALCULATOR KEYS

Name: Sine key

Function: Instructs the calculator to find the sine of the displayed value (angle → value).

Example: $\sin 37° = ?$

Press keys: ⟨3⟩ ⟨7⟩ ⟨SIN⟩

Answer: 0.601815

Name: Arcsine ($\sin^{-1}$) or inverse sine sequence.

Function: Calculates the smallest angle whose sine is in the display (value → angle).

Example: invsin 0.666 = ?

Press keys: ⟨.⟩ ⟨6⟩ ⟨6⟩ ⟨6⟩ ⟨INV⟩ ⟨SIN⟩

Answer: 41.759°

■ INTEGRATED MATH APPLICATION: SURVEYING

On the set of plans shown in Figure 6–17(a), angle theta is equal to 41°, and the opposite side is equal to 20 centimeters in length. Calculate the length of the hypotenuse.

■ *Solution:*

Inserting these values in our formula, we obtain the following:

$$\sin \theta = \frac{O}{H}$$

$$\sin 41° = \frac{20 \text{ cm}}{H}$$

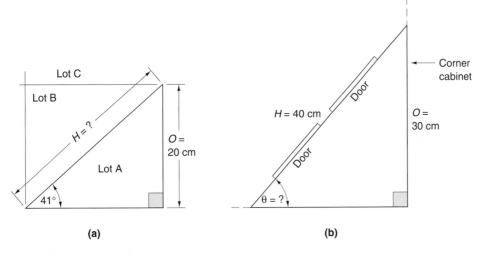

(a) (b)

FIGURE 6–17 **Sine of Theta.**

By looking up 41° in a sine trigonometry table or by using a scientific calculator that has all the trigonometry tables stored permanently in its memory, you will find that the sine of 41° is 0.656. This value describes the fact that when $\theta = 41°$, the opposite side will be 0.656, or 65.6%, as long as the hypotenuse side. By inserting this value into our formula and transposing the formula according to the rules of algebra, we can determine the length of the hypotenuse.

$$\sin 41° = \frac{20\,\text{cm}}{H}$$

Calculator sequence: $\boxed{4}\,\boxed{1}\,\boxed{\text{SIN}}$

$$0.656 = \frac{20\,\text{cm}}{H}$$

$$0.656 \times H = \frac{20\,\text{cm} \times \cancel{H}}{\cancel{H}}$$

Multiply both sides by H.

$$\frac{\cancel{0.656} \times H}{\cancel{0.656}} = \frac{20\,\text{cm}}{0.656}$$

Divide both sides by 0.656.

$$H = \frac{20\,\text{cm}}{0.656} = 30.5\,\text{cm}$$

■   **INTEGRATED MATH APPLICATION: DRAFTING**

Figure 6–17(b) illustrates another example; however, in this case the lengths of sides H and O are known but θ is not.

■   *Solution:*

$$\sin \theta = \frac{O}{H}$$

$$\sin \theta = \frac{30\,\text{cm}}{40\,\text{cm}}$$

$$= 0.75$$

The ratio of side O to side H is 0.75, or 75%, which means that the opposite side is 75%, or 0.75, as long as the hypotenuse. To calculate angle θ we must isolate it on one side of the equation. To achieve this we must multiply both sides of the equation by **arcsine**, or **inverse sine** (invsin), which does the opposite of sine.

Arcsine or Inverse Sine

The inverse function to the sine. (If y is the sine of θ, then θ is the arcsine of y.)

$$\sin \theta = 0.75$$

$$\cancel{\text{invsin}}\,(\cancel{\sin}\,\theta) = \text{invsin}\,0.75$$

Take the inverse sine of 0.75.

$$\theta = \text{invsin}\,0.75$$

Calculator sequence: $\boxed{.}\,\boxed{7}\,\boxed{5}\,\boxed{\text{INV}}\,\boxed{\text{SIN}}$

$$= 48.6°$$

In summary, therefore, the sine trig functions take an angle θ and give you a number x. The inverse sine (arcsin) trig functions take a number x and give you an angle θ. In both cases, the number x is the ratio of the opposite side to the hypotenuse.

Sine: angle $\theta \rightarrow$ number x

Inverse sine: number $x \rightarrow$ angle θ

■   **INTEGRATED MATH APPLICATION: MACHINING**

Referring to the part shown, determine:

a. the length of the taper (H).

b. Angle θ.

CHAPTER 6 / GEOMETRY AND TRIGONOMETRY

■ *Solution:*

   a. $H = \sqrt{6.2^2 + 3.4^2} = \sqrt{38.44 + 11.56} = \sqrt{50} = 7.07$ centimeters

   b. $\theta = \text{invsin}\dfrac{O}{H} = \text{invsin}\dfrac{3.4\ \text{cm}}{7.07\ \text{cm}} = 28.7°$

6–3–2 *Cosine of Theta (Cos θ)*

Sine is a comparison between the opposite side and the hypotenuse, and **cosine** is a comparison between the adjacent side and the hypotenuse.

$$\text{Cosine of theta } (\cos \theta) = \frac{\text{adjacent } (A)}{\text{hypotenuse } (H)}$$

CALCULATOR KEYS

Name: Cosine key

Function: Instructs the calculator to find the cosine of the displayed value (angle → value).

Example: $\cos 26° = ?$

Press keys: | 2 | 6 | cos |

Answer: 0.89879

Name: Arccosine $(\cos^{-1})$ or inverse cosine sequence.

Function: Calculates the smallest angle whose cosine is in the display (value → angle).

Example: $\text{invcos } 0.234 = ?$

Press keys: | . | 2 | 3 | 4 | INV | cos |

Answer: 76.467

■ **INTEGRATED MATH APPLICATION: WELDING**

Figure 6–18(a) illustrates a metal support frame that is to be welded. With this right triangle the angle θ and the length of the hypotenuse are known. From this information, calculate the length of the adjacent side.

(a)

(b)

FIGURE 6–18 **Cosine of Theta.**

■ *Solution:*

$$\cos\theta = \frac{A}{H}$$

$$\cos 30° = \frac{A}{40\,cm}$$

Calculator sequence: [3] [0] [cos]

$$0.866 = \frac{A}{40\,cm}$$

Looking up the cosine of 30° you will obtain the fraction 0.866. This value states that when $\theta = 30°$, the adjacent side will always be 0.866, or 86.6%, as long as the hypotenuse. By transposing this equation we can calculate the length of the adjacent side:

$$0.866 = \frac{A}{40\,cm}$$

$$40 \times 0.866 = \frac{A}{40\,cm} \times 40 \quad \text{Multiply both sides by 40.}$$

$$A = 0.866 \times 40$$

$$= 34.64\,cm$$

■ INTEGRATED MATH APPLICATION: CONSTRUCTION

Calculate the angle θ in Figure 6–18(b) if $A = 120$ centimeters and $H = 180$ centimeters.

■ *Solution:*

$$\cos\theta = \frac{A}{H}$$

$$= \frac{120\,cm}{180\,cm}$$

$$= 0.667 \quad \text{\textit{A} is 66.7\% as long as \textit{H}.}$$

$$\text{invcos}(\cos\theta) = \text{invcos}\,0.667 \quad \text{Multiply both sides by invcos.}$$

$$\theta = \text{invcos}\,0.667 \quad \textit{Calculator sequence:} \; [0][.][6][6][7][\text{INV}][\text{COS}]$$

$$= 48.2$$

The inverse cosine trig function performs the reverse operation of the cosine function:

Cosine: angle $\theta \rightarrow$ number x

Inverse cosine: number $x \rightarrow$ angle θ

6–3–3 *Tangent of Theta (Tan θ)*

Tangent

The trigonometric function that for an acute angle is the ratio between the leg opposite the angle when it is considered part of a right triangle, and the leg adjacent.

Tangent is a comparison between the opposite side of a right triangle and the adjacent side.

$$\text{Tangent of theta } (\tan\theta) = \frac{\text{opposite } (O)}{\text{adjacent } (A)}$$

CHAPTER 6 / GEOMETRY AND TRIGONOMETRY

CALCULATOR KEYS

Name: Tangent key

Function: Instructs the calculator to find the tangent of the displayed value (angle → value).

Example: $\sin 73° = ?$

Press keys: [7] [3] [TAN]

Answer: 3.2709

Name: Arctangent ($\tan^{-1}$) or inverse tangent sequence.

Function: Calculates the smallest angle whose tangent is in the display (value → angle).

Example: invtan $0.95 = ?$

Press keys: [.] [9] [5] [INV] [TAN]

Answer: 43.5312

■ INTEGRATED MATH APPLICATION: CARPENTRY

Figure 6–19(a) shows a wooden shelf support. With this right triangle $\theta = 65°$ and the opposite side is 43 centimeters. Calculate the length of the adjacent side.

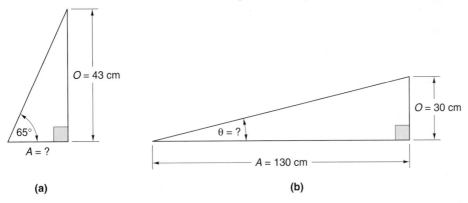

(a)   (b)

FIGURE 6–19   Tangent of Theta.

■ *Solution:*

$$\tan \theta = \frac{O}{A}$$

$$\tan 65° = \frac{43 \text{ cm}}{A}$$ *Calculator sequence:* [6] [5] [TAN]

$$2.14 = \frac{43 \text{ cm}}{A}$$ Whenever $\theta = 65°$, the opposite side will be 2.14 times longer than the adjacent side.

$$2.14 \times A = \frac{43 \text{ cm}}{A} \times A$$ Multiply both sides by A.

$$\frac{2.14 \times A}{2.14} = \frac{43 \text{ cm}}{2.14}$$ Divide both sides by 2.14.

$$A = \frac{43 \text{ cm}}{2.14}$$

$$= 20.1 \text{ cm}$$

■ INTEGRATED MATH APPLICATION: CONSTRUCTION

Calculate the angle θ in Figure 6–19(b) for the concrete handicapped ramp if $O = 30$ centimeters and $A = 130$ centimeters.

■ *Solution:*

$$\tan \theta = \frac{O}{A}$$

$$= \frac{30 \text{ cm}}{130 \text{ cm}}$$

$$= 0.231 \qquad \text{\textit{O} is 0.231 or 23.1\% as long as \textit{A}.}$$

$$\text{invtan} \left(\tan \theta \right) = \text{invtan } 0.231 \qquad \text{Multiply both sides by invtan.}$$

$$\theta = \text{invtan } 0.231 \qquad \textit{Calculator sequence:} \boxed{0} \boxed{.} \boxed{2} \boxed{3} \boxed{1} \boxed{\text{INV}} \boxed{\text{TAN}}$$

$$= 12.99 \text{ or } 13°$$

■ INTEGRATED MATH APPLICATION: ELECTRONICS

Some components in electronics circuits, such as resistors, consume power from a source and dissipate it in the form of heat. This power consumption is known as *resistive power* or *true power*. Other components, like capacitors, take power from the source at one time and then return it to the circuit at another time. This power consumption is known as *reactive power* or *imaginary power*.

When capacitors and resistors are together in a circuit, they consume power at different times, and so you cannot simply add together the resistive and reactive power to find the third. The resulting apparent power is obtained by vectorially adding the resistive and reactive power.

Referring to the right triangle shown, determine:

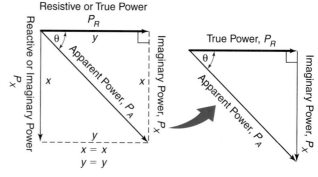

a. P_A when $P_R = 12$ watts and $P_X = 14$ watts.
b. Angle θ, using sine.
c. Since all three side lengths are known, show how the angle can be determined using cosine and tangent.

■ *Solution:*

a. $P_A = \sqrt{P_R^2 + P_X^2} = \sqrt{12^2 + 14^2} = \sqrt{144 + 196} = \sqrt{340} = 18.44$ watts

b.   Opposite $= P_X = 12$
Hypotenuse $= P_A = 18.44$ (All three sides are known, so you
Adjacent $= P_R = 14$ can use either sine, cosine, or tangent.)

$$\sin \theta = \frac{O}{H}$$

$$\theta = \text{invsin } \frac{O}{H} = \text{invsin } \frac{12}{18.44} = 40.6°$$

c. $\theta = \text{invcos } \dfrac{A}{H} = \text{invcos } \dfrac{14}{18.44} = 40.6°$

$$\theta = \text{invtan } \frac{O}{A} = \text{invtan } \frac{12}{14} = 40.6°$$

6–3–4   *Summary*

As you have seen, trigonometry involves the study of the relationships among the three sides (O, H, A) of a right triangle and also the relationships among the sides of the right triangle and the number of degrees contained in the angle theta (θ).

If the lengths of two sides of a right triangle are known, and the length of the third side is needed, remember that

$$H^2 = O^2 + A^2$$

If the angle θ is known along with the length of one side, or if angle θ is needed and the lengths of the two sides are known, one of the three formulas can be chosen based on which variables are known and what is needed.

$$\sin\theta = \frac{O}{H} \quad \cos\theta = \frac{A}{H} \quad \tan\theta = \frac{O}{A}$$

When I was introduced to trigonometry, my mathematics professor spent 15 minutes having the whole class practice what he described as an old Asian war cry that went like this: "SOH CAH TOA." After he explained that it wasn't a war cry but in fact a memory aid to help us remember that SOH was in fact $\sin\theta = O/H$, CAH was $\cos\theta = A/H$, and TOA was $\tan\theta = O/A$, we understood the method in his madness.

■ INTEGRATED MATH APPLICATION: ELECTRICAL

The **sine wave** is the most common type of electrical waveform. It is the natural output of a generator that converts a mechanical input, in the form of a rotating shaft, into an electrical output in the form of a sine wave. In fact, for one cycle of the input shaft, the generator will produce one sinusoidal AC voltage waveform, as shown in Figure 6–20. When the input shaft of the generator is at 0°, the AC output is 0 volts. As the shaft is rotated through 360° the AC output voltage will rise to a maximum or peak positive voltage at 90°, fall back to 0 volts at 180°, then reach a maximum or peak negative voltage at 270°, and finally return to 0 volts at 360°. If this AC voltage is applied across a closed circuit, it produces a current that continually reverses or alternates in direction. (It switches from being a positive voltage to a negative voltage, back to a positive voltage, and so on.) The sine wave is the most common type of waveform shape, and why the name *sine* was given to this wave needs to be explained further.

Sine Wave

A waveform that represents periodic oscillations in which the amplitude of displacement at each point is proportional to the sine of the phase angle of the displacement and that is visualized as a sine curve.

FIGURE 6–20   Degrees of a Sine Wave.

Figure 6–21(a) shows the correlation between the 360° of a circle and the 360° of a sine wave. Within the circle a triangle has been drawn to represent the right-angle triangle formed at 30°. The hypotenuse side will always remain the same length throughout 360° and will equal the maximum voltage or peak of the sine wave. The opposite side of the triangle is equal to the amplitude vector of the sine wave at 30°. To calculate the amplitude of the op-

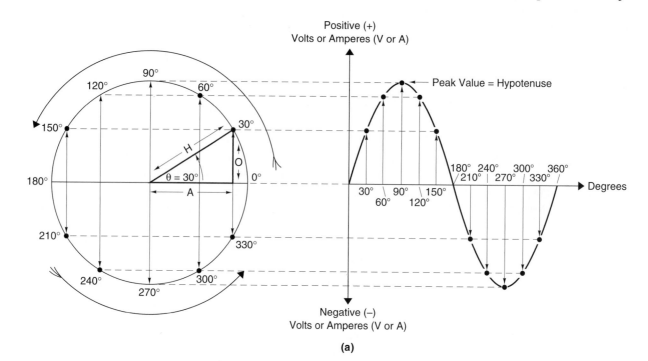

(a)

Angle	Sine
0°	0.000
15	0.259
30	0.500
45	0.707
60	0.866
75	0.966
90	1.000

(b)

At 90°, Amplitude = 1.000 × 10 V = 10 V
At 75°, Amplitude = 0.966 × 10 V = 9.66 V
At 60°, Amplitude = 0.866 × 10 V = 8.66 V
At 45°, Amplitude = 0.707 × 10 V = 7.07 V
At 30°, Amplitude = 0.500 × 10 V = 5.00 V
At 15°, Amplitude = 0.259 × 10 V = 2.59 V
At 0°, Amplitude = 0.000 × 10 V = 0 V

(c)

FIGURE 6–21 Meaning of Sine.

posite side, and therefore the amplitude of the sine wave at 30°, we use the sine of theta formula discussed previously:

$$\sin \theta = \frac{\text{opposite}}{\text{hypotenuse}}$$

$$\sin 30° = \frac{O}{H}$$

$$0.500 = \frac{O}{H}$$

Calculator sequence: [3] [0] [SIN]

This equation tells us that at 30°, the opposite side is 0.500 or half the size of the hypotenuse. At 30°, therefore, the amplitude or magnitude of the sine wave will be 0.5 (50%) of the peak value.

Figure 6–21(b) lists the sine values at 15° increments. Figure 6–21(c) shows an example of a 10 volt peak sine wave. At 15°, a sine wave will always be at 0.259 (sine of 15°) of the peak value, which for a 10 volt sine wave will be 2.59 volts (0.259×10 volts $=$ 2.59 volts). At 30°, a sine wave will have increased to 0.500 (sine of 30°) of the peak value. At 45°, a sine wave will be at 0.707 of the peak, and so on. The sine wave is called a sine wave because it changes in amplitude at the same rate as the sine trigonometric function.

■ INTEGRATED MATH APPLICATION: MANUFACTURING

In the automated manufacturing process shown, a connecting conveyor is needed to transport parts from one system to another. Determine the length of the conveyor needed.

θ = 18°

22 ft

■ **Solution:**

$$\text{Hypotenuse} = ?$$

$$\text{Adjacent} = 22 \text{ feet}$$

$$\text{Angle } \theta = 18°$$

$$\cos 18° = \frac{A}{H} = \frac{22'}{H}$$

$$0.95 = \frac{22}{A}, H = \frac{22}{0.95} = 23.2 \text{ feet}$$

SELF-TEST EVALUATION POINT FOR SECTION 6–3

Now that you have completed this section, you should be able to:

■ **Objective 9.** *Define the trigonometric terms opposite, adjacent, hypotenuse, theta, sine, cosine, and tangent.*

■ **Objective 10.** *Demonstrate how the sine, cosine, and tangent trigonometric functions can be used to*

calculate (1) the length of an unknown side of a right-angle triangle if the length of another side and the angle theta are known or (2) the angle theta of a right-angle triangle if the lengths of two sides of the triangle are known.

Use the following questions to test your understanding of Section 6–3.

1. Calculate the length of the unknown side in the following triangles.

a.

H = ?

O = 35 mm

36°

b.

H = 160 km

38°

A = ?

c.

72°

O = 163 cm

A = ?

2. Calculate the angle θ for the following right-angle triangles.

a.

$H = 120$ mi $O = 38$ mi $\theta = ?$

b.

$H = 25$ ft $\theta = ?$ $A = 17$ ft

c.

$O = 51$ cm $\theta = ?$ $A = 69$ cm

6–4   SOLID FIGURES

As mentioned previously, a plane figure is a flat surface that has only two dimensions—length and width. A **solid figure,** on the other hand, is a three-dimensional figure having length, width, and height.

6–4–1   *Prisms*

A **prism** is a solid figure made up of plane figure faces and has at least one pair of parallel surfaces. Figure 6–22 details the facts and formulas for the prism. The visual breakdown in Figure 6–22(a) shows how a cube is made up of an upper base, lower base, and a set of sides. The **lateral surface area (L)** of the cube is the area of the sides only, as shown in Figure 6–22(b). The **total surface area (S)** is the lateral surface area of the cube plus the area of the two bases, as shown in Figure 6–22(c). The holding capacity or **volume** of the cube measures the space inside the prism and is described visually in Figure 6–22(d). Figure 6–22(e) shows examples of other prism types.

■ **INTEGRATED MATH APPLICATION: PAINTING**

If 1 gallon of paint covers 300 square feet, how many gallons would be needed to paint the office building shown?

Front   Side   Back   Side

20 ft

|← 20 ft →|← 16 ft →|

Doors 7 × 3 ft
Windows 4 × 4 ft

■ *Solution:*

$$\text{Lateral surface area} = P \times h$$
$$= 72 \times 20$$
$$= 1440 \text{ square feet}$$
$$7 \text{ windows} \times (4 \times 4) = 112 \text{ square feet}$$
$$2 \text{ doors} \times (7 \times 3) = 42 \text{ square feet}$$
$$1440 - 112 - 42 = 1286 \text{ square feet}$$
$$1286 \text{ square feet} \div 300 \text{ square feet/gallon} = 4.3 \text{ gallons}$$

Solid Figure
A 3-dimensional figure having length, width, and height.

Prism
A solid figure made up of plane figure faces; having at least one pair of parallel surfaces.

Lateral Surface Area
The side area of a solid figure.

Total Surface Area
The lateral surface area of a solid figure plus the area of its two bases.

Volume
The space within a solid figure, or its holding capacity.

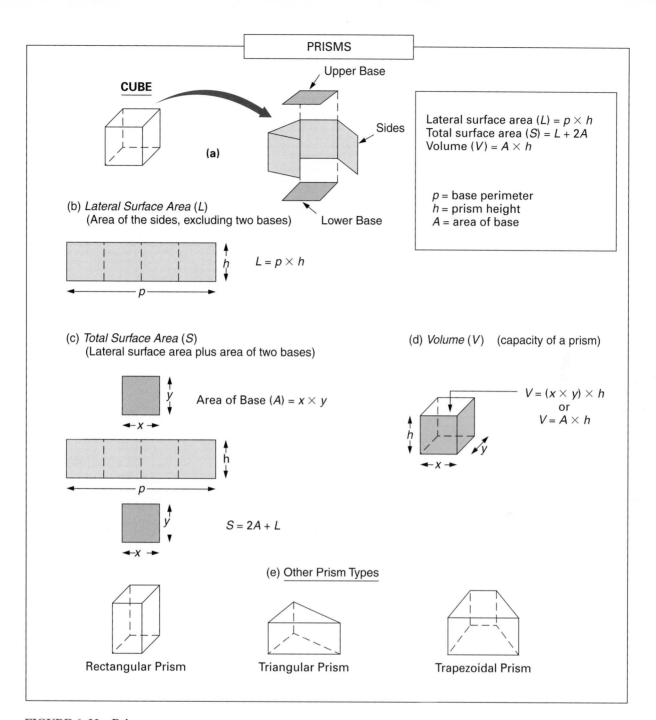

FIGURE 6–22 Prisms.

6–4–2 *Cylinders*

A **cylinder** is a solid figure with curved side walls extending between two identical circular bases. Figure 6–23 details the facts and formulas for the cylinder. As with prisms, plane figure formulas are used to calculate the cylinder's lateral surface area and base area, and as expected, the volume of a cylinder is the product of its base area and height.

Cylinder

A solid figure with curved side walls extending between two identical circular bases.

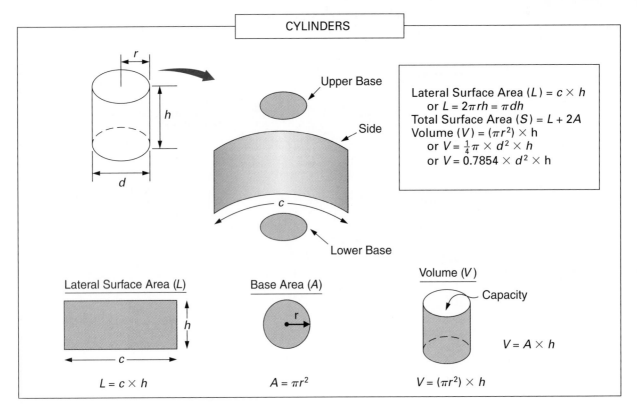

Lateral Surface Area $(L) = c \times h$
or $L = 2\pi rh = \pi dh$
Total Surface Area $(S) = L + 2A$
Volume $(V) = (\pi r^2) \times h$
or $V = \frac{1}{4}\pi \times d^2 \times h$
or $V = 0.7854 \times d^2 \times h$

Upper Base

Side

Lower Base

Lateral Surface Area (L)

$L = c \times h$

Base Area (A)

$A = \pi r^2$

Volume (V)

Capacity

$V = A \times h$

$V = (\pi r^2) \times h$

FIGURE 6–23   Cylinders.

■   **INTEGRATED MATH APPLICATION: LANDSCAPING**

To prepare for topsoil and sod, a circular section measuring 35 feet in diameter will have $1\frac{1}{2}$ feet of dirt removed. How many cubic feet is this?

■   *Solution:*

Cylinder volume:

$$
\begin{aligned}
V &= 0.7854 \times d^2 \times h \\
&= 0.7854 \times 35^2 \times 1.5 \\
&= 0.7854 \times 1225 \times 1.5 \\
&= 1443 \text{ cubic feet}
\end{aligned}
$$

6–4–3   *Spheres*

Sphere

A globular solid figure in which all points on the surface are at equal distance from the center.

A **sphere** is a globular solid figure in which all points on the surface are at an equal distance from the center. Figure 6–24 lists the formulas for a sphere along with a visual breakdown for a sphere's surface area and volume.

■   **INTEGRATED MATH APPLICATION: MACHINING**

If aluminium has a density of 0.093 pound per cubic inch, what is the weight of the machined capsule shown?

4 in.

2 in.

$$\text{Surface Area } (S) = \pi d^2 \text{ or } S = 4\pi r^2$$
$$\text{Volume } (V) = \frac{1}{6} \pi \times d^3 \text{ or } V = \frac{4\pi r^3}{3}$$
$$\text{or } V = 0.524 \times d^3$$

Surface Area

Surface area $(s) = c \times d$ (Since $c = \pi \times d$)
$$= (\pi \times d) \times d$$
$$= \pi \times d^2$$

Volume (V)

Cube's volume $= d \times d \times d = d^3$
Sphere occupies 52.4% of cube.

FIGURE 6–24   **Spheres.**

■   *Solution:*

Two $\frac{1}{2}$ spheres $= 1$ sphere

Volume of sphere $= 0.524 \times d^3$
$$= 0.524 \times 2^3$$
$$= 0.524 \times 8 = 4.19 \text{ cubic inches}$$

Volume of cylinder $= 0.7854 \times d^2 \times h$
$$= 0.7854 \times 2^2 \times 4$$
$$= 12.6 \text{ cubic inches}$$

Total volume $= 4.19 + 12.6 = 16.79 \text{ cubic inches}$

Weight $= $ density $\times$ volume
$$= 0.093 \text{ pound/cubic inch} \times 16.79 \text{ cubic inches} = 1.56 \text{ pounds}$$

6–4–4   *Pyramids and Cones*

Figure 6–25(a) lists the formulas for a **pyramid,** which has only one base and three or more lateral surfaces that taper up to a single point called an **apex.** Figure 6–25(a) lists the formulas for a **cone,** which has a circular base and smooth surface that extends up to an apex. Figures 6–25(b) and 6–26(b) list the formulas for the frustum of a pyramid and the frustum of a cone. A **frustum** is the base section of a solid pyramid or cone and is formed by cutting off the top of the solid figure so that the upper base and lower base are parallel to one another.

Pyramid

A solid figure having only one base and 3 or more lateral surfaces that taper up to a single point.

Apex

The uppermost point or tip of a solid figure.

Cone

A solid figure having a circular base and smooth surface that extends up to an apex.

Frustum

The base section of a solid pyramid or cone.

(a) PYRAMID

Lateral Surface Area $(L) = \frac{1}{2}ps$

Volume $(V) = \frac{1}{3}Ah$

p = perimeter of base s = slant height
A = area of base h = height

Lateral Surface Area (*L*)

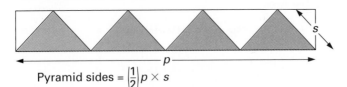

Pyramid sides $= \left|\frac{1}{2}\right|p \times s$

Volume (*V*)

Cube Volume $= A \times h$.
Pyramid occupies $\frac{1}{3}$ of cube.

(b) FRUSTUM OF PYRAMID

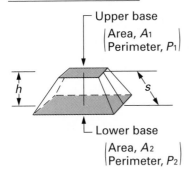

Upper base
$\begin{pmatrix}\text{Area, } A_1 \\ \text{Perimeter, } P_1\end{pmatrix}$

Lower base
$\begin{pmatrix}\text{Area, } A_2 \\ \text{Perimeter, } P_2\end{pmatrix}$

Lateral surface area $(L) = \frac{1}{2}(P_1 + P_2)s$

Volume $(V) = \frac{1}{3}h(A_1 + A_2 + \sqrt{A_1 \times A_2})$

FIGURE 6–25   Pyramids and Frustums of Pyramids.

■   **INTEGRATED MATH APPLICATION: ROOFING**

A roof's base measures 40 meters square and has a slant height of 24 meters. If roofing materials cost \$5.75 per square meter, how much will it cost to roof the building?

■   *Solution:*

Lateral surface area $= \left(\frac{1}{2}\right)p \times s$

$= \left(\frac{1}{2}\right)160 \times 24$

$= \left(\frac{1}{2}\right) \times 3840$

$= 1920$ square meters

Perimeter $= 4 \times 40$ meters
$= 160$ meters

Total cost $= 1920 \times \$5.75 = \$11,040$

(a) CONE

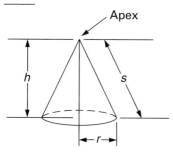

Apex

h s

r

Lateral Surface Area $(L) = \pi r s$ or $L = \dfrac{1}{2}\pi d s$

Volume $(V) = \dfrac{1}{3}\pi r^2 h$ or $V = \dfrac{1}{12}\pi d^2 h$

r = base radius h = height
s = slant height d = base diameter

(b) FRUSTUM OF CONE

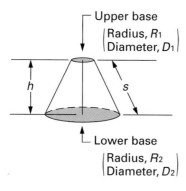

Upper base
$\left(\begin{array}{l}\text{Radius, } R_1 \\ \text{Diameter, } D_1\end{array}\right)$

h s

Lower base
$\left(\begin{array}{l}\text{Radius, } R_2 \\ \text{Diameter, } D_2\end{array}\right)$

Lateral Surface Area $(L) = \pi s\,(r_1 + r_2)$

$\quad$ or $L = \dfrac{1}{2}\pi s\,(d_1 + d_2)$

Volume $(V) = \dfrac{1}{3}\pi h\,(r_1^2 + r_2^2 + (r_1 r_2))$

$\quad$ or $V = \dfrac{1}{12}\pi h\,(d_1^2 + d_2^2 + (d_1 d_2))$

FIGURE 6–26   Cones and Frustums of Cones.

SELF-TEST EVALUATION POINT FOR SECTION 6–4

Now that you have completed this section, you should be able to:

■ **Objective 11.** *Define the term* solid figure, *and identify the different types.*

■ **Objective 12.** *List the formulas for determining the lateral surface area, total surface area, and volume of solid figures.*

Use the following questions to test your understanding of Section 6–4:

1. What is the difference between a plane figure and a solid figure?

2. What is the difference between lateral surface area and total surface area?

3. Sketch the following solid figures and list the formulas for lateral surface area, total surface area and volume.
   a. Prism
   b. Cylinder
   c. Sphere
   d. Pyramid
   e. Cone

4. What is a frustum?

REVIEW QUESTIONS

Multiple Choice Questions

1. What instrument can be used to measure an angle?
   a. Compass
   b. Protractor
   c. Ruler
   d. None of the above

2. A right angle is equal to:
   a. 45°
   b. 100°
   c. 180°
   d. 90°

3. A protractor can be used to:

   a. Measure an angle   **c.** Both (a) and (b)
   b. Draw an angle   **d.** None of the above

4. A _____ is a quadrilateral parallelogram, that has four sides that are all of equal length.

   a. Square   **c.** Trapezoid
   b. Rectangle   **d.** Prism

5. Trigonometry is the study of:

   a. Triangles   **c.** Trigonometric functions
   b. Angles   **d.** All the above

6. The Pythagorean theorem states that the _____ of the length of the hypotenuse of a right-angle triangle equals the sum of the _____ of the lengths of the other two sides.

   a. Square, square roots   **c.** Square, squares
   b. Square root, squares   **d.** Square root, square roots

7. A vector is an arrow whose length is used to represent the _____ of a quantity and whose point is used to indicate the same quantity's _____.

   a. Magnitude, size   **c.** Direction, size
   b. Magnitude, direction   **d.** Direction, magnitude

8. If two vectors in a vector diagram are not working together, or are out of phase with each other, they must be _____ to obtain a _____ vector.

   a. Added, resultant
   b. Multiplied by one another, parallel
   c. Vectorially added, resultant
   d. Both (a) and (b)

9. The _____ of a right-angle triangle is always the longest side.

   a. Opposite side
   b. Adjacent side
   c. Hypotenuse
   d. Angle theta

10. The _____ of a right-angle triangle is always across from the angle theta.

    a. Opposite side
    b. Adjacent side
    c. Hypotenuse
    d. Angle theta

11. The _____ of a right-angle triangle is always formed between the hypotenuse and the adjacent side.

    a. Opposite side   **c.** Hypotenuse
    b. Adjacent side   **d.** Angle theta

12. Which trigonometric function can be used to determine the length of the hypotenuse if the angle theta and the length of the opposite side are known?

    a. Tangent
    b. Sine
    c. Cosine

13. Which trigonometric function can be used to determine the angle theta when the lengths of the opposite and adjacent sides are known?

    a. Tangent
    b. Sine
    c. Cosine

14. Which trigonometric function can be used to determine the length of the adjacent side when the angle theta and the length of the hypotenuse are known?

    a. Tangent
    b. Sine
    c. Cosine

15. A plane figure has _____ dimensions, while a solid figure has _____ dimensions.

    a. 3, 4
    b. 2, 4
    c. 2, 3
    d. 3, 2

Communication Skill Questions

16. Define the following terms: (6–1 to 6–4)

    a. Geometry   **c.** Plane figure
    b. Trigonometry   **d.** Solid figure

17. Describe how to use a protractor to draw and measure an angle.   (6–1)

18. Sketch the following plane figures and list their associated formulas for calculating perimeter and area: (6–2)

    a. Square   **f.** Right triangle
    b. Rectangle   **g.** Scalene triangle
    c. Rhombus   **h.** Equilateral triangle
    d. Parallelogram   **i.** Isosceles triangle
    e. Trapezoid   **j.** Circle

19. What are the number of degrees in each vertex of a: (6–2–5)

    a. Square   **d.** Hexagon
    b. Equilateral triangle   **e.** Octagon
    c. Pentagon

20. Describe the purpose of Pythagoras's theorem.   (6–2–2)

21. Demonstrate how the sine, cosine and tangent trigonometric functions can be used to calculate: (6–3)

    a. The length of an unknown side of a right triangle when the length of another side is known and the angle theta is known
    b. The angle theta of a right triangle if the lengths of two sides of the triangle are known

22. Sketch the following solid figures and list their associated formulas: (6–4)

    a. Prism   **d.** Pyramid
    b. Cylinder   **e.** Cone
    c. Sphere

23. Why is it necessary to have inverse trigonometric functions?   (6–3)

24. What is a sine wave?   (6–3)

25. What is a frustum?   (6–4)

Practice Problems

26. Sketch the following:

    a. A line segment
    b. Two parallel lines
    c. A right angle
    d. An acute angle

    e. Two perpendicular lines
    f. An angle of 45°
    g. A vertical line
    h. An obtuse angle

27. Calculate the length of the hypotenuse (C side) of the following right triangles.

    a. $A = 20\,\text{mi}, B = 53\,\text{mi}$
    b. $A = 2\,\text{km}, B = 3\,\text{km}$
    c. $A = 4\,\text{in.}, B = 3\,\text{in.}$
    d. $A = 12\,\text{mm}, B = 12\,\text{mm}$

28. Calculate the length or magnitude of the resultant vectors in the following vector diagrams.

a.

b.

c.

d.

29. Calculate the value of the following trigonometric functions.

    a. sin 0°
    b. sin 30°
    c. sin 45°
    d. sin 60°
    e. sin 90°
    f. cos 0°
    g. cos 30°
    h. cos 45°

    i. cos 60°
    j. cos 90°
    k. tan 0°
    l. tan 30°
    m. tan 45°
    n. tan 60°
    o. tan 90°

30. Calculate angle θ from the function given.

    a. $\sin \theta = 0.707, \theta = ?$
    b. $\sin \theta = 0.233, \theta = ?$
    c. $\cos \theta = 0.707, \theta = ?$
    d. $\cos \theta = 0.839, \theta = ?$
    e. $\tan \theta = 1.25, \theta = ?$
    f. $\tan \theta = 0.866, \theta = ?$

31. Name each of the triangle's sides, and calculate the unknown values.

a.

b.

c.

32. If a cylinder has a radius of 3 centimeters and a height of 10 centimeters, calculate its:

    a. Lateral surface area
    b. Total surface area
    c. Volume

33. If a sphere has a diameter of 2 inches, what would be its surface area and volume?

34. What is the lateral surface area of a rectangular prism that has a base perimeter of 40 centimeters and a height of 12 centimeters?

35. Calculate the lateral surface area of a pyramid that has a base perimeter of 24 inches and a slant height of 8 inches.

Web Site Questions

Go to the Web site http://www.prenhall.com/cook, select the textbook *Introductory Mathematics*, select this chapter, and then follow the instructions when answering the multiple-choice practice problems.

Logarithms and Graphs

The First Pocket Calculator

During the seventeenth century, European thinkers were obsessed with any device that could help them with mathematical calculation. Scottish mathematician John Napier decided to meet this need, and in 1614 he published his new discovery of logarithms. In this book, consisting mostly of tediously computed tables, Napier stated that a logarithm is the exponent of a base number. For example:

The common logarithm (base 10) of 100 is 2 ($100 = 10^2$).

The common logarithm of 10 is 1 ($10 = 10^1$).

The common logarithm of 27 is 1.43136 ($27 = 10^{1.43136}$).

The common logarithm of 6 is 0.77815 ($6 = 10^{0.77815}$).

Any number, no matter how large or small, can be represented by or converted to a logarithm. Napier also outlined how the multiplication of two numbers could be achieved by simply adding the numbers' logarithms. For example, if the logarithm of 2 (which is 0.30103) is added to the logarithm of 4 (which is 0.60206), the result will be 0.90309, which is the logarithm of the number 8 ($0.30103 + 0.60206 = 0.90309$, $2 \times 4 = 8$). Therefore, the multiplication of two large numbers can be achieved by looking up the logarithms of the two numbers in a log table, adding them together, and then finding the number that corresponds to the sum in an antilog (reverse log) table. In this example, the antilog of 0.90309 is 8.

Napier's table of logarithms was used by William Oughtred, who developed, just 10 years after Napier's death in 1617, a handy mechanical device that could be used for rapid calculation. This device, considered the first pocket calculator, was the slide rule.

Outline and Objectives

VIGNETTE: THE FIRST POCKET CALCULATOR

INTRODUCTION

7–1   COMMON LOGARITHMS

Objective 1: Define the following terms:
   a. Common logarithm
   b. Exponential expression
   c. Logarithmic expression
   d. Natural logarithm

Objective 2: Describe how to use the calculator's stored table of common logarithms to find the log of any number.

Objective 3: Explain how logarithms can be used to simplify multiplication and division.

Objective 4: Describe how to use the calculator's stored table of common antilogarithms to convert the log answer to a number.

7–1–1 The Calculator's Table of Common Logarithms

7–1–2 Numbers That Are Not Multiples of Ten

7–1–3 Using Logarithms to Simplify Multiplication

7–1–4 Using Logarithms to Simplify Division

7–2   GRAPHS

Objective 5: Define the purpose of a graph.

Objective 6: Describe how to interpret the data displayed in a graph.

Objective 7: Explain the differences among a:
   a. Line graph
   b. Bar graph
   c. Circle graph or pie chart

Objective 8: Show how data collected from an experiment are plotted on a graph.

Objective 9: Describe the difference between a logarithmic response and a linear response.

Objective 10: Explain how the graphing calculator can be used to find the solution to a problem.

7–2–1 Reading Graphs

7–2–2 Types of Graphs

7–2–3 Plotting a Graph

7–2–4 The Graphing Calculator

Logarithm

The exponent that indicates the power to which a number is raised to produce a given number.

In the first half of this chapter, we will discuss **logarithms.** Although this seems like a new topic, it is simply a continuation of our discussion of exponents from Chapter 4. For example, you know that a distance of 1 kilometer is equal to 1×10^3 meters (or 1000 meters). The power of 10, exponent 3 in this example, is the logarithm of 1000 (log 1000 = 3). Logarithms can be used to simplify multiplication to addition, division to subtraction, raising to a power to multiplication, and extracting square roots to division. Today, multiplication, division, raising to a power, and extracting square roots can be accomplished more easily with a calculator, and so the use of logarithms to simplify these operations has decreased. Logarithms have found new applications, however, in many areas of science and technology, and so it is important to have a clear understanding of them.

In the second half of this chapter, we will study graph interpretation, types, and plotting. A graph is a diagram used frequently in all areas of science and technology to show the relationship between two or more factors. Graphs actually make it easier for us to understand information because they convert long lists of recorded data into an easy-to-understand visual representation.

7–1   COMMON LOGARITHMS

Common Logarithm

A logarithm whose base is 10.

Stated simply, a **common logarithm** is a base 10 exponent. So far, you have already had considerable experience with base 10 exponents, or as they are officially called, *power of 10.* For example, you know that a computer having 1 megabyte of memory means that the memory can store 1×10^6 bytes (or 1,000,000 bytes), and that a distance of 1 kilometer is equal to 1×10^3 meters (or 1000 meters). The power-of-10 exponent used in these examples is the logarithm; for the first example, *the common logarithm (base 10) of 1,000,000 is 6,* and for the second example, *the common logarithm (base 10) of 1000 is 3.* If we were to state the first of these examples in mathematical form, it would appear as follows:

$$
\begin{array}{cc}
\text{Exponent} & \text{Logarithm} \\
\swarrow & \swarrow \\
10^6 = 1{,}000{,}000 \qquad & \log_{10} 1{,}000{,}000 = 6 \\
\nearrow & \nearrow \quad \nwarrow \\
\text{Number} & \text{Base} \quad \text{Number}
\end{array}
$$

The first expression ($10^6 = 1,000,000$) is an *exponential expression* because the quantity is expressed in *exponential form.* The second expression ($\log_{10} 1,000,000 = 6$) is a *logarithmic expression* because the quantity is expressed in *logarithmic form.*

Natural Logarithm

A logarithm with *e* as a base, in which *e* is a transcendental number having a value to eight decimal places of 2.71828183.

There are two basic types of logarithms: **natural logarithms,** which use a base of 2.71828, and *common logarithms,* which use a base of 10. The base of a logarithm should be indicated when there is any chance of confusion; however, when using only common logs, it is not necessary to indicate or say the base because it is assumed to be 10. Therefore, $\log_{10} 1,000,000 = 6$ can be simplified to log 1,000,000 = 6.

As another example: What is the log of 1000? To find the answer, we simply convert 1000 to its power-of-10 equivalent, which in exponential form is 10^3. Because the common logarithm of any number is its base 10 exponent, in logarithmic form, log 1000 = 3. Here is a list of logs for numbers that are greater than 1:

Exponential Form	Logarithmic Form
$10^0 = 1$	$\log 1 = 0$
$10^1 = 10$	$\log 10 = 1$
$10^2 = 100$	$\log 100 = 2$
$10^3 = 1000$	$\log 1000 = 3$
$10^4 = 10,000$	$\log 10,000 = 4$
$10^5 = 100,000$	$\log 100,000 = 5$
$10^6 = 1,000,000$	$\log 1,000,000 = 6$

What do we do when we are dealing with numbers that are a fraction of 1, for example, 0.001 and 0.000001? The answer is to follow exactly the same procedure. First, express the number or value in exponential form ($10^{-3} = 0.001$). Then, because the logarithm is the base 10 exponent, express the number in logarithmic form ($\log 0.001 = -3$). Here is a list of logs for numbers that are a fraction of 1:

Exponential Form	Logarithmic Form
$10^0 = 1$	$\log 1 = 0$
$10^{-1} = 0.1$	$\log 0.1 = -1$
$10^{-2} = 0.01$	$\log 0.01 = -2$
$10^{-3} = 0.001$	$\log 0.001 = -3$
$10^{-4} = 0.0001$	$\log 0.0001 = -4$
$10^{-5} = 0.00001$	$\log 0.00001 = -5$
$10^{-6} = 0.000001$	$\log 0.000001 = -6$

7–1–1   *The Calculator's Table of Common Logarithms*

Your calculator contains a table of common logarithms. It operates as follows.

CALCULATOR KEYS

Name: Common logarithm key

Function: Calculates the common logarithm (base 10) of the displayed number.

Example: $\log 35.5 = ?$

Press keys: $\boxed{3}\,\boxed{5}\,\boxed{.}\,\boxed{5}\,\boxed{\text{LOG}}$

Answer: 1.5502283

Therefore, $35.5 = 10^{1.5502283}$.

■ **EXAMPLE**

Using a calculator, determine the log of 1000.

■ *Solution:*

To determine the log of 1000 using a calculator, you input the value 1000 and then press the log calculator key, as follows:

$$\text{Log } 1000 = ?$$

Press keys: $\boxed{1}\ \boxed{0}\ \boxed{0}\ \boxed{0}\ \boxed{\text{LOG}}$

Answer: 3

■ **EXAMPLE**

Determine the log of 0.00001.

■ *Solution:*

To determine the log of 0.00001 using the calculator, you simply input the value 0.00001 and then press the log calculator key, as follows:

$$\text{Log } 0.00001 = ?$$

Press keys: $\boxed{0}\ \boxed{.}\ \boxed{0}\ \boxed{0}\ \boxed{0}\ \boxed{0}\ \boxed{1}\ \boxed{\text{LOG}}$

Answer: -5

7–1–2   *Numbers That Are Not Multiples of 10*

Only numbers that are powers of 10, for example, 0.001, 0.1, 100, and 1,000,000,000, have an exponent or logarithm that is a whole number. In the following examples, you will see that any number between two multiples of 10 will have a decimal fraction. Using your calculator, check the logs (which have been rounded off to 4 decimal places) for the following numbers:

Logarithmic Form	Exponential Form
$\log 0.000089 = -4.0506$	$10^{-4.0506} = 0.000089$
$\log 0.28 = -0.5528$	$10^{-0.5528} = 0.28$

Because the example numbers are a fraction of 1, the log or exponent will be negative.

$\log 2.5 = 0.3979$	$10^{0.3979} = 2.5$
$\log 6 = 0.7782$	$10^{0.7782} = 6$
$\log 9.75 = 0.9890$	$10^{0.9890} = 9.75$

Because the log of 1 is 0, and the log of 10 is 1, all the example numbers between 1 and 10 will have logs or exponents between 0 and 1.

$\log 12 = 1.0792$	$10^{1.0792} = 12$
$\log 76.4 = 1.8831$	$10^{1.8831} = 76.4$

Because the log of 10 is 1 and the log of 100 is 2, all the example numbers between 10 and 100 will have logs or exponents between 1 and 2.

$\log 135.54 = 2.132$	$10^{2.132} = 135.54$
$\log 873 = 2.941$	$10^{2.941} = 873$

Because the log of 100 is 2 and the log of 1000 is 3, all the example numbers between 100 and 1000 will have logs or exponents between 2 and 3.

7–1–3 *Using Logarithms to Simplify Multiplication*

Now that we know what logarithms are and how to find them, let us see what we can do with them. As mentioned previously, one reason for converting numbers to logarithms is to simplify calculations. For example, using logarithms, *we can reduce the multiplication of two numbers to simple addition.* This advantage was first shown in this chapter's opening vignette; however, let us study the steps involved in a little more detail. To find the product of two numbers using logs, you have to follow these four steps:

Step 1: Find the log of each number.

Step 2: Change each number to base 10.

Step 3: Add the base 10 exponents.

Step 4: Convert the log answer back to a number using the antilog.

■ INTEGRATED MATH APPLICATION: CONSTRUCTION

Let us use logarithms to determine the total number of nails in inventory, if there are 26 boxes, and each box holds 789 nails.

■ *Solution:*

Step 1: Find the log of each number:

$$\log 26 = 1.41497$$
$$\log 789 = 2.89708$$

Step 2: Change each number to base 10. Because the log of a number is the base 10 exponent, we have the following:

$$26 = 10^{1.41497}$$
$$789 = 10^{2.89708}$$

Step 3: The answer to 26×789 can be obtained by simply adding the exponents from step 2.

$$10^{1.41497} \times 10^{2.89708} = 10^{1.41497+2.89708} = 10^{4.31205}$$

Note: This law has always applied to exponents; for example, $100 \times 1000 = 10^2 \times 10^3 = 100{,}000 = 10^5$. When like bases are multiplied, the exponents or number of zeros are added.

Step 4: Once you have converted all your numbers to logarithms and finished all your calculations, you then have to convert the log answer back to a number using a **common antilogarithm** table. This reverse log table is also stored in your calculator and operates as follows.

Antilogarithm

The number corresponding to a given logarithm.

CALCULATOR KEYS

Name: Common antilogarithm (10^x) key

Function: Calculates the common antilogarithm of the displayed value.

Example: $10^{1.5502283}$

Press keys: $\boxed{1}\ \boxed{.}\ \boxed{5}\ \boxed{5}\ \boxed{0}\ \boxed{2}\ \boxed{2}\ \boxed{8}\ \boxed{3}\ \boxed{10^x}$

Answer: 35.5

Therefore, $10^{1.5502283} = 35.5$.

7–1–4 *Using Logarithms to Simplify Division*

Using logarithms, *we can reduce the division of one number by another number to simple subtraction.* The steps for calculating the result of a division using logs are:

Step 1: Find the log of the dividend and divisor.

Step 2: Change each number to base 10.

Step 3: Subtract the log of the divisor from the log of the dividend.

Step 4: Convert the log answer back to a number using the antilog.

■ **INTEGRATED MATH APPLICATION: OFFICE MANAGEMENT**

Inventory has 8992.5 quarts of oil. If a service station uses 165 quarts on average per week, how many weeks will this inventory last? Let us use logarithms to calculate the quotient of $8992.5 \div 165$.

■ *Solution:*

Step 1: Find the log of both the dividend (8992.5) and the divisor (165):

$$\log 8992.5 = 3.95388$$
$$\log 165 = 2.21748$$

Step 2: Change each number to base 10. Because the log of a number is the base 10 exponent, we have the following:

$$8992.5 = 10^{3.95388}$$
$$165 = 10^{2.21748}$$

Step 3: The answer to $8992.5 \div 165$ can be obtained by simply subtracting the log or exponent of the divisor from the log of the dividend.

$$10^{3.95388} \div 10^{2.21748} = 10^{3.95388-2.21748} = 10^{1.7364}$$

Note: This law has always applied to exponents; for example, $10{,}000 \div 10 = 10^4 \div 10^1 = 10^{4-1} = 10^3 = 1000$. When like bases are divided, the exponents are subtracted.

Step 4: Convert the log answer back to a number using the antilog table in your calculator.

$$\text{antilog } 1.7364 = 54.5 \left(54\tfrac{1}{2} \text{ weeks}\right)$$

Using the calculator to double-check this answer, we find that $8992.5 \div 165 = 54.5$.

SELF-TEST EVALUATION POINT FOR SECTION 7–1

Now that you have completed this section, you should be able to:

■ *Objective 1.* *Define the terms* common logarithm, exponential expression, logarithmic expression, *and* natural logarithm.

■ *Objective 2.* *Describe how to use the calculator's stored table of common logarithms to find the log of any number.*

■ *Objective 3.* *Explain how logarithms can be used to simplify multiplication and division.*

Use the following questions to test your understanding of Section 7–1.

1. A common logarithm is a base 10 _____.

2. The logarithm of 10,000 is _____.
   a. 2
   b. 40
   c. 4
   d. 100

3. Determine the log of each of the following.
   a. 10
   b. 100
   c. 25
   d. 150
   e. 10,500
   f. 3746
   g. 0.35
   h. 1.75

4. Convert the following logs to their numerical value.
   a. 1.75
   b. 2.35
   c. 5
   d. 1
   e. 5.5
   f. 0.25
   g. 0.1
   h. 10

7–2 GRAPHS

A **graph** is a diagram showing the relationship between two or more factors. As an example, the graph in Figure 7–1 shows how new technologies, such as electric light, radio, the telephone, and television, were adopted in homes in the United States. This type of graph is called a **line graph** because the data are represented by points joined by line segments. The vertical scale in this graph indicates U.S. households in percent (0% to 100%), while the horizontal scale indicates the year (1900 to 1980). The four plotted lines in this graph show how quickly these new technologies found their way into the home. The data used to create the four plotted lines were obtained by recording what percentage of homes in the United States had electric light, radio, telephone, and television every year from 1900 to 1980.

7–2–1 Reading Graphs

If the yearly facts for the graph shown in Figure 7–1 were listed in columns instead of illustrated graphically, it would be difficult to see the characteristics of each individual line graph or to compare the characteristics of one technology with those of another. To explain this point, let us try an example.

Graph

A diagram (as a series of one or more points, lines, line segments, curves, or areas) that represents the variation of a variable in comparison with that of one or more other variables.

Line Graph

A graph in which points representing values of a variable are connected by a broken line.

FIGURE 7–1 A Graph Showing How New Technologies were Adopted in the United States.

From the graph shown in Figure 7–1, determine the following:

  a. Approximately what percentage of U.S. households had radio in 1930?
  b. Approximately what percentage of U.S. households had telephones in 1960?
  c. Which technology was the slowest to be adopted into the home, and which was the fastest?

■ *Solution:*

  a. To find this point on the graph, move along the horizontal scale until you reach 1930, and then proceed directly up until you connect with the plotted radio line graph (shown as a dashed line in Figure 7–1). Comparing this point with the vertical scale, you can see that it corresponds to about 40%. In 1930, therefore, approximately 40% of U.S. households had radio.
  b. To find this point on the graph, move along the horizontal scale until you reach 1960, and then proceed directly up until you connect with the plotted telephone line graph. Comparing this point with the vertical scale, you can see that it corresponds to about 80%. In 1960, therefore, approximately 80% of U.S. households had a telephone.
  c. Comparing the four line graphs shown in Figure 7–1, you can see that the telephone seemed to be the slowest technology to be adopted, while the television was the fastest.

Questions (a) and (b) in the previous example show how a graph can be used to analyze the characteristics of each individual line graph. Question (c) showed how a graph can be used to compare the characteristics of several line graphs.

■ **INTEGRATED MATH APPLICATION: RETAIL**

From the graph shown in Figure 7–2, determine the following:

  a. Which camera type sells the most, and approximately how many were sold in 1992?
  b. Which camera type came on and then went off the market?
  c. Compare the sales of 35 mm cameras in 1984 with sales in 1994.

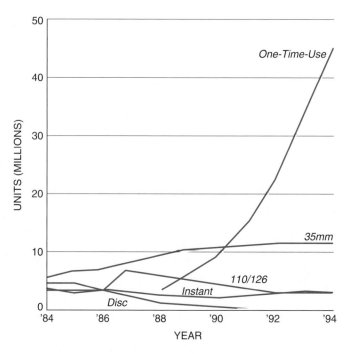

FIGURE 7–2 A Graph Showing Camera Sales.

Solution:

a. Disposable or one-time-use cameras sell the most, and in 1992 approximately 22 million units were sold.

b. Disc cameras.

c. In 1984 approximately 5 million 35 mm cameras were sold, whereas in 1994 approximately 11 million 35 mm cameras were sold.

7–2–2   *Types of Graphs*

There are three types of graphs: line graphs, bar graphs, and circle graphs. *Line graphs*, like the examples shown in Figures 7–1 and 7–2, are ideal for displaying historical data in which a factor is constantly changing.

   Bar graphs use parallel bars of different sizes to compare one variable with others. For example, Figure 7–3(a) shows how a bar graph can be used to compare the sales records of four computer salespersons for the month of May.

> **Bar Graph or Bar Chart**
> A graphic means of quantitative comparison by rectangles with lengths proportional to the measure of the data being compared.

■ **INTEGRATED MATH APPLICATION: OFFICE MANAGEMENT**

From the graph shown in Figure 7–3(a), determine the following:

a. How many sales representatives are there?

b. Which sales rep sold the most computers in the month of May, and which sales rep sold the least?

c. How many computers were sold in May?

d. Calculate the number of computers Ann sold as a percentage of the total sold.

■ *Solution:*

a. There are four bars in this bar graph, one for each of the four sales reps.

b. Ann has the largest bar in the graph because she sold 23 computers, while Pete has the smallest bar because he sold only 12 computers.

c. Ann sold 23, Jim sold 15, Pete sold 12, and Sally sold 15. Therefore, a total of 65 computers were sold in May.

d. To calculate the number of computers Ann sold as a percentage of the total sold, use the following formula;

$$\text{Percentage} = \frac{\text{units sold by rep}}{\text{total units sold}} \times 100$$

$$\text{Percentage} = \frac{23}{65} \times 100 \qquad (23 \div 65 = 0.35)$$

$$= 0.35 \times 100 = 35\%$$

FIGURE 7–3   Bar Graphs and Circle Graphs.

Pie Chart or
Circle Graph

A circular chart cut by radii
into segments illustrating
relative magnitudes.

Figure 7–3(b) shows an example of a **circle graph,** which uses different size segments within a circle to compare one variable with others. These circle graphs are often called **pie charts** because the circle resembles a pie, and the segments look like different slices in the pie. The size of the segments within a circle graph is usually given as a percentage, as shown in the example in Figure 7–3(b), with the whole circle equal to 100%. Like the bar graph in Figure 7–3(a), the circle graph in Figure 7–3(b) compares the sales records of four computer salespersons for the month of May. Because the data represented in these two graphs are the same, the only difference between these two graphs is the presentation format.

■   **INTEGRATED MATH APPLICATION: OFFICE MANAGEMENT**

From the graph shown in Figure 7–3(b), determine the following:

a. How many sales representatives are there?
b. Which sales rep sold the most computers in the month of May, and which sales rep sold the least?
c. If Ann sold 23 units, Jim sold 15 units, Pete sold 12 units, and Sally sold 15 units, calculate each sales rep's percentage of total sales.
d. If a total of 87 computers were sold in the month of June, and Jim again sold 23%, how many did he sell?

■   *Solution:*

a. There are four segments in this circle graph, one for each of the four sales reps.
b. The largest segment is assigned to Ann because she sold 23 computers, and the smallest segment is assigned to Pete, who sold only 12 computers.
c. Each sales rep's percentage is calculated as follows:

$$\text{Ann's percentage} = \frac{\text{units sold by Ann}}{\text{total units sold}} \times 100 = \frac{23}{65} \times 100 = 0.35 \times 100 = 35\%$$

$$\text{Jim's percentage} = \frac{\text{units sold by Jim}}{\text{total units sold}} \times 100 = \frac{15}{65} \times 100 = 0.23 \times 100 = 23\%$$

$$\text{Pete's percentage} = \frac{\text{units sold by Pete}}{\text{total units sold}} \times 100 = \frac{12}{65} \times 100 = 0.19 \times 100 = 19\%$$

$$\text{Sally's percentage} = \frac{\text{units sold by Sally}}{\text{total units sold}} \times 100 = \frac{15}{65} \times 100 = 0.23 \times 100 = 23\%$$

d. $23\% \times 87 = 0.23 \times 87 = 20$ Jim sold 20 computers in June.

7–2–3   *Plotting a Graph*

As the expression goes, "a picture is worth a thousand words," and graphs are pictures that can convert a thousand lines of recorded data into an easy-to-understand visual representation. To understand the graph better, let us collect data from an experiment and then plot these data on a graph.

It has been proven through experimentation that the human ear responds logarithmically to changes in sound level. To explain this in more detail, refer to the illustration in Figure 7–4(a), which shows a tone generator connected to a speaker. As the volume control of the generator is increased the power out of the speaker is gradually increased from 1 watt to 100 watts. The listener is asked to signal the points at which a change in volume is detected. These points are listed in the table in Figure 7–4(b). Studying this table, you can see that the power has to be increased from its start point of 1 watt (1.0 W) to approximately 1.26 W before the ear can detect an increase in volume. As the power is increased further a change in volume is detected at 1.58 W, 2 W, 2.5 W, 3.17 W, and so on, as shown in the table in Figure 7–4(b).

FIGURE 7–4   The Human Ear's Logarithmic Response to Changes in Sound Level.

217

Studying the table in Figure 7–4(b), you can see that the human ear is rather slow to respond to increased stimulation. For example, even when the sound intensity is doubled to twice its original value (1 W to 2 W), it is not interpreted by the listener as being twice as loud. In fact, the listener seems to detect only a small change in loudness. To make the tone sound twice as loud to the listener, the power must be increased 10 times (1 W to 10 W), and to make the tone sound three times as loud, the power must be increased 100 times (1 W to 100 W). Let us see how these data can be put to some practical purpose.

■ **INTEGRATED MATH APPLICATION: PERSONAL FINANCE**

You are considering the purchase of a new amplifier for your home music system and cannot decide whether to buy a 50 watt or a 100 watt unit. If the 100 watt unit is twice the price, will you get twice the benefit?

■ *Solution:*

Although the 100 watt amplifier delivers twice the output power, your ears will perceive this as only a small difference in volume. For your ears to detect twice the volume, therefore, you would need an amplifier that can deliver 10 times the power. Compared with a 50 watt amplifier, a 500 watt amplifier will sound as though you are getting twice your money's worth.

Logarithmic Response

Having or being a response or output that is logarithmic.

The graph in Figure 7–4(c) shows the **logarithmic response** of the human ear as a plot of the information from Figure 7–4(b). At point 10, the sound is sensed by the human ear as being twice (2 times) as loud, although power has actually increased from 1 W to 10 W (10 times). At point 20, the sound is sensed as being three times (3 times) as loud, although power has actually increased from 1 watt to 100 watts (100 times). This response is called *logarithmic* because a multiple-of-10 change was needed in the input sound level (power had to be increased from 1 watt to 10 watts to 100 watts) to be detected as a change in the sensed output sound (twice as much, or three times as much, respectively).

Linear Response

Having or being a response or output that is directly proportional to the input.

The logarithmic response of the human ear is very different from the **linear response** of the microphone, which is also shown in the graph in Figure 7–4(c). By definition, a linear change at the input causes a proportional change at the output. With a linear response, each change in the sound level applied to the microphone input will result in a proportional change in the microphone's output signal. For example, when the sound level is increased from 20 watts to 30 watts, the increase on the horizontal scale will be from point 4 to point 6 (a 10 watt change at the input results in a 2-point change at the output). Similarly, when the sound level is increased from 70 watts to 80 watts, the increase on the horizontal scale will be from point 14 to point 16 (a 10 watt change at the input will again result in a 2-point change at the output).

7–2–4 *The Graphing Calculator*

Up to this point we have been using a standard scientific calculator to perform mathematical operations. Graphing calculators possess all the functions of the scientific calculator, with the added ability that they can be used to graph functions. An example of a typical graphing calculator is shown in Figure 7–5.

The keyboard is divided into four zones: *graphing keys* access the calculator's interactive graphing features, *editing keys* allow you to edit expressions and values, *advanced function keys* display menus that access the advanced functions, and *scientific calculator keys* access the standard capabilities discussed previously.

To explain why we would need graphing capabilities, let us go straight to an application example.

FIGURE 7–5   **A Graphing Calculator.**

GRAPHING KEYS

EDITING KEYS

ADVANCED
FUNCTION KEYS

SCIENTIFIC
CALCULATOR KEYS

■ **INTEGRATED MATH APPLICATION: SHIPPING**

Figure 7–6 shows a 20-cm $\times$ 25-cm sheet of cardboard with two $X \times X$ corners cut out of the left side, and two $X \times$ 12.5-cm rectangles cut out of the right side. The cardboard can then be folded into a box with a lid that will have a volume that is equal to:

$$V = A \times B \times X$$

Volume = length $\times$ width $\times$ height

where $A = 20 - 2X$

$B = \dfrac{25}{2} - X$

If a small value of X is chosen, the base of the box will be large, but the sides or height of the box will be small. On the other hand, if a large value of X is chosen, the base of the box will be small, but the sides or height of the box will be large. The question, therefore, is, What value of X will give the box the maximum value of volume?

FIGURE 7–6   **A Cardboard Box.**

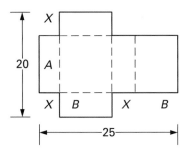

■ *Solution:*

If $V = A \times B \times X$, and $A = 20 - 2X$, and $B = \frac{25}{2} - X$, then

$$V = (20 - 2X) \times \left(\frac{25}{2} - X\right) \times (X)$$

To enter this equation into the graphing calculator, you would use the "$Y =$ Editor" function and input the formula as shown in Figure 7–7(a). Now, selecting the calculator's "Table" function will determine the volume of the box ($Y1$) for every possible value of X, as shown in Figure 7–7(b). Scanning down the table shown in Figure 7–7(c), you can see that the maximum value of $Y1$ or volume is 410.26, which occurs when $X = 3.7$. Pressing the

(a)

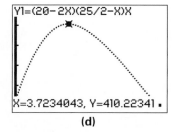

(b)

(c)

(d)

FIGURE 7–7   Display Results for Box Volume.

calculator's "Graph" function will display the plotted data in graphical rather than tabular form, as seen in Figure 7–7(d).

This ability of the graphing calculator to display every possible output result for every possible input value can be extremely helpful in the analysis of a problem.

SELF-TEST EVALUATION POINT FOR SECTION 7–2

Now that you have completed this section, you should be able to:

■   **Objective 5.**   *Define the purpose of a graph.*

■   **Objective 6.**   *Describe how to interpret the data displayed in a graph.*

■   **Objective 7.**   *Explain the differences among a line graph, a bar graph, and a circle graph or pie chart.*

■   **Objective 8.**   *Show how the data collected from an experiment are plotted on a graph.*

■   **Objective 9.**   *Describe the difference between a logarithmic response and a linear response.*

■   **Objective 10.**   *Explain how the graphing calculator can be used to find the solution to a problem.*

Use the following questions to test your understanding of Section 7–2.

  1. Plot a line graph to display the following data:

The Number of Indoor and Outdoor Movie Screens

YEAR	INDOOR	OUTDOOR
1975	12,000	4000
1980	14,000	4000
1985	18,000	3000
1990	23,000	2000

  2. Draw a circle graph and bar graph to indicate the following data: Each 24-hour day Tom works for 8 hours, studies for 3 hours, showers and dresses for 2 hours, eats for 1.5 hours, sleeps for 8 hours, and exercises for 1.5 hours.

  3. What is the difference between a standard scientific calculator and a graphing calculator?

  4. Describe the purpose for each of the following graphing calculator steps:
    a. Y = Editor
    b. Table
    c. Graph

REVIEW QUESTIONS

Multiple Choice Questions

1. Logarithms were first discovered by:
   - **a.** Carl Gauss
   - **b.** John Napier
   - **c.** Isaac Newton
   - **d.** Blaise Pascal

2. Natural logarithms use a base of _____, while common logarithms use a base of _____.
   - **a.** 10, 100
   - **b.** 10, 2.71828
   - **c.** 2.71828, 100
   - **d.** 2.71828, 10

3. What is the log of 10,000?
   - **a.** 9.21
   - **b.** 6
   - **c.** 4
   - **d.** 3

4. What is the log of 0.32?
   - **a.** −0.5
   - **b.** −1.1
   - **c.** −1.5
   - **d.** −0.15

5. What is the common antilogarithm of 10^6?
   - **a.** 1,000,000
   - **b.** 1000.000
   - **c.** 1000
   - **d.** 100,000

6. A _____ is a diagram showing the relationship between two or more factors.
   - **a.** Scale
   - **b.** Logarithm
   - **c.** Graph
   - **d.** Exponent

7. In a _____ _____, parallel columns of different sizes are used to compare one variable with others.
   - **a.** Line graph
   - **b.** Bar graph
   - **c.** Circle graph
   - **d.** All the above

8. In a _____ _____, the data are represented by points joined by line segments.
   - **a.** Line graph
   - **b.** Bar graph
   - **c.** Circle graph
   - **d.** All the above

9. In a _____ _____, different size segments within a circle are used to compare one variable with others.
   - **a.** Line graph
   - **b.** Bar graph
   - **c.** Pie chart
   - **d.** All the above

10. With a _____ response, a multiple-of-10 change at the input causes a change at the output, whereas with a _____ response, a change at the input causes a proportional change at the output.
    - **a.** Linear, logarithmic
    - **b.** Logarithmic, linear
    - **c.** Logarithmic, exponential
    - **d.** Linear, exponential

Communication Skill Questions

11. What is a common logarithm? (7–1)
12. What is the difference between a natural logarithm and a common logarithm? (7–1)
13. What is a common antilogarithm? (7–1)
14. State the four-step procedure you would use to simplify multiplication using logarithms. (7–1)
15. Arbitrarily choose values, and then describe the following:
    - **a.** Multiplication using logarithms
    - **b.** Division using logarithms

16. What is a graph? (7–2)
17. List some of the different types of graphs. (7–2)
18. Sketch a diagram and describe the different types of graphs. (7–2)
19. What is a linear response? (7–2)
20. What is a logarithmic response? (7–2)

Practice Problems

21. Determine the log of the following:
    - **a.** 10
    - **b.** 10,000
    - **c.** 1,234,567
    - **d.** 50.5
    - **e.** 1.5
    - **f.** 0.25
    - **g.** 0.5
    - **h.** 0.75
    - **i.** 0.125
    - **j.** 0.95
    - **k.** 12.25
    - **l.** 75.75

22. Determine the antilog of the following:
    - **a.** 1
    - **b.** 2
    - **c.** 3
    - **d.** 4
    - **e.** 5
    - **f.** 1.5
    - **g.** 2.5
    - **h.** 1.25
    - **i.** 0.125
    - **j.** 0.75
    - **k.** 0.33
    - **l.** 0.95

23. What steps should be followed to simplify multiplication to addition using logarithms?

24. Apply the four-step procedure from Problem 23 to the following:

   a. $2 \times 2 = ?$ **c.** $1.25 \times 150 = ?$
   b. $244 \times 30 = ?$ **d.** $0.5 \times 0.33 = ?$

25. What steps should be followed to simplify division to subtraction using logarithms?

26. Apply the four-step procedure from Problem 25 to the following:

   a. $12 \div 2 = ?$
   b. $244 \div 2 = ?$
   c. $250 \div 0.25 = ?$
   d. $1255 \div 0.5 = ?$

27. Identify the types of graphs shown in Figure 7–8.

(a)

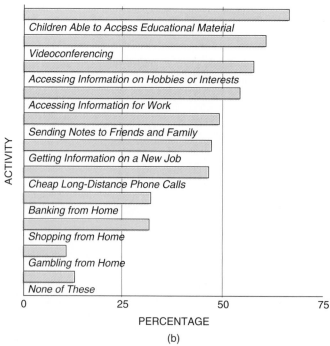

(b)

FIGURE 7–8   Graphs (a) Music Types. (b) Why People Connect to the Internet.

28. Refer to Figure 7–8(a), and answer the following questions:
   a. Which music type has remained the most popular?
   b. Which music type is the least popular?
   c. Which music type became less popular as country music had an increase in popularity in 1992 and 1993?
   d. What was the approximate market share for each of the music types in 1995?

29. Refer to Figure 7–8(b), and answer the following questions:
   a. What is the main reason for the average person to connect to the Internet?
   b. In order, what are the top five reasons why people want to go online?

30. Figure 7–9 shows an electronic device called a *potentiometer*. This device has two terminals (*A* and *B*), a circular resistive track, and a wiper that can be placed at any point on the resistive track by the turn of a control shaft. (Potentiometers are often used as volume controls for music systems and televisions.) There are two basic types of potentiometers: those with linear resistive tracks, and those with tapered or nonlinear resistive tracks. Figure 7–9 lists the resistance values that were measured between terminals *A* and *B* for both a linear and a tapered potentiometer. When the control shaft was fully counterclockwise (CCW), the resistance between terminals *A* and *B* for both the linear and tapered potentiometer was 0 ohms (0 Ω). When the control shaft was turned one-quarter clockwise ($\frac{1}{4}$ CW), the resistance between terminals *A* and *B* for the linear potentiometer was 250 Ω, and for the tapered potentiometer it was 350 Ω. When the control shaft was turned half clockwise ($\frac{1}{2}$ CW), the resistance between terminals *A* and *B* for the linear potentiometer was 500 Ω, and for the tapered potentiometer it was 625 Ω. When the control shaft was turned three-quarters clockwise ($\frac{3}{4}$ CW), the resistance between terminals *A* and *B* for the linear potentiometer was 750 Ω, and for the tapered potentiometer it was 900 Ω. Finally, when the control shaft was turned fully clockwise (CW), the resistance between terminals *A* and *B* for the linear potentiometer was 1000 Ω (1 kΩ), and for the tapered potentiometer it was also 1000 Ω. Sketch a "resistance versus control shaft position" line graph, and then plot a linear line and a tapered line.

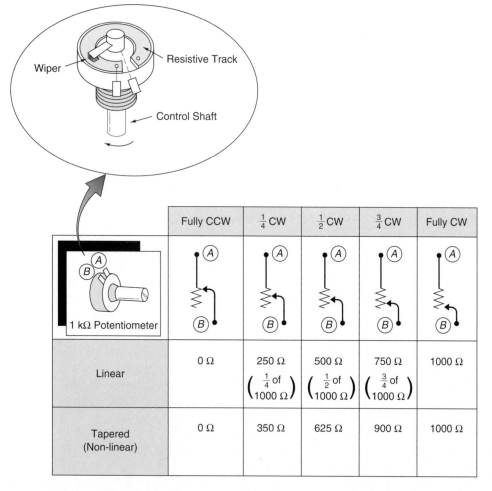

	Fully CCW	$\frac{1}{4}$ CW	$\frac{1}{2}$ CW	$\frac{3}{4}$ CW	Fully CW
1 kΩ Potentiometer					
Linear	0 Ω	250 Ω $\left(\frac{1}{4} \text{ of } 1000\,\Omega\right)$	500 Ω $\left(\frac{1}{2} \text{ of } 1000\,\Omega\right)$	750 Ω $\left(\frac{3}{4} \text{ of } 1000\,\Omega\right)$	1000 Ω
Tapered (Non-linear)	0 Ω	350 Ω	625 Ω	900 Ω	1000 Ω

FIGURE 7–9   A Linear and a Tapered Resistive Track Within a Potentiometer.

Web Site Questions

Go to the Web site http://www.prenhall.com/cook, select the textbook *Introductory Mathematics*, select this chapter, and then follow the instructions when answering the multiple-choice practice problems.

Computer Math

Moon Walk

In July of 1969 almost everyone throughout the world was caught at one stage or another gazing up to the stars and wondering. For U.S. astronaut Neil Armstrong, the age-old human dream of walking on the moon was close to becoming a reality. On earth, millions of people and thousands of newspapers and magazines waited anxiously to celebrate a successful moon landing. Then, the rather brief but eloquent message from Armstrong was transmitted 240,000 miles to Houston, Texas, where it was immediately retransmitted to a waiting world. This message was: "That's one small step for man, one giant leap for mankind." These words were received by many via television, but for magazines and newspapers this entire mission, including the speech from Armstrong, was converted into a special code made up of ON–OFF pulses that traveled from computer to computer. Every letter of every word was converted into a code that used the two symbols of the binary number system, 0 and 1. This code is still used extensively by all modern computers and is called *American Standard Code for Information Interchange*, or ASCII (pronounced "askey").

It is only fitting that these codes made up of zeros and ones conveyed the finale to this historic mission, since they played such an important role throughout. Commands encoded into zeros and ones were used to control almost everything, from triggering the takeoff to keeping the spacecraft at the proper angle for reentry into the earth's atmosphere.

No matter what its size or application, a digital electronic computer is quite simply a system that manages the flow of information in the form of zeros and ones. Referring to the ASCII code table in Figure 8–2 on p. 228, see if you can decode the following famous Armstrong message that reads from left to right:

```
0100010  1010100  1101000  1100001  1110100  0100111  1110011  0100000  1101111  1101110
1100101  0100000  1110011  1101101  1100001  1101100  1101100  0100000  1110011  1110100
1100101  1110000  0100000  1100110  1101111  1110010  0100000  1100001  0100000  1101101
1100001  1101110  0101100  0100000  1101111  1101110  1100101  0100000  1100111  1101101
1100001  1101110  1110100  0100000  1101100  1100101  1100001  1110000  0100000  1100110
1101111  1110010  0100000  1101101  1100001  1101110  1101011  1101001  1101110  1100100
0101110  0100010  0100000  0100000  0101101  1001110  1100101  1101001  1101100  0100000
1000001  1110010  1101101  1110011  1110100  1110010  1101111  1101110  1100111  0101100
0100000  1000001  1110000  1101111  1101100  1101100  1101111  0100000  0110001  0110001
```

Outline and Objectives

VIGNETTE: MOON WALK

INTRODUCTION

8–1   ANALOG AND DIGITAL DATA AND DEVICES

Objective 1: Define the differences between analog and digital data and devices.

Objective 2: Explain why analog circuits are also referred to as linear circuits.

Objective 3: Describe the function of the digital ASCII code.

Objective 4: Explain why the 10-state digital system was replaced by the two-state digital system.

8–1–1   Analog Data and Devices

8–1–2   Digital Data and Devices

8–2   ANALOG AND DIGITAL SIGNAL CONVERSION

Objective 5: List the three basic blocks within a computer.

Objective 6: Describe how analog and digital devices are connected to a computer.

Objective 7: Define the function of the analog-to-digital converter (ADC) and the digital-to-analog converter (DAC).

8–3   THE BINARY NUMBER SYSTEM

Objective 8: Explain the positional weight, the reset and carry action, and conversion for binary.

8–3–1   Positional Weight

8–3–2   Reset and Carry

8–3–3   Converting Binary Numbers to Decimal Numbers

8–3–4   Converting Decimal Numbers to Binary Numbers

8–3–5   Converting Information Signals

8–4   THE HEXADECIMAL NUMBER SYSTEM

Objective 9: Explain the positional weight, the reset and carry action, and conversion for hexadecimal.

8–4–1   Converting Hexadecimal Numbers to Decimal Numbers

8–4–2   Converting Decimal Numbers to Hexadecimal Numbers

8–4–3   Converting between Binary and Hexadecimal

8–5   BINARY CODES

Objective 10: Describe the following binary codes: binary coded decimal (BCD) and the American Standard Code for Information Interchange (ASCII).

8–5–1   The Binary Coded Decimal (BCD) Code

8–5–2   The American Standard Code for Information Interchange (ASCII)

8–6   LOGIC GATES AND BOOLEAN EXPRESSIONS

Objective 11: Give the Boolean expressions for each of the seven basic logic gates.

Objective 12: Describe how to write an equivalent Boolean expression for a logic circuit and how to sketch an equivalent logic circuit for a Boolean expression.

8–6–1   The NOT Gate and Its Boolean Expression

8–6–2   The OR Gate and Its Boolean Expression

8–6–3   The AND Gate and Its Boolean Expression

8–6–4   The NOR Gate and Its Boolean Expression

8–6–5 The NAND Gate and Its Boolean Expression

8–6–6 The XOR Gate and Its Boolean Expression

8–6–7 The XNOR Gate and Its Boolean Expression

8–7 BINARY ARITHMETIC

Objective 13: Demonstrate how to add binary numbers.

Objective 14: Explain how the two's complement number system can be used to represent positive and negative numbers.

Objective 15: Demonstrate how two's complement numbers enable an adder circuit to add, subtract, multiply, and divide binary numbers.

8–7–1 Binary Addition

8–7–2 Representing Positive and Negative Numbers

8–7–3 Two's Complement Arithmetic

Introduction

Since World War II no branch of science has contributed more to the development of the modern world than electronics. It has stimulated dramatic advances in the fields of communication, computing, consumer products, industrial automation, test and measurement, and health care. It has now become the largest single industry in the world, exceeding the automobile and oil industries, with annual sales of electronic systems exceeding $2 trillion.

One of the most important trends in this huge industry has been a gradual shift from *analog electronics* to *digital electronics*. This movement began in the 1960s and is almost complete today. In fact, a recent statistic stated that, on average, 90% of the circuitry within electronic systems is now digital and only 10% is analog. This digitalization of the electronics industry is bringing sectors that were once separate closer together. For example, two of the largest sectors or branches of electronics are *computing* and *communications*. Being able to communicate with each other using the common language of digital has enabled computers and communications to interlink so that computers can now function within communication-based networks, and communications networks can now function through computer-based systems. Industry experts call this merging *convergence* and predict that digital electronics will continue to unite the industry and stimulate progress in practically every field of human endeavor.

Needless to say, this part of the book you are about to read involving digital electronic mathematics and concepts is an essential element in your study of electronics.

8–1 ANALOG AND DIGITAL DATA AND DEVICES

Before we begin, let's first review analog electronics and then compare it with digital so that we can clearly see the differences between the two.

8–1–1 *Analog Data and Devices*

Figure 8–1(a) shows an electronic circuit designed to amplify speech information detected by a microphone. One of the easiest ways to represent data or information is to have a voltage change in direct proportion to the information it is representing. In the example in Figure 8–1(a), the *pitch and loudness* of the sound waves applied to the microphone should

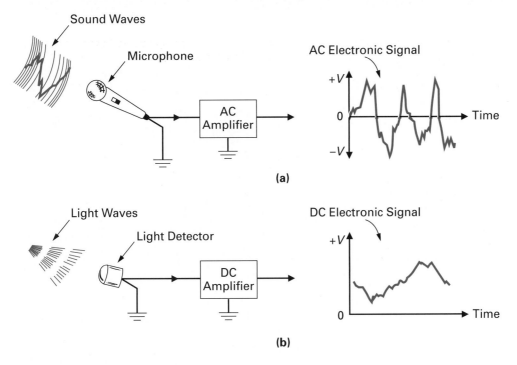

(a)

(b)

FIGURE 8–1   Analog Data and Devices.

control the *frequency and amplitude* of the voltage signal from the microphone. The output voltage signal from the microphone is said to be an analog of the input speech signal. The word **analog** means "similar to," and in Figure 8–1(a) the electronic signal produced by the microphone is an analog (or similar) to the speech signal, since a change in speech "loudness or pitch" will cause a corresponding change in signal voltage "amplitude or frequency."

In Figure 8–1(b), a light detector or solar cell converts light energy into an electronic signal. This information signal represents the amount of light present, since changes in voltage amplitude result in a change in light-level intensity. Once again, the output electronic signal is an analog (or similar) to the sensed light level at the input.

Figure 8–1 therefore indicates two analog circuits. The microphone in Figure 8–1(a) generates an **AC analog signal** that is amplified by an **AC amplifier circuit.** The microphone is considered an **analog device or component** and the amplifier an **analog circuit.** The light detector in Figure 8–1(b) would also be an analog component; however, in this example it generates a **DC analog signal** that is amplified by a **DC amplifier circuit.**

Both of the information signals in Figure 8–1 vary smoothly and continuously, in accordance with the natural quantities they represent (sound and light). Analog circuits are often called **linear circuits,** since linear by definition is *the variation of an output in direct proportion to the input.* This linear circuit response is evident in Figure 8–1, where you can see that in both the DC and AC circuit, the output signal voltage varies in direct proportion to the sound or light signal input.

8–1–2   *Digital Data and Devices*

In **digital electronic circuits,** information is first converted into a coded group of pulses. To explain this concept, let's take a closer look at the example shown in Figure 8–2(a). This code consists of a series of HIGH and LOW voltages, in which the HIGH voltages are called "1s" (ones) and the LOW voltages are called "0s" (zeros). Figure 8–2(b) lists the ASCII code, which is one example of a digital code. Referring to Figure 8–2(a), you will notice that the "1101001" information or data stream code corresponds to lowercase *i* in the ASCII table shown highlighted in Figure 8–2(b). Computer keyboards are one of many devices that

Analog
The representation of physical properties by a proportionally varying signal.

AC Analog Signal
An analog signal that alternates positive and negative.

AC Amplifier
An amplifier designed to increase the magnitude of an AC signal.

Analog Device or Component
Advice or component that makes up a circuit designed to manage analog signals.

Analog Circuit
A circuit designed to manage analog signals.

DC Analog Signal
An analog signal that is always either positive or negative.

DC Amplifier
An amplifier designed to increase the magnitude of a DC signal.

Linear Circuit
A circuit in which the output varies in direct proportion to the input.

Digital Electronic Circuit
A circuit designed to manage digital information signals.

(a)

(c)

The American Standard Code for Information Interchange (ASCII)

Char	b7	b6	b5	b4	b3	b2	b1		Char	b7	b6	b5	b4	b3	b2	b1
SP	0	1	0	0	0	0	0		P	1	0	1	0	0	0	0
!	0	1	0	0	0	0	1		Q	1	0	1	0	0	0	1
"	0	1	0	0	0	1	0		R	1	0	1	0	0	1	0
#	0	1	0	0	0	1	1		S	1	0	1	0	0	1	1
$	0	1	0	0	1	0	0		T	1	0	1	0	1	0	0
%	0	1	0	0	1	0	1		U	1	0	1	0	1	0	1
&	0	1	0	0	1	1	0		V	1	0	1	0	1	1	0
'	0	1	0	0	1	1	1		W	1	0	1	0	1	1	1
(	0	1	0	1	0	0	0		X	1	0	1	1	0	0	0
)	0	1	0	1	0	0	1		Y	1	0	1	1	0	0	1
*	0	1	0	1	0	1	0		Z	1	0	1	1	0	1	0
+	0	1	0	1	0	1	1		[	1	0	1	1	0	1	1
,	0	1	0	1	1	0	0		/	1	0	1	1	1	0	0
–	0	1	0	1	1	0	1		]	1	0	1	1	1	0	1
.	0	1	0	1	1	1	0		^	1	0	1	1	1	1	0
/	0	1	0	1	1	1	1		_	1	0	1	1	1	1	1
0	0	1	1	0	0	0	0		'	1	1	0	0	0	0	0
1	0	1	1	0	0	0	1		a	1	1	0	0	0	0	1
2	0	1	1	0	0	1	0		b	1	1	0	0	0	1	0
3	0	1	1	0	0	1	1		c	1	1	0	0	0	1	1
4	0	1	1	0	1	0	0		d	1	1	0	0	1	0	0
5	0	1	1	0	1	0	1		e	1	1	0	0	1	0	1
6	0	1	1	0	1	1	0		f	1	1	0	0	1	1	0
7	0	1	1	0	1	1	1		g	1	1	0	0	1	1	1
8	0	1	1	1	0	0	0		h	1	1	0	1	0	0	0
9	0	1	1	1	0	0	1		i	1	1	0	1	0	0	1
:	0	1	1	1	0	1	0		j	1	1	0	1	0	1	0
;	0	1	1	1	0	1	1		k	1	1	0	1	0	1	1
<	0	1	1	1	1	0	0		l	1	1	0	1	1	0	0
=	0	1	1	1	1	0	1		m	1	1	0	1	1	0	1
>	0	1	1	1	1	1	0		n	1	1	0	1	1	1	0
?	0	1	1	1	1	1	1		o	1	1	0	1	1	1	1
@	1	0	0	0	0	0	0		p	1	1	1	0	0	0	0
A	1	0	0	0	0	0	1		q	1	1	1	0	0	0	1
B	1	0	0	0	0	1	0		r	1	1	1	0	0	1	0
C	1	0	0	0	0	1	1		s	1	1	1	0	0	1	1
D	1	0	0	0	1	0	0		t	1	1	1	0	1	0	0
E	1	0	0	0	1	0	1		u	1	1	1	0	1	0	1
F	1	0	0	0	1	1	0		v	1	1	1	0	1	1	0
G	1	0	0	0	1	1	1		w	1	1	1	0	1	1	1
H	1	0	0	1	0	0	0		x	1	1	1	1	0	0	0
I	1	0	0	1	0	0	1		y	1	1	1	1	0	0	1
J	1	0	0	1	0	1	0		z	1	1	1	1	0	1	0
K	1	0	0	1	0	1	1		(	1	1	1	1	0	1	1
L	1	0	0	1	1	0	0		¦	1	1	1	1	1	0	0
M	1	0	0	1	1	0	1		)	1	1	1	1	1	0	1
N	1	0	0	1	1	1	0		~	1	1	1	1	1	1	0
O	1	0	0	1	1	1	1		DEL	1	1	1	1	1	1	1

(b)

FIGURE 8–2 Two-State Digital Information. (a) Information or Data Code for *i*. (b) ASCII Code. (c) Keyboard ASCII Generator.

make use of the digital ASCII code. In Figure 8–2(c), you can see how the lowercase *i* ASCII code is generated whenever the *i* key is pressed, encoding the information *i* into a group of pulses (1101001).

The next question you may have is: Why do we go to all this trouble to encode our data or information into these two-state codes? The answer can best be explained by examining history. The early digital systems constructed in the 1950s made use of a decimal code that used 10 levels or voltages, with each of these voltages corresponding to one of the 10 digits in the decimal number system (0 = 0 V, 1 = 1 V, 2 = 2 V, 3 = 3 V, up to 9 = 9 V). The circuits that had to manage these decimal codes, however, were very complex since they had to generate one of 10 voltages and sense the difference among all 10 voltage levels. This complexity led to inaccuracy, since some circuits would periodically confuse one voltage level for a different voltage level. *The solution to the problem of circuit complexity and inaccuracy was solved by adopting a two-state system instead of a 10-state system.* Using a two-state or two-digit system, you can generate codes for any number, letter, or symbol, as we have seen in the ASCII table in Figure 8–2. The electronic circuits that manage these two-state codes are less complex, since they have to generate and sense only either a HIGH or a LOW voltage. In addition, two-state circuits are much more accurate, since there is little room for error between the two extremes of ON and OFF, or HIGH voltage and LOW voltage.

Abandoning the 10-state system and adopting the two-state system for the advantages of circuit simplicity and accuracy meant that we were no longer dealing with the decimal number system. As a result, having only two digits (0 and 1) means that we are now operating in the two-state number system, which is called **binary.** Figure 8–3(a) shows how the familiar decimal system has 10 digits or levels labeled 0 through 9, and Figure 8–3(b) shows how we could electronically represent each decimal digit. With the base 2, or binary number system, we have only two digits in the number scale, and therefore only the digits 0 (zero) and 1 (one) exist in binary, as shown in Figure 8–3(c). These two states are typically represented in an electronic circuit as two different values of voltage (binary 0 = LOW voltage; binary 1 = HIGH voltage), as shown in Figure 8–3(d).

Using combinations of **binary digits** (abbreviated **bits**) we can represent information as a binary code. This code is called a **digital signal** because it is an *information signal* that makes use of *binary digits*. Today, almost all information, from your telephone conversations to the music on your compact discs, is **digitized** or converted to binary data form.

You will probably hear the terms *analog* and *digital* used to describe the difference between the two types of meters shown in Figure 8–4. On the **analog readout multimeter** shown in Figure 8–4(a), the amount of pointer deflection across the scale is an analog (or similar) to the magnitude of the electrical property being measured. On the other hand, with the **digital readout multimeter** shown in Figure 8–4(b), the magnitude of the electrical property being measured is displayed using digits, which in this case are decimal digits.

Binary
Having only two alternatives, two-state.

Binary Digit
Abbreviated bit; it is either a 0 or 1 from the binary number system.

Bit
Binary digit.

Digital Signal
An electronic signal made up of binary digits.

Digitize
To convert an analog signal to a digital signal.

Analog Readout Multimeter
A multimeter that uses a movement on a calibrated scale to indicate a value.

Digital Readout Multimeter
A multimeter that uses digits to indicate a value.

SELF-TEST EVALUATION POINT FOR SECTION 8–1

Now that you have completed this section, you should be able to:

■ *Objective 1. Define the differences between analog and digital data and devices.*

■ *Objective 2. Explain why analog circuits are also referred to as linear circuits.*

■ *Objective 3. Describe the function of the digital ASCII code.*

■ *Objective 4. Explain why the 10-state digital system was replaced by the two-state digital system.*

Use the following questions to test your understanding of Section 8–1:

1. What does the word *analog* mean?
2. What is an analog signal?
3. What is the ASCII code?
4. What were the two key reasons for adopting the two-state binary system in place of the 10-state decimal system?
5. How many digits does the binary number system have?
6. Describe the difference between an analog readout watch and a digital readout watch.

FIGURE 8–3 **Electronically Representing the Digits of a Number System.**

8–2   ANALOG AND DIGITAL SIGNAL CONVERSION

Whenever you have two different forms, like analog and digital, there is always a need to convert back and forth between the two. Computers are used at the heart of almost every electronic system today because of their ability to quickly process and store large amounts of data, to make systems more versatile, and to perform many functions. Many of the input signals applied to a computer for storage or processing are analog in nature, and therefore a data conversion circuit is needed to interface these analog inputs with a digital system. Similarly, a data conversion circuit may also be needed to interface the digital computer system with an analog output device.

FIGURE 8–4 Analog and Digital Readouts.

Figure 8–5 shows a simplified block diagram of a computer. No matter what the application, a computer has three basic blocks: a microprocessor unit, a memory unit, and an input/output unit.

Getting information into the computer is the job of input devices, and as you can see in Figure 8–5, there are two input paths into a computer: one path from analog input devices, and the other path from digital input devices. Information in the form of light, sound, heat, pressure, or any other real-world quantity is analog in nature and has an infinite number of input levels. Sensors or transducers of this type, such as photoelectric cells for light or microphones for sound, will generate analog signals. Since the computer operates on information in only digital or binary form, a translation or conversion is needed if the computer is going to be able to understand or interpret an analog signal. This signal processing is achieved by an **analog-to-digital converter (ADC),** which transforms the varying analog input voltage into equivalent digital codes that the computer can understand. Once the analog information has been encoded into digital form, the information can then enter the computer through an electronic doorway called the **input port.** Some input devices, such as a computer keyboard, automatically generate a digital code for each key that is pressed (ASCII codes). The codes generated by these digital input devices can be connected directly to the computer, without having to pass through a converter.

In contrast, the digital output information from a computer's **output port** can be used to drive either an analog device or a digital device. If an analog output is desired, as in the case of a voice signal for a speaker, the digital computer's output will have to be converted into a corresponding equivalent analog signal by a **digital-to-analog converter (DAC).** If only a digital output is desired, as in the case of a printer, which translates the binary codes into printed characters, the digital computer output can be connected directly to the output device without the need for a converter.

Analog-to-Digital Converter
A circuit that converts analog input signals into equivalent digital output signals.

Input Port
An electronic doorway used for receiving information into a computer.

Output Port
An electronic doorway used for transmitting information out of a computer.

Digital-to-Analog Converter
A circuit that converts digital input signals into equivalent analog output signals.

SELF-TEST EVALUATION POINT FOR SECTION 8–2

Now that you have completed this section, you should be able to:

■ **Objective 5.** *List the three basic blocks within a computer.*

■ **Objective 6.** *Describe how analog and digital devices are connected to a computer.*

■ **Objective 7.** *Define the function of the analog-to-digital converter (ADC) and the digital-to-analog converter (DAC).*

Use the following questions to test your understanding of Section 8–2:

1. What are the three basic blocks of a computer?

2. Give the full names of the following acronyms:
   a. ADC   **b.** DAC

3. An ADC will convert a/an _____ input into an equivalent _____ output.

4. A DAC will convert a/an _____ input into an equivalent _____ output.

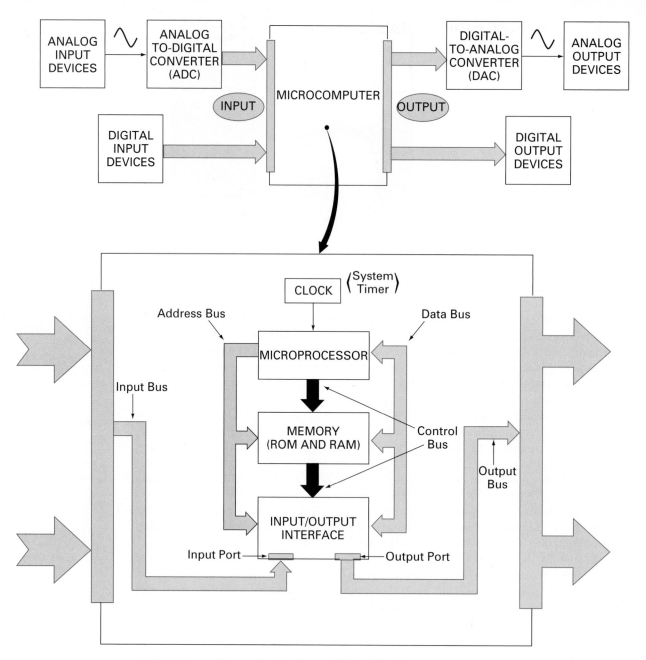

FIGURE 8–5   Connecting Analog and Digital Devices to a Computer.

8–3   THE BINARY NUMBER SYSTEM

Binary Point
A symbol used to separate the whole from the fraction in a binary number.

As in the decimal system, the value of a binary digit is determined by its position relative to the other digits. In the decimal system, each position to the left of the decimal point increases by a power of 10. Similarly, in the binary number system, since it is a base 2 (bi) number system, each place to the left of the **binary point** increases by a power of 2. Figure 8–6 shows how columns of the binary and decimal number systems have different weights.

FIGURE 8–6 **A Comparison of Decimal and Binary.**

DECIMAL

10^2	10^1	10^0
100	10	1
		0
		1
		2
		3
		4
		5
		6
		7
		8
		9
	1	0
	1	1
	1	2

BINARY

2^3	2^2	2^1	2^0
8	4	2	1
			0
			1
		1	0
		1	1
	1	0	0
	1	0	1
	1	1	0
	1	1	1
1	0	0	0
1	0	0	1
1	0	1	0
1	0	1	1
1	1	0	0

For example, with binary, the columns are weighted so that 2^0 is one, 2^1 is two, 2^2 is four, 2^3 is eight ($2 \times 2 \times 2 = 8$), and so on.

As we know, the base or radix of a number system also indicates the maximum number of digits used by the number system. The base 2 binary number system uses only the first two digits on the number scale, 0 and 1. The 0s and 1s in binary are called binary digits, or *bits*, for short.

8–3–1 *Positional Weight*

As in the decimal system, each column in binary carries its own weight, as seen in Figure 8–7(a). With the decimal number system, each position to the left increases 10 times. With binary, the weight of each column to the left increases 2 times. The first column therefore has a weight of 1, the second column has a weight of 2, the third column 4, the fifth 8, and so on. The value or quantity of a binary number is determined by the digit in each column and the positional weight of the column. For example, in Figure 8–7(a), the binary number 101101_2 is equal in decimal to 45_{10}, since we have 1×32, 1×8, 1×4, and 1×1 ($32 + 8 + 4 + 1 = 45$). The 0s in this example are not multiplied by their weights (16 and 2), since they are "0" and of no value. The leftmost binary digit is called the most significant bit (MSB), since it carries the most weight, and the rightmost digit is called the least significant bit (LSB), since it carries the least weight. Applying these definitions to the example in Figure 8–7(a), we see that the 1 in the thirty-twos column is the MSB, and the 1 in the units column is the LSB.

8–3–2 *Reset and Carry*

The reset and carry action occurs in binary in exactly the same way as does in decimal; however, since binary has only two digits, a column will reach its maximum digit much sooner, and therefore the reset and carry action in the binary number system will occur much more frequently.

Referring to Figure 8–7(b), you can see that the binary counter begins with 0 and advances to 1. At this stage, the units column has reached its maximum, and therefore the next count forces the units column to reset and carry into the next column, producing a count of 0010_2 (2_{10}). The units column then advances to a count of 0011_2 (3_{10}). At this stage, both the units column and twos column have reached their maximums. As the count advances

2^6	2^5	2^4	2^3	2^2	2^1	2^0
	32	16	8	4	2	1
	1	0	1	1	0	1

MSB   LSB

$$\begin{aligned}
\text{Value of number} &= (1 \times 2^5) + (0 \times 2^4) + (1 \times 2^3) + (1 \times 2^2) + (1 \times 2^1) + (1 \times 2^0) \\
&= (1 \times 32) + (0 \times 16) + (1 \times 8) + (1 \times 4) + (0 \times 2) + (1 \times 1) \\
&= 32 + 0 + 8 + 4 + 0 + 1 \\
&= 45_{10}
\end{aligned}$$

(a)

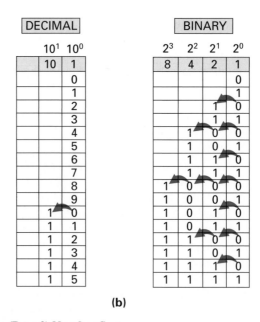

(b)

FIGURE 8–7   **The Binary (Base 2) Number System.**

by 1 it will cause the units column to reset and carry a 1 into the twos column, which will also have to reset and carry a 1 into the fours column. This will result in a final count of 0100_2 (4_{10}). The count will then continue to 0101_2 (5_{10}), 0110_2 (6_{10}), 0111_2 (7_{10}), and then 1000_2 (8_{10}), which, as you can see in Figure 8–7(b), is a result of three reset and carries.

Comparing the binary reset and carry with the decimal reset and carry in Figure 8–7(b), you can see that since the binary number system uses only two digits, binary numbers quickly turn into multidigit figures. For example, a decimal eight (8) uses only one digit, whereas a binary eight (1000) uses four digits.

8–3–3   *Converting Binary Numbers to Decimal Numbers*

Binary numbers can easily be converted to their decimal equivalent by simply adding together all the column weights that contain a binary 1, as we did previously in Figure 8–7(a).

■ **INTEGRATED MATH APPLICATION:**

Convert the following binary numbers to their decimal equivalents.

   a. 1010
   b. 101101

■ *Solution:*

  a. Binary column weights   32   16 8 4 2 1
       Binary number   1 0 1 0
       Decimal equivalent

$$= (1 \times 8) + (0 \times 4) + (1 \times 2) + (0 \times 1)$$
$$= 8 + 0 + 2 + 0 = 10_{10}$$

  b. Binary column weights   32   16 8 4 2 1
       Binary number   1 0 1 1 0 1
       Decimal equivalent

$$= (1 \times 32) + (0 \times 16) + (1 \times 8) + (1 \times 4)$$
$$+ (0 \times 2) + (1 \times 1)$$
$$= 32 + 0 + 8 + 4 + 0 + 1 = 45_{10}$$

8–3–4   *Converting Decimal Numbers to Binary Numbers*

To convert a decimal number to its binary equivalent, continuously subtract the largest possible power of two until the decimal number is reduced to zero, placing a binary 1 in columns that are used and a binary 0 in the columns that are not used. To see how simple this process is, refer to Figure 8–8, which shows how decimal 53 can be converted to its binary equivalent. As you can see in this example, the first largest power of two that can be subtracted from decimal 53 is 32, and therefore a 1 is placed in the thirty-twos column, and 21 remains. The largest power of two that can be subtracted from the remainder 21 is 16, and therefore a 1 is placed in the sixteens column, and 5 remains. The next largest power of two that can be subtracted from 5 is 4, and therefore a 1 is placed in the fours column, and 1 remains. The final 1 is placed in the units column, and therefore decimal 53 is represented in binary as 110101, which indicates that the value is the sum of 1×32, 1×16, 1×4, and 1×1.

■ **INTEGRATED MATH APPLICATION:**

Convert the following decimal numbers to their binary equivalents.

   a. 25
   b. 55

■ *Solution:*

  a. Binary column weights   32   16 8 4 2 1
       Binary number   1 1 0 0 1
       Decimal equivalent $25 - 16 = 9, 9 - 8 = 1, 1 - 1 = 0$

  b. Binary column weights   32   16 8 4 2 1
       Binary number   1 1 0 1 1 1
       Decimal equivalent $55 - 32 = 23, 23 - 16 = 7, 7 - 4 = 3,$
                                 $3 - 2 = 1, 1 - 1 = 0$

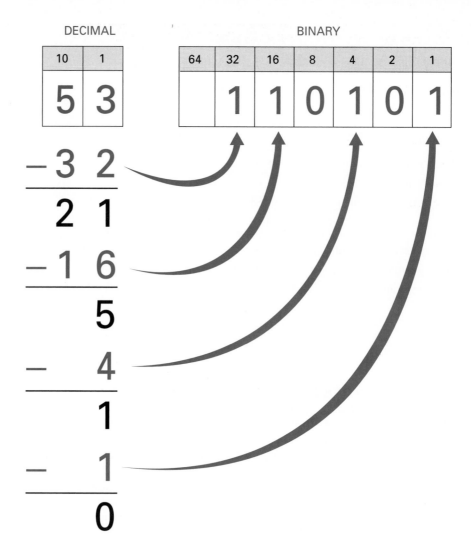

FIGURE 8–8 Decimal-to-Binary Conversion.

8–3–5 *Converting Information Signals*

Figure 8–9(a) shows how analog data can be converted into digital data, and Figure 8–9(b) shows how digital data can be converted into analog data. To give this process purpose, let us consider how the music information stored on a compact disc (CD) is first recorded and then played back. Like all sound, music is made up of waves of compressed air. When these waves strike the diaphragm of a microphone, an analog voltage signal is generated. In the past, this analog data was recorded on magnetic tapes or as a grooved track on a record. These data storage devices were susceptible to wear and tear, temperature, noise, and age. Using digital recording techniques, binary codes are stored on a compact disc to achieve near-perfect fidelity to live sound.

Figure 8–9(a) shows how an ADC is used during the recording process to convert the analog music input into a series of digital output codes. These digital codes are used to control the light beam of a recording laser so that it will engrave the binary 0s and 1s onto a CD in the form of pits or spaces. The ADC is triggered into operation by a sampling pulse that causes it to measure the input voltage of the analog signal at that particular time and generate an equivalent digital output code. For example, on the active edge of sampling pulse 1, the analog input voltage is at 2 V, so the binary code 010 (decimal 2) is generated at the output of the ADC. On the active edge of sampling pulse 2, the analog voltage has risen to 4 V,

(a)

(b)

FIGURE 8–9   Converting Information Signals.

so the binary code 100 (decimal 4) is generated at the output of the ADC. On the active edge of sampling pulses 3, 4, and 5, the binary codes 110 (decimal 6), 111 (decimal 7), and 110 (decimal 6) are generated at the output, representing the analog voltages 6 V, 7 V, and 6 V, respectively, and so on.

    Figure 8–9(b) shows how a DAC is used during the music playback process to convert the digital codes stored on a CD to an analog output signal. Another laser is used during playback to read the pits and spaces on the compact disc as 0s and 1s. These codes are then applied to the DAC, which converts the digital input codes into discrete voltages. The DAC is triggered into operation by a strobe pulse that causes it to convert the code currently being applied at the input into an equivalent output voltage.

For example, on the active edge of strobe pulse 1, the digital code 010 (decimal 2) is being applied to the DAC, so it will generate 2 V at its output. On the active edge of strobe pulse 2, the digital code 100 (decimal 4) is being applied to the DAC, so it will generate 4 V at its output. On the active edge of strobe pulse 3, the digital code 110 (decimal 6) is being applied to the DAC, so it will generate 6 V at its output. On the active edge of strobe pulse 4, the digital code 111 (decimal 7) is being applied to the DAC, so it will generate 7 V at its output. On the active edge of strobe pulse 5, the digital code 110 (decimal 6) is being applied to the DAC, so it will generate 6 V at its output, and so on. If the output of a DAC is then applied to a low-pass filter, the discrete voltage steps can be blended into a smooth wave that closely approximates the original analog wave, as shown by the dashed line in Figure 8–9(b).

SELF-TEST EVALUATION POINT FOR SECTION 8–3

Now that you have completed this section, you should be able to:

■ **Objective 8.** *Explain the positional weight, the reset and carry action, and conversion for binary.*

Use the following questions to test your understanding of Section 8–3:

1. What is the decimal equivalent of 11010_2?
2. Convert 23_{10} to its binary equivalent.
3. What are the full names for the acronyms LSB and MSB?
4. Convert 110_{10} to its binary equivalent.

8–4 THE HEXADECIMAL NUMBER SYSTEM

If digital electronic circuits operate using binary numbers and we operate using decimal numbers, why is there any need for us to have any other system? The hexadecimal, or hex, system is used as a sort of shorthand for large strings of binary numbers, as will be explained in this section. To begin, let us examine the basics of this number system and then look at its application.

Hexadecimal

A base 16 number system.

Hexadecimal means "sixteen," and this number system has 16 different digits, as shown in Figure 8–10(a), which shows a comparison between decimal, hexadecimal, and binary. Looking at the first 10 digits in the decimal and hexadecimal columns, you can see that there is no difference between the two columns; however, beyond 9, hexadecimal makes use of the letters A, B, C, D, E, and F. Digits 0 through 9 and letters A through F make up the 16 total digits of the hexadecimal number system. Comparing hexadecimal with decimal once again, you can see that $A_{16} = 10_{10}$, $B_{16} = 11_{10}$, $C_{16} = 12_{10}$, $D_{16} = 13_{10}$, $E_{16} = 14_{10}$ and $F_{16} = 15_{10}$. Having these extra digits means that a column reset and carry will not occur until the count has reached the last and largest digit, F. The hexadecimal column in Figure 8–10(a) shows how reset and carry will occur whenever a column reaches its maximum digit, F, and this action is further illustrated in the examples in Figure 8–10(b).

8–4–1 *Converting Hexadecimal Numbers to Decimal Numbers*

To find the decimal equivalent of a hexadecimal number, simply multiply each hexadecimal digit by its positional weight. For example, referring to Figure 8–11, you can see that the positional weight of each of the hexadecimal columns, as expected, is progressively 16 times

DECIMAL			HEXADECIMAL				BINARY					
10^2	10^1	10^0	16^3	16^2	16^1	16^0	2^5	2^4	2^3	2^2	2^1	2^0
100	10	1	4096	256	16	1	32	16	8	4	2	1
		0				0	0	0	0	0	0	0
		1				1	0	0	0	0	0	1
		2				2	0	0	0	0	1	0
		3				3	0	0	0	0	1	1
		4				4	0	0	0	1	0	0
		5				5	0	0	0	1	0	1
		6				6	0	0	0	1	1	0
		7				7	0	0	0	1	1	1
		8				8	0	0	1	0	0	0
		9				9	0	0	1	0	0	1
	1	0				A	0	0	1	0	1	0
	1	1				B	0	0	1	0	1	1
	1	2				C	0	0	1	1	0	0
	1	3				D	0	0	1	1	0	1
	1	4				E	0	0	1	1	1	0
	1	5				F	0	0	1	1	1	1
	1	6			1	0	0	1	0	0	0	0
	1	7			1	1	0	1	0	0	0	1
	1	8			1	2	0	1	0	0	1	0
	1	9			1	3	0	1	0	0	1	1
	2	0			1	4	0	1	0	1	0	0
	2	1			1	5	0	1	0	1	0	1
	2	2			1	6	0	1	0	1	1	0
	2	3			1	7	0	1	0	1	1	1
	2	4			1	8	0	1	1	0	0	0
	2	5			1	9	0	1	1	0	0	1
	2	6			1	A	0	1	1	0	1	0
	2	7			1	B	0	1	1	0	1	1
	2	8			1	C	0	1	1	1	0	0
	2	9			1	D	0	1	1	1	0	1
	3	0			1	E	0	1	1	1	1	0
	3	1			1	F	0	1	1	1	1	1
	3	2			2	0	1	0	0	0	0	0
	3	3			2	1	1	0	0	0	0	1
	3	4			2	2	1	0	0	0	1	0

(a)

A9	DE7	78	F9	10C
AA	DE8	79	FA	10D
AB	DE9	7A	FB	10E
AC	DEF	7B	FC	10F
AD	DF0	7C	FD	110
AE	DF1	7D	FE	111
AF	DF2	7E	FF	112
B0	DF3	7F	100	
B1		80	101	
		81		

(b)

FIGURE 8–10 **Hexadecimal Reset and Carry (a) Number System Comparison (b) Hex Counting Examples.**

larger as you move from right to left. The hexadecimal number 4C therefore indicates that the value is $4 \times 16 (4 \times 16 = 64)$ and $C \times 1$. Since C is equal to 12 in decimal (12×1), the result of $12 + 64$ is 76, so hexadecimal 4C is equivalent to decimal 76.

16^4	16^3	16^2	16^1	16^0
65,536	4096	256	16	1
			4	C

$$\text{Value of Number} = (4 \times 16) + (C \times 1)$$
$$= (4 \times 16) + (12 \times 1)$$
$$= 64 + 12 = 76_{10}$$

FIGURE 8–11 Positional Weight of the Hexadecimal Number System.

■ **INTEGRATED MATH APPLICATION:**

Convert hexadecimal 8BF to its decimal equivalent.

■ *Solution:*

Hexadecimal column weights   4096   256   16   1

                                      8   B   F

Decimal equivalent
$$= (8 \times 256) + (B \times 16) + (F \times 1)$$
$$= 2048 + (11 \times 16) + (15 \times 1)$$
$$= 2048 + 176 + 15$$
$$= 2239$$

8–4–2 *Converting Decimal Numbers to Hexadecimal Numbers*

Decimal-to-hexadecimal conversion is achieved in the same way as decimal to binary. First, subtract the largest possible power of 16, and then keep subtracting the largest possible power of 16 from the remainder. Each time a subtraction takes place, add a 1 to the respective column until the decimal value has been reduced to zero. Figure 8–12 illustrates this procedure with an example showing how decimal 425 is converted to its hexadecimal equivalent. To begin, the largest possible power of sixteen (256) is subtracted once from 425, and therefore a 1 is placed in the 256s column, leaving a remainder of 169. The next largest power of 16 is 16, which can be subtracted 10 times from 169, and therefore the hexadecimal equivalent of 10, which is A, is placed in the sixteens column, leaving a remainder of 9. Since nine 1s can be subtracted from the remainder of 9, the units column is advanced nine times, giving us our final hexadecimal result, 1A9.

■ **INTEGRATED MATH APPLICATION:**

Convert decimal 4525 to its hexadecimal equivalent.

■ *Solution:*

Hexadecimal column weights   4096   256   16   1

                                      1   1   A   D

Decimal equivalent
$$= 4525 - 4096 = 429, 429 - 256 = 173,$$
$$173 - 16 - 16 - 16 - 16 - 16 - 16 -$$
$$16 - 16 - 16 - 16 = 13, 13 - 1 -$$
$$1 - 1 - 1 - 1 - 1 - 1 - 1 -$$
$$1 - 1 - 1 - 1 = 0$$

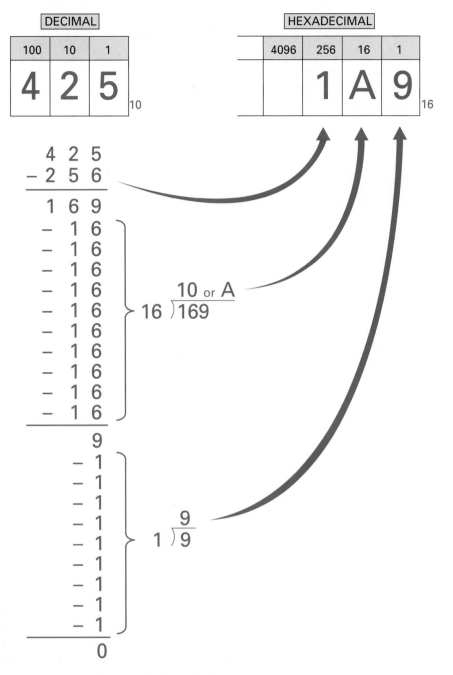

FIGURE 8–12   **Decimal-to-Hexadecimal Conversion.**

8–4–3   *Converting between Binary and Hexadecimal*

As mentioned at the beginning of this section, hexadecimal is used as a shorthand for representing large groups of binary digits. To illustrate this, Figure 8–13(a) shows how a 16-bit binary number, which is more commonly called a 16-bit binary **word,** can be represented by 4 hexadecimal digits. To explain this, Figure 8–13(b) shows how a 4-bit binary word can have any value from 0_{10} (0000_2) to 15_{10} (1111_2), and since hexadecimal has the same number of digits (0 through F), we can use one hexadecimal digit to represent 4 binary bits. As you can see in Figure 8–13(a), it is much easier to work with a number like $AE73_{16}$ than with 1010111001110011_2.

To convert from hexadecimal to binary, we simply do the opposite, as shown in Figure 8–14. Since each hexadecimal digit represents 4 binary digits, a 4-digit hexadecimal number will convert to a 16-bit binary word.

Word

An ordered set of characters that is treated as a unit.

16-Bit Binary Word

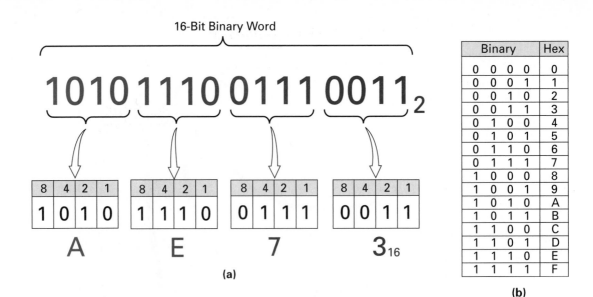

Binary	Hex
0 0 0 0	0
0 0 0 1	1
0 0 1 0	2
0 0 1 1	3
0 1 0 0	4
0 1 0 1	5
0 1 1 0	6
0 1 1 1	7
1 0 0 0	8
1 0 0 1	9
1 0 1 0	A
1 0 1 1	B
1 1 0 0	C
1 1 0 1	D
1 1 1 0	E
1 1 1 1	F

(a)

(b)

FIGURE 8–13   Representing Binary Numbers in Hexadecimal.

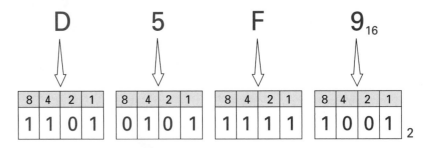

FIGURE 8–14   Hexadecimal-to-Binary Conversion.

■   INTEGRATED MATH APPLICATION:

Convert hexadecimal 2BF9 to its binary equivalent.

■   *Solution:*

Hexadecimal number	2	B	F	9
Binary equivalent	0010	1011	1111	1001

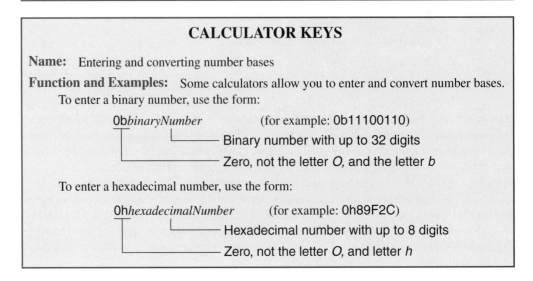

CALCULATOR KEYS

Name: Entering and converting number bases

Function and Examples: Some calculators allow you to enter and convert number bases.
To enter a binary number, use the form:

0b*binaryNumber*   (for example: 0b11100110)

————— Binary number with up to 32 digits

————— Zero, not the letter *O*, and the letter *b*

To enter a hexadecimal number, use the form:

0h*hexadecimalNumber*   (for example: 0h89F2C)

————— Hexadecimal number with up to 8 digits

————— Zero, not the letter *O*, and letter *h*

If you enter without the 0b or 0h prefix, such as 11, it is always treated as a decimal number. If you omit the 0h prefix on a hexadecimal number containing A–F, all or part of the entry is treated as a variable.

Use the ▶ conversion operator.

integerExpression ▶ Bin
integerExpression ▶ Dec
integerExpression ▶ Hex

For ▶, press 2nd [▶]. Also, you can select base conversions from the MATH/Base menu.

For example, to convert 256 from decimal to binary:

256 ▶ Bin

For a binary or hex entry, you must use the 0b or 0h prefix.

■ 256 ▶ Bin	0b100000000
■ 0b101110 ▶ Hex	0h2E
0b101110 ▶ hex	
MATH RAD AUTO FUNC	2738

To convert 101110 from binary to hexidecimal:

0b101110 ▶ Hex

Result uses the 0b or 0h prefix to identify the base.

■ **INTEGRATED MATH APPLICATION:**

Convert binary 110011100001 to its hexadecimal equivalent.

■ *Solution:*

Binary number	1100	1110	0001
Hexadecimal equivalent	C	E	1

SELF-TEST EVALUATION POINT FOR SECTION 8–4

Now that you have completed this section, you should be able to:

■ **Objective 9.**   *Explain the positional weight, the reset and carry action, and conversion for hexadecimal.*

Use the following questions to test your understanding of Section 8–4:

1. What is the base of each?
   a. Decimal number system
   b. Binary number system
   c. Hexadecimal number system
2. What are the decimal and hexadecimal equivalents of 10101_2?
3. Convert $1011\ 1111\ 0111\ 1010_2$ to its hexadecimal equivalent.
4. What are the binary and hexadecimal equivalents of 33_{10}?

8–5   BINARY CODES

The process of converting a decimal number to its binary equivalent is called *binary coding*. The result of the conversion is a binary number or code that is called **pure binary.** There are, however, binary codes used in digital circuits other than pure binary, and in this section we will examine some of the most frequently used.

Pure Binary

Uncoded binary.

Binary Coded Decimal
(BCD)

A code in which each
decimal digit is
represented by a group of
4 binary bits.

8–5–1   *The Binary Coded Decimal (BCD) Code*

No matter how familiar you become with binary, it will always be less convenient to work with than the decimal number system. For example, it will always take a short time to convert 1111000_2 to 120_{10}. Designers realized this disadvantage early on and developed a binary code that had decimal characteristics and that was appropriately named **binary coded decimal (BCD).** Being a binary code, it has the advantages of a two-state system, and since it has a decimal format, it is also much easier for an operator to interface via a decimal keypad or decimal display with systems such as pocket calculators and wristwatches.

The BCD code expresses each decimal digit as a 4-bit word, as shown in Figure 8–15(a). In this example, decimal 1753 converts to a BCD code of 0001 0111 0101

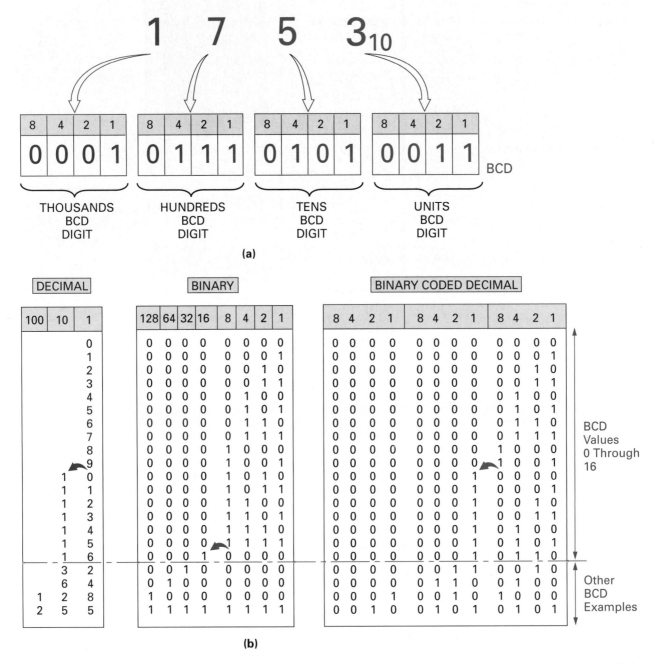

(a)

(b)

FIGURE 8–15   Binary Coded Decimal (BCD) (a) Decimal-to-BCD Conversion Example (b) A Comparison of Decimal, Binary, and BCD.

0011, with the first 4-bit code ($0001_2 = 1_{10}$) representing the 1 in the thousands column, the second 4-bit code ($0111_2 = 7_{10}$) representing the 7 in the hundreds column, the third 4-bit code ($0101_2 = 5_{10}$) representing the 5 in the tens column, and the fourth 4-bit code ($0011_2 = 3_{10}$) representing the 3 in the units column. As can be seen in Figure 8–15(a), the subscript BCD is often used after a BCD code to distinguish it from a pure binary number ($1753_{10} = 0001\ 0111\ 0101\ 0011_{BCD}$).

Figure 8–15(b) compares decimal, binary, and BCD. As you can see, the reset and carry action occurs in BCD at the same time it does in decimal. This is because BCD was designed to have only ten 4-bit binary codes, 0000, 0001, 0010, 0011, 0100, 0101, 0110, 0111, 1000, and 1001 (0 through 9), to make it easy to convert between this binary code and decimal. Binary codes 1010, 1011, 1100, 1101, 1110, and 1111 (A_{16} through F_{16}), are invalid codes that are not used in BCD because they are not used in decimal.

■ **INTEGRATED MATH APPLICATION:**

Convert the BCD code 0101 1000 0111 0000 to decimal, and the decimal number 369 to BCD.

■ *Solution:*

BCD code	0101	1000	0111	0000
Decimal equivalent	5	8	7	0
Decimal number	3	6	9	
BCD code	0011	0110	1001	

8–5–2 *The American Standard Code for Information Interchange (ASCII)*

To this point we have discussed only how we can encode numbers into binary codes. Digital electronic computers must also be able to generate and recognize binary codes that represent letters of the alphabet and symbols. The ASCII code is the most widely used **alphanumeric code** (alphabet and numeral code). As I write this book the ASCII codes for each of these letters is being generated by my computer keyboard, and the computer is decoding these ASCII codes and then displaying the alphanumeric equivalent (letter, number, or symbol) on my computer screen.

Figure 8–16 lists all the 7-bit ASCII codes and the full names for the abbreviations used. This diagram also shows how each of the 7-bit ASCII codes is made up of a 4-bit group that indicates the row of the table and a 3-bit group that indicates the column of the table. For example, the uppercase letter K is in column 100 (4_{10}) and in row 1011 (11_{10}) and is therefore represented by the ASCII code 1001011.

Alphanumeric Code

A code used to represent the letters of the alphabet and decimal numbers.

■ **INTEGRATED MATH APPLICATION:**

List the ASCII codes for the message "Digital."

■ *Solution:*

$$D = 1000100$$
$$i = 1101001$$
$$g = 1100111$$
$$i = 1101001$$
$$t = 1110100$$
$$a = 1100001$$
$$l = 1101100$$

Column		0	1	2	3	4	5	6	7
Row Bits 4321 765 →		000	001	010	011	**100**	101	110	111
0	0000	NUL	DLE	SP	0	@	P	\	p
1	0001	SOH	DC1	!	1	A	Q	a	q
2	0010	STX	DC2	"	2	B	R	b	r
3	0011	ETX	DC3	#	3	C	S	c	s
4	0100	EOT	DC4	$	4	D	T	d	t
5	0101	ENQ	NAK	%	5	E	U	e	u
6	0110	ACK	SYN	&	6	F	V	f	v
7	0111	BEL	ETB	'	7	G	W	g	w
8	1000	BS	CAN	(	8	H	X	h	x
9	1001	HT	EM	)	9	I	Y	i	y
10	1010	LF	SUB	*	:	J	Z	j	z
11	**1011**	VT	ESC	+	;	**K**	[	k	{
12	1100	FF	FS	,	<	L	\	l	¦
13	1101	CR	GS	-	=	M	]	m	}
14	1110	SO	RS	.	>	N	⌒	n	~
15	1111	SI	US	/	?	O	—	o	DEL

Example: Upper Case "K" = Column 4   (100), Row 11 (1011)

Bits	7 6 5	4 3 2 1
ASCII Code	1 0 0	1 0 1 1

NUL Null	FF Form Feed	CAN Cancel
SOH Start of Heading	CR Carriage Return	EM End of Medium
STX Start of Text	SO Shift Out	SUB Substitute
ETX End of Text	SI Shift In	ESC Escape
EOT End of Transmission	SP Space (blank)	FS File Separator
ENQ Enquiry	DLE Data Link Escape	GS Group Separator
ACK Acknowledge	DC1 Device Control 1	RS Record Separator
BEL Bell (audible signal)	DC2 Device Control 2	US Unit Separator
BS Backspace	DC3 Device Control 3	DEL Delete
HT Horizontal Tabulation	DC4 Device Control 4	
(punched card skip)	NAK Negative Acknowledge	(ASCII
LF Line Feed	SYN Synchronous Idle	Abbreviations)
VT Vertical Tabulation	ETB End of Transmission Block	

FIGURE 8–16   The ASCII Code.

SELF-TEST EVALUATION POINT FOR SECTION 8–5

Now that you have completed this section, you should be able to:

■ **Objective 10.**   *Describe the following binary codes: binary coded decimal (BCD) and the American Standard Code for Information Interchange (ASCII).*

Use the following questions to test your understanding of Section 8–5:

1. Convert 7629_{10} to BCD.

2. What does the following string of ASCII codes mean?
   1000010   1100001   1100010   1100010
   1100001   1100111   1100101

8–6   LOGIC GATES AND BOOLEAN EXPRESSIONS

Logic gates are used as the basic building blocks for all digital circuits, and their purpose is to control the movement of binary data and binary instructions. Logic gates, and all other digital electronics circuits, are often referred to as **hardware** circuits. By definition, the hardware of a digital electronic system includes all the electronic, magnetic, and mechanical devices of a digital system. In contrast, the **software** of a digital electronic system includes the binary data (like pure binary and ASCII codes) and binary instructions that are processed by the digital electronic system hardware. To use an analogy, we could say that a compact disc player is hardware, and the music information stored on a compact disc and processed by the player is software. Just as a CD player is useless without CDs, a digital system's hardware is useless without software. In other words, the information on the CD determines what music is played and similarly, the digital software determines the actions of the digital electronic hardware.

A logic gate accept inputs in the form of HIGH or LOW voltage, judges these input combinations based on a predetermined set of rules, and then produces a single output in the form of a HIGH or LOW voltage. The term *logic* is used because the output is predictable or logical, and the term *gate* is used because only certain input combinations will "unlock the gate". For example, any HIGH input to an OR gate will unlock the gate and allow the HIGH at the input to pass through to the output.

The operation of each of the seven basic logic gates can be described with a Boolean expression. To explain this in more detail, let us first consider the most basic of all the logic gates, the NOT gate.

8–6–1   *The NOT Gate and Its Boolean Expression*

In Figure 8–17 you can see the NOT gate, and because of the inversion that occurs between input and output, Y is always equal to the opposite, or complement, of input A.

$$Y = \text{NOT } A$$

Therefore, if the input is 0, the output will be 1.

$$Y = \text{NOT } 0 = 1$$

If, on the other hand, the A input is 1, the output will be 0.

$$Y = \text{NOT } 1 = 0$$

In Boolean algebra, this inversion of the A input is indicated with a bar over the letter A as follows:

$$Y = \overline{A} \text{ (pronounced "}Y\text{ equals not }A\text{")}$$

Using this Boolean expression, we can easily calculate the output Y for either of the two input A conditions. For example, if $A = 0$,

$$Y = \overline{A} = \overline{0} = 1$$

If, on the other hand, the A input is 1,

$$Y = \overline{A} = \overline{1} = 0$$

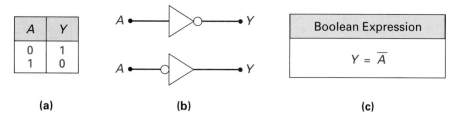

A	Y
0	1
1	0

(a)   **(b)**   **(c)**

FIGURE 8–17   The Boolean Expression for a NOT Gate.

Logic Gate Circuits

Circuits containing predictable gate functions that either open or close their outputs.

Hardware

A term used to describe the electronic, magnetic, and mechanical devices of a digital system.

Software

A term used to describe the binary instructions and data processed by digital electronic system hardware.

8–6–2   The OR Gate and Its Boolean Expression

The operation of the OR gate can also be described with a Boolean expression, as seen in Figure 8–18(a). In Boolean algebra, a + sign is used to indicate the OR operation.

$$Y = A + B \text{ (pronounced "Y equals A or B")}$$

The expression $Y = A + B$ (Boolean equation) is the same as $Y = A$ OR B (word equation). If you find yourself wanting to say "plus" instead of "OR," don't worry; it will require some practice to get used to using this traditional symbol in a new application. Just as the word "wind" can be used in two different ways ("a gusty wind" or "to wind a clock"), so can the + symbol be used to mean either addition or the OR operation. With practice you will get more comfortable with this symbol's dual role.

  The truth table for a two-input OR gate has been repeated in Figure 8–18(b). Using the Boolean expression for this gate, we can determine the output Y by substituting all the possible 1 and 0 input combinations for the input variables A and B. Studying this truth table you can see that the output of the OR gate is a 1 when either A OR B is 1. George Boole described the OR function as *Boolean addition*, which is why the "plus" (addition) symbol was used. Boolean addition, or *logical addition*, is different from normal addition, and the OR gate is said to produce at its output the *logical sum* of the inputs.

8–6–3   The AND Gate and Its Boolean Expression

In Boolean algebra, the AND operation is indicated with a ·, and therefore the Boolean expression for the AND gate shown in Figure 8–19(a) would be

$$Y = A \cdot B \text{ (pronounced "Y equals A and B")}$$

This · is also used to indicate multiplication, and therefore like the + symbol, the · symbol has two meanings. In multiplication, the · is often dropped; however, the meaning is still known to be multiplication. For example:

$$V = I \times R \text{ is the same as } V = I \cdot R \text{ or } V = IR$$

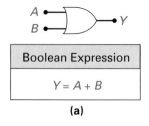

A	B	Y = A + B
0	0	Y = 0 + 0 = **0**
0	1	Y = 0 + 1 = **1**
1	0	Y = 1 + 0 = **1**
1	1	Y = 1 + 1 = **1**

The OR gate delivers at its output the logical sum (+) of its inputs. (logical sum = OR function)

(a)   **(b)**

FIGURE 8–18   The Boolean Expression for an OR Gate.

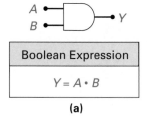

A	B	Y = A · B
0	0	Y = 0 · 0 = **0**
0	1	Y = 0 · 1 = **0**
1	0	Y = 1 · 0 = **0**
1	1	Y = 1 · 1 = **1**

The AND gate delivers at its output the logical product (•) of its inputs. (logical product = AND function)

(a)   **(b)**

FIGURE 8–19   The Boolean Expression for an AND Gate.

In Boolean algebra the same situation occurs, in that

$$Y = A \cdot B \text{ is the same as } Y = AB$$

The truth table for a two-input AND gate is shown in Figure 8–19(b). Using the Boolean expression, we can calculate the output Y by substituting all the possible 1 and 0 input combinations for the input variables A and B. Looking at the truth table, you can see that the output of the AND gate is a 1 only when both A and B are 1. George Boole described the AND function as *Boolean multiplication* or *logical multiplication*, which is why the period (multiplication) symbol was used. Boolean multiplication is different from standard multiplication as you have seen, and the AND gate is said to produce at its output the *logical product* of its inputs.

■ **INTEGRATED MATH APPLICATION:**

What would be the Boolean expression for the circuit shown in Figure 8–20, and what would be the output if $A = 0$ and $B = 1$?

FIGURE 8–20   Single Bubble Input AND Gate.

■ *Solution:*

The A input in Figure 8–20 is first inverted before it is ANDed with the B input. The Boolean equation therefore would be

$$Y = \overline{A} \cdot B$$

Using this expression, we can substitute the A and B for the 0 and 1 inputs, and then determine the output Y.

$$Y = \overline{A} \cdot B = \overline{0} \cdot 1 = 1 \cdot 1 = 1$$

(Y equals NOT A and B = NOT 0 AND 1 = 1 AND 1 = 1)

■ **INTEGRATED MATH APPLICATION:**

Give the Boolean expression and truth table for the logic circuit shown in Figure 8–21(a).

(a)

A	B	C	$Y = AB + C$
0	0	0	$Y = (0 \cdot 0) + 0 = 0 + 0 = 0$
0	0	1	$Y = (0 \cdot 0) + 1 = 0 + 1 = 1$
0	1	0	$Y = (0 \cdot 1) + 0 = 0 + 0 = 0$
0	1	1	$Y = (0 \cdot 1) + 1 = 0 + 1 = 1$
1	0	0	$Y = (1 \cdot 0) + 0 = 0 + 0 = 0$
1	0	1	$Y = (1 \cdot 0) + 1 = 0 + 1 = 1$
1	1	0	$Y = (1 \cdot 1) + 0 = 1 + 0 = 1$
1	1	1	$Y = (1 \cdot 1) + 1 = 1 + 1 = 1$

(b)

FIGURE 8–21   (*A* AND *B*) OR (*C*) Logic Circuit.

Solution:

First, there are three inputs or variables to this circuit: A, B, and C. Inputs A and B are ANDed together $(A \cdot B)$ and the result is then ORed with input $C(+C)$. The resulting equation is

$$Y = (A \cdot B) + C \quad \text{or} \quad Y = AB + C$$

As you can see in the truth table in Figure 8–21(b), the Y output is HIGH when A AND B are HIGH, OR if C is HIGH.

In the two previous examples, we have gone from a logic circuit to a Boolean equation. Now let us reverse the procedure and see how easy it is to go from a Boolean equation to a logic circuit.

■ INTEGRATED MATH APPLICATION:

Sketch the logic-gate circuit for the following Boolean equation.

$$Y = AB + CD$$

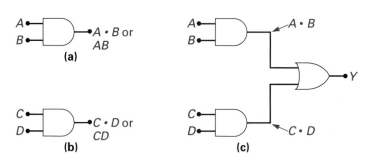

FIGURE 8–22   (*A* AND *B*) OR (*C* AND *D*) Logic Circuit.

■ Solution:

Studying the equation, we can see that there are four variables: A, B, C, and D. Input A is ANDed with input B, giving $A \cdot B$ or AB, as seen in Figure 8–22(a). Input C is ANDed with input D, giving CD, as seen in Figure 8–22(b). The ANDed AB output and the ANDed CD output are then ORed to produce a final output Y, as seen in Figure 8–22(c).

8-6-4   *The NOR Gate and Its Boolean Expression*

The Boolean expression for the two-input NOR gate shown in Figure 8–23(a) is

$$Y = \overline{A + B} \text{ (pronounced "} Y \text{ equals not } A \text{ or } B\text{")}$$

Boolean Expression
$Y = \overline{A + B}$

(a)

A	B	$Y = \overline{A + B}$
0	0	$Y = \overline{0 + 0} = \overline{0} = 1$
0	1	$Y = \overline{0 + 1} = \overline{1} = 0$
1	0	$Y = \overline{1 + 0} = \overline{1} = 0$
1	1	$Y = \overline{1 + 1} = \overline{1} = 0$

(b)

FIGURE 8–23   The Boolean Expression for a NOR Gate.

CHAPTER 8 / COMPUTER MATH

A	B	Y
0	0	1
0	1	0
1	0	0
1	1	0

$$Y = \overline{A} \cdot \overline{B}$$

(a)

A	B	Y
0	0	1
0	1	0
1	0	0
1	1	0

$$Y = \overline{A + B}$$

(b)

FIGURE 8–24 DeMorgan's First Theorem.

To explain why this expression describes a NOR gate, let us examine each part in detail. In this equation the two input variables A and B are first ORed (indicated by the $A + B$ part of the equation), and then the result is complemented, or inverted (indicated by the bar over the whole OR output expression).

The truth table in Figure 8–23(b) tests the Boolean expression for a NOR gate for all possible input combinations.

While studying Boolean algebra Augustus DeMorgan discovered two important theorems. The first theorem stated that a bubbled (inverting) input AND gate was logically equivalent to a NOR gate. These two logic gates are shown in Figure 8–24(a) and (b), respectively, with their associated truth tables. Comparing the two truth tables you can see how any combination of inputs to either gate circuit will result in the same output, making the two circuits interchangeable.

■ **INTEGRATED MATH APPLICATION:**

Give the two Boolean equations for the two logic-gate circuits shown in Figure 8–24(a) and (b).

■ *Solution:*

With the bubbled input AND gate shown in Figure 8–24(a), you can see that input A is first inverted by a bubble (NOT gate), giving $\overline{A}$. Input B is also complemented by a bubble (NOT gate), giving $\overline{B}$. These two complemented inputs are then ANDed, giving the following final function:

$$Y = \overline{A} \cdot \overline{B}$$

With the NOR gate shown in Figure 8–24(b), inputs A and B are first ORed and then the result is complemented, giving

$$Y = \overline{A + B}$$

DeMorgan's first theorem, therefore, is as follows:

DeMorgan's First Theorem

A bubbled input AND gate is logically equivalent to a NOR gate.

DeMorgan's first theorem: $\overline{A} \cdot \overline{B} = \overline{A + B}$

Studying the differences between the left equation and the right equation, you can see that there are actually two basic changes: the line is either broken or solid, and the logical sign is different. Therefore, if we wanted to convert the left equation into the right equation, all we would have to do is follow this simple procedure:

$$Y = \overline{A} \cdot \overline{B} \rightarrow Y = \overline{A + B} \qquad \begin{matrix} \text{Mend} \\ \text{the} \\ \text{line, and} \\ \text{change} \\ \text{the sign.} \end{matrix}$$

To convert the right equation into the left equation, all we would have to do is follow this simple procedure:

$$\begin{matrix} \text{Break} \\ \text{the} \\ \text{line, and} \\ \text{change} \\ \text{the sign.} \end{matrix} \qquad Y = \overline{A} \cdot \overline{B} \leftarrow Y = \overline{A + B}$$

This ability to interchange an AND function with an OR function and to interchange a bubbled input with a bubbled output will come in handy when we are trying to simplify logic circuits, as you will see in this chapter. For now, remember the rules on how to "De-Morganize" an equation.

> *Mend the line, and change the sign.*
> *Break the line, and change the sign.*

8–6–5 *The NAND Gate and Its Boolean Expression*

The Boolean expression for the two-input NAND gate shown in Figure 8–25(a) is

> $Y = \overline{A \cdot B}$ or $Y = \overline{AB}$ (pronounced "Y equals not A and B")

This expression states that the two inputs A and B are first ANDed (indicated by the $A \cdot B$ part of the equation), and then the result is complemented (indicated by the bar over the AND expression).

The truth table shown in Figure 8–25(b) tests this Boolean equation for all possible input combinations.

■ **INTEGRATED MATH APPLICATION:**

DeMorganize the following equation, and then sketch the logically equivalent circuits with their truth tables and Boolean expressions.

$$Y = \overline{A} + \overline{B}$$

Boolean Expression
$Y = \overline{AB}$

(a)

A	B	$Y = \overline{A \cdot B}$
0	0	$Y = \overline{0 \cdot 0} = \overline{0} = 1$
0	1	$Y = \overline{0 \cdot 1} = \overline{0} = 1$
1	0	$Y = \overline{1 \cdot 0} = \overline{0} = 1$
1	1	$Y = \overline{1 \cdot 1} = \overline{1} = 0$

(b)

FIGURE 8–25 **The Boolean Expression for a NAND Gate.**

■ *Solution:*

To apply DeMorgan's theorem to the equation $Y = \overline{A} + \overline{B}$, simply apply the rule "mend the line, and change the sign," as follows:

$$\overline{A} + \overline{B} = \overline{A \cdot B}$$

The circuits for these two equations can be seen in Figure 8–26(a) and (b), and if you study the truth tables you can see that the outputs of both logic gates are identical for all input combinations.

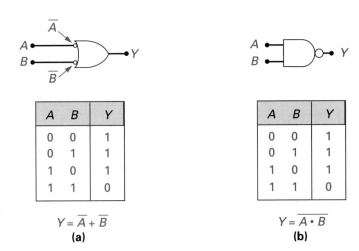

A	B	Y
0	0	1
0	1	1
1	0	1
1	1	0

$Y = \overline{A} + \overline{B}$

(a)

A	B	Y
0	0	1
0	1	1
1	0	1
1	1	0

$Y = \overline{A \cdot B}$

(b)

FIGURE 8–26   DeMorgan's Second Theorem.

This observation—that a bubbled input OR gate is interchangeable with a bubbled output AND gate, or NAND gate—was first made by DeMorgan and is referred to as **DeMorgan's second theorem.**

> DeMorgan's second theorem: $\overline{A} + \overline{B} = \overline{A \cdot B}$

DeMorgan's Second Theorem

A bubbled input OR gate is interchangeable with a NAND gate.

8-6-6   *The XOR Gate and Its Boolean Expression*

In Boolean algebra, the $\oplus$ symbol is used to describe the exclusive-OR action. This means that the Boolean expression for the two-input XOR gate shown in Figure 8–27(a) is

> $Y = A \oplus B$ (pronounced "Y equals A exclusive-OR B")

The truth table in Figure 8–27(b) shows how this Boolean expression is applied to all possible input combinations.

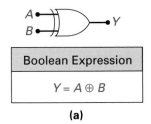

Boolean Expression

$Y = A \oplus B$

(a)

A	B	$Y = A \oplus B$
0	0	$Y = 0 \oplus 0 = 0$
0	1	$Y = 0 \oplus 1 = 1$
1	0	$Y = 1 \oplus 0 = 1$
1	1	$Y = 1 \oplus 1 = 0$

(b)

FIGURE 8–27   The Boolean Expression for an XOR Gate.

Give the Boolean expression and truth table for the circuit shown in Figure 8–28(a).

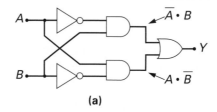

(a)

A B	$Y = \overline{A}B + A\overline{B}$
0 0	$Y = \overline{(0 \cdot 0)} + (0 \cdot \overline{0}) = (1 \cdot 0) + (0 \cdot 1) = 0 + 0 = 0$
0 1	$Y = \overline{(0 \cdot 1)} + (0 \cdot \overline{1}) = (1 \cdot 1) + (0 \cdot 0) = 1 + 0 = 1$
1 0	$Y = \overline{(1 \cdot 0)} + (1 \cdot \overline{0}) = (0 \cdot 0) + (1 \cdot 1) = 0 + 1 = 1$
1 1	$Y = \overline{(1 \cdot 1)} + (1 \cdot \overline{1}) = (0 \cdot 1) + (1 \cdot 0) = 0 + 0 = 0$

(b)

FIGURE 8–28   **(NOT *A* AND *B*) or (*A* AND NOT *B*).**

■   *Solution:*

Looking at the circuit in Figure 8–28(a), you can see that the upper AND gate has an inverted *A* input $(\overline{A})$ and a *B* input, giving a result of $\overline{A} \cdot B$. This $\overline{A} \cdot B$ output is then ORed with the input from the lower AND gate, which has an inverted *B* input and *A* input, giving a result of $A \cdot \overline{B}$. The final equation therefore is

$$Y = (\overline{A} \cdot B) + (A \cdot \overline{B}) \quad \text{or} \quad Y = \overline{A}B + A\overline{B}$$

The truth table for this logic circuit that combines five logic gates is shown in Figure 8–28(b). Looking at the input combinations and the output, you can see that the output is 1 only when an odd number of 1s appears at the input. This circuit is therefore acting as an XOR gate, and so

$$\overline{A}B + A\overline{B} = A \oplus B$$

8-6-7   *The XNOR Gate and Its Boolean Expression*

The Boolean expression for the two-input XNOR gate shown in Figure 8–29(a) will be

$$Y = \overline{A \oplus B} \text{ (pronounced "} Y \text{ equals not } A \text{ exclusive-or } B\text{")}$$

From this equation, you can see that the two inputs *A* and *B* are first XORed and then the result is complemented.

The truth table in Figure 8–29(b) tests this Boolean expression for all possible input combinations.

Boolean Expression	
$Y = \overline{A \oplus B}$	

(a)

A	B	$Y = \overline{A \oplus B}$
0	0	$Y = \overline{0 \oplus 0} = \overline{0} = 1$
0	1	$Y = \overline{0 \oplus 1} = \overline{1} = 0$
1	0	$Y = \overline{1 \oplus 0} = \overline{1} = 0$
1	1	$Y = \overline{1 \oplus 1} = \overline{0} = 1$

(b)

FIGURE 8–29   **The Boolean Expression for an XNOR Gate.**

Give the Boolean expression and truth table for the circuit shown in Figure 8–30(a).

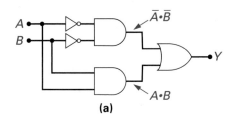

(a)

A	B	$Y = \bar{A}\bar{B} + AB$
0	0	$Y = (\bar{0}\cdot\bar{0}) + (0\cdot0) = (1\cdot1) + (0\cdot0) = 1 + 0 = \mathbf{1}$
0	1	$Y = (\bar{0}\cdot\bar{1}) + (0\cdot1) = (1\cdot0) + (0\cdot1) = 0 + 0 = \mathbf{0}$
1	0	$Y = (\bar{1}\cdot\bar{0}) + (1\cdot0) = (0\cdot1) + (1\cdot0) = 0 + 0 = \mathbf{0}$
1	1	$Y = (\bar{1}\cdot\bar{1}) + (1\cdot1) = (0\cdot0) + (1\cdot1) = 0 + 1 = \mathbf{1}$

(b)

FIGURE 8–30 **(NOT A AND NOT B) or (A AND B).**

■ *Solution:*

Studying the circuit in Figure 8–30(a), you can see that the upper AND gate has an inverted A input and an inverted B input, giving $\bar{A}\cdot\bar{B}$ at its output. The lower AND gate simply ANDs the A and B input, giving $A\cdot B$ at its output. The two AND gate outputs are then ORed, so the final equation for this circuit will be

$$Y = \bar{A}\cdot\bar{B} + A\cdot B \quad \text{or} \quad Y = \bar{A}\bar{B} + AB$$

The truth table for this circuit is shown in Figure 8–30(b). Comparing this truth table with an XNOR gate's truth table, you can see that the two are equivalent, since a 1 output is present only when an even number of 1s are applied to the input; therefore,

$$\bar{A}\bar{B} + AB = \overline{A \oplus B}$$

SELF-TEST EVALUATION POINT FOR SECTION 8–6

Now that you have completed this section, you should be able to:

■ **Objective 11.** *Give the Boolean expressions for each of the seven basic logic gates.*

■ **Objective 12.** *Describe how to write an equivalent Boolean expression for a logic circuit and how to sketch an equivalent logic circuit for a Boolean expression.*

Use the following questions to test your understanding of Section 8–6:

1. Which logic gate would be needed to perform each of the following Boolean expressions?
   a. AB **c.** $N + M$ **e.** $\overline{S + T}$
   b. $A \oplus B$ **d.** $\overline{X \cdot Y}$ **f.** $\overline{GH}$
2. State DeMorgan's first theorem using Boolean expressions.
3. Apply DeMorgan's theorem to the following.
   a. $\bar{A} + \bar{B}$ **b.** $Y = (\overline{A + B}) \cdot C$
4. Sketch the logic gates or circuits for the following expressions.
   a. ABC **c.** $(\bar{A} \cdot \bar{B}) + \bar{C}$
   b. $\overline{L + D}$ **d.** $A \cdot (\overline{L \oplus D})$

8–7 BINARY ARITHMETIC

At the heart of every digital electronic computer system is a circuit called a microprocessor unit (MPU). To perform its function, the MPU contains several internal circuits, one of which is called the arithmetic–logic unit, or ALU. This unit is the number-crunching circuit of every digital system, and as its name implies, it can be controlled to perform either arithmetic operations (such as addition and subtraction) or logic operations (such as AND and OR).

In order to understand the operation of an ALU, we must begin by discussing how binary numbers can represent any value, whether it is large, small, positive, or negative. We must also discuss how these numbers can be manipulated to perform arithmetic operations such as addition, subtraction, multiplication, and division.

8–7–1 *Binary Addition*

Figure 8–31(a) illustrates how the decimal values 29,164 and 63,729 are added together. First the units 4 and 9 are added, producing a sum of 13, or 3 carry 1. The carry is then added to the 6 and 2 in the tens column, producing a sum of 9 with no carry. This process is then

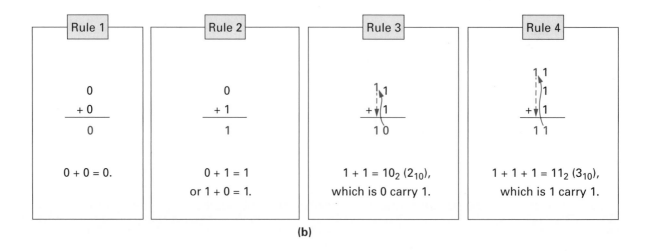

	(10,000)	(1,000)	(100)	(10)	(1)
Carry	0	1	0	0	1
Addend:		2	9,	1	6	4
Augend:	+	6	3,	7	2	9
Sum:		9	2,	8	9	3

4 + 9 = 13 or 3 1 carry

(a)

Rule 1	Rule 2	Rule 3	Rule 4
0 + 0 ___ 0	0 + 1 ___ 1	1 1 + 1 ___ 1 0	1 1 1 + 1 ___ 1 1
0 + 0 = 0.	0 + 1 = 1 or 1 + 0 = 1.	$1 + 1 = 10_2 \ (2_{10})$, which is 0 carry 1.	$1 + 1 + 1 = 11_2 \ (3_{10})$, which is 1 carry 1.

(b)

	(32)	(16)	(8)	(4)	(2)	(1)		(10)	(1)
Carry:	1	1	0	0	0			1	
Augend:		1	1	0	1	0_2	=	2	6_{10}
Addend:	+	1	1	1	0	0_2		+ 2	8_{10}
Sum:	1	1	0	1	1	0_2		5	4_{10}

(c)

FIGURE 8–31 Binary Addition.

repeated for each column, moving right to left, until the total sum of the addend, augend, and their carries has been obtained.

In the addition of two decimal values, many combinations are available, since decimal has 10 different digits. With binary, however, only 2 digits are available, and therefore only four basic combinations are possible. These four rules of binary addition are shown in Figure 8–31(b). To help explain these rules, let us apply them to the example in Figure 8–31(c), in which we will add 11010 (decimal 26) to 11100 (decimal 28). As in decimal addition, we will begin at the right in the ones column and proceed to the left. In the ones column $0 + 0 = 0$ with a carry of 0 (Rule 1). In the twos column $1 + 0$ and the 0 carry from the ones column $= 1$ with no carry (Rule 2). In the fours column we have $0 + 1$ plus a carry of 0 from the twos column. This results in a fours column total of 1, with a carry of 0 (Rule 2). In the eights column we have $1 + 1$ plus a 0 carry from the fours column, producing a total of 10_2 ($1 + 1 =$ decimal 2, which is 10 in binary). In the eights column this 10 (decimal 2) is written down as a total of 0 with a carry of 1 (Rule 3). In the sixteens column, we have $1 + 1$, plus a carry of 1 from the eights column, producing a total of 11_2 ($1 + 1 + 1 =$ decimal 3, which is 11 in binary). In the sixteens column, this 11 (decimal 3) is written down as a total of 1 with a carry of 1 (Rule 4). Since there is only a carry of 1 in the thirty-twos column, the total in this column will be 1.

Converting the binary addend and augend to their decimal equivalents as seen in Figure 8–31(c), we see that 11010 (decimal 26) plus 11100 (decimal 28) will yield a sum or total of 110110 (decimal 54).

■ **INTEGRATED MATH APPLICATION:**

Find the sum of the following binary numbers.

    a. $1011 + 1101 = ?$ **b.** $1000110 + 1100111 = ?$

■ *Solution:*

a.

Carry:		1	1	1	1	
Addend:		1	0	1	1_2	(11_{10})
Augend:	+	1	1	0	1_2	(13_{10})
Sum:	1	1	0	0	0_2	(24_{10})

b.

Carry:		1	0	0	0	1	1	0
Addend:		1	0	0	0	1	1	0_2	(70_{10})
Augend:	+	1	1	0	0	1	1	1_2	(103_{10})
Sum:	1	0	1	0	1	1	0	1_2	(173_{10})

8–7–2   *Representing Positive and Negative Numbers*

Many digital systems, such as calculators and computers, are used to perform mathematical functions on a wide range of numbers. Some of these numbers are positive and negative signed numbers; therefore, a binary code is needed to represent these numbers.

1. The Sign and Magnitude Number System

The **sign and magnitude number system** is a binary code system used to represent positive and negative numbers. A sign–magnitude number contains a *sign bit* (0 for positive, 1 for negative) followed by the *magnitude bits*.

As an example, Figure 8–32(a) shows some 4-bit sign–magnitude positive numbers. All these sign–magnitude binary numbers—0001, 0010, 0011, and 0100—have an MSB, or sign bit, of 0, and therefore they are all positive numbers. The remaining 3 bits in these 4-bit words indicate the magnitude of the number based on the standard 421 binary column weight ($001_2 = 1_{10}, 010_2 = 2_{10}, 011_2 = 3_{10}, 100_2 = 4_{10}$).

Sign and Magnitude Number System

A binary code used to represent positive and negative numbers in which the MSB signifies sign and the following bits indicate magnitude.

Sign Bit	Magnitude Bits
0 = +	
1 = −	= Binary Equivalent

4-Bit Sign-Magnitude Numbers

$$0\ 0\ 0\ 1_2 \quad = \quad +1_{10}$$
$$0\ 0\ 1\ 0_2 \quad = \quad +2_{10}$$
$$0\ 0\ 1\ 1_2 \quad = \quad +3_{10}$$
$$0\ 1\ 0\ 0_2 \quad = \quad +4_{10}$$

Sign Bit   Magnitude Bits

(a)

$$1\ 1\ 0\ 0_2 \quad = \quad -4_{10}$$
$$1\ 1\ 0\ 1_2 \quad = \quad -5_{10}$$
$$1\ 1\ 1\ 0_2 \quad = \quad -6_{10}$$
$$1\ 1\ 1\ 1_2 \quad = \quad -7_{10}$$

Sign Bit   Magnitude Bits

(b)

8-Bit Sign-Magnitude Numbers

$$0\ 0\ 0\ 0\ 1\ 1\ 1\ 1_2 \quad = \quad +15_{10}$$
$$0\ 1\ 0\ 1\ 0\ 0\ 0\ 0_2 \quad = \quad +80_{10}$$
$$1\ 0\ 0\ 0\ 1\ 0\ 1\ 0_2 \quad = \quad -10_{10}$$
$$1\ 1\ 1\ 1\ 1\ 1\ 1\ 1_2 \quad = \quad -127_{10}$$

(c)

16-Bit Sign-Magnitude Numbers

$$0\ 0\ 0\ 0\ \ 0\ 0\ 0\ 0\ \ 0\ 0\ 0\ 0\ \ 0\ 0\ 0\ 1_2 \quad = \quad +1_{10}$$
$$0\ 0\ 0\ 0\ \ 0\ 0\ 0\ 0\ \ 1\ 0\ 1\ 0\ \ 0\ 1\ 0\ 0_2 \quad = \quad +164_{10}$$
$$1\ 0\ 0\ 0\ \ 0\ 0\ 0\ 0\ \ 0\ 0\ 1\ 1\ \ 0\ 0\ 1\ 0_2 \quad = \quad -50_{10}$$
$$1\ 0\ 0\ 0\ \ 1\ 1\ 1\ 1\ \ 0\ 0\ 0\ 0\ \ 0\ 0\ 0\ 0_2 \quad = \quad -3840_{10}$$

(d)

FIGURE 8–32 The Sign–Magnitude Number System.

As another example, Figure 8–32(b) shows some 4-bit sign–magnitude negative numbers. All of these sign–magnitude binary numbers—1100, 1101, 1110, and 1111—have an MSB, or sign bit, of 1, and therefore they are all negative numbers. The remaining 3 bits in these 4-bit words indicate the magnitude of the number based on the standard 421 binary column weight $(100_2 = 4_{10}, 101_2 = 5_{10}, 110_2 = 6_{10}, 111_2 = 7_{10})$.

If you need to represent larger decimal numbers, you simply use more bits, as shown in Figure 8–32(c) and (d). The principle still remains the same in that the MSB indicates the sign of the number, and the remaining bits indicate the magnitude of the number.

The sign–magnitude number system is an ideal example of how binary numbers could be coded to represent positive and negative numbers. This code system, however, requires complex digital hardware for addition and subtraction, and is therefore seldom used.

■ **INTEGRATED MATH APPLICATION:**

What are the decimal equivalents of the following sign–magnitude number system codes?

   a. 0111   **b.** 1010   **c.** 0001 1111   **d.** 1000 0001

■ *Solution:*

   a. +7   **b.** −2   **c.** +31   **d.** −1

■ **INTEGRATED MATH APPLICATION:**

What are the 8-bit sign–magnitude number system codes for the following decimal values?

   a. +8   **b.** −65   **c.** +127   **d.** −127

■ *Solution:*

   a. 0000 1000   **b.** 1100 0001   **c.** 0111 1111   **d.** 1111 1111

2. The One's Complement (1's Complement) Number System

The **one's complement number system** became popular in early digital systems. It also used an MSB sign bit that was either a 0 (indicating a positive number) or a 1 (indicating a negative number).

Figure 8–33(a) shows some examples of 8-bit one's complement positive numbers. As you can see, one's complement positive numbers are no different from sign–magnitude positive numbers. The only difference between the one's complement number system and the sign–magnitude number system is the way in which negative numbers are represented. To understand the code used to represent negative numbers, we must first understand the meaning of the term *one's complement*. To one's complement a binary number means to invert or change all the 1s in the binary word to 0s, and all the 0s in the binary word to 1s. Performing the one's complement operation on a positive number will change the number from a one's complement positive number to a one's complement negative number. For example, by one's complementing the number 00000101, or +5, we will get 11111010, which is −5. In Figure 8–33(b), you can see the negative one's complement codes of the positive one's complement codes shown in Figure 8–33(a).

The next question you may have is, How can we determine the decimal value of a negative one's complement code? The answer is to do the complete opposite, or complement, of the positive one's complement code. This means count the 1s in a positive one's complement code, and count the 0s in a negative one's complement code. For example, with a positive one's complement code (MSB = 0), the value is determined by simply adding together all the column weights that contain a 1 (00000110 has a 1 in the fours column and twos column, so this number is 4 + 2, or +6). On the other hand, for a negative one's complement code (MSB = 1), the value is determined by doing the opposite and adding together all the column weights that contain a 0 (11110101 has a 0 in the eights column and twos column, so this number is 8 + 2, or −10).

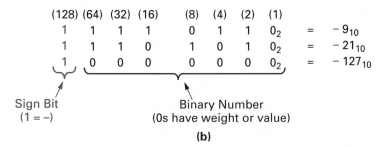

FIGURE 8–33   The One's Complement Number System.

■ **INTEGRATED MATH APPLICATION:**

What are the decimal equivalents of the following one's complement number system codes?

    a. 0010 0001   **b.** 1101 1101

■   *Solution:*

    a. MSB = 0 (number is positive, so add together all the column weights that contain a 1). The number 0010 0001 has a 1 in the thirty-twos column and a one in the units column, so 0010 0001 = +33.
    b. MSB = 1 (number is negative, so add together all the column weights that contain a 0). The number 1101 1101 has a 0 in the thirty-twos column and a 0 in the twos column, so 1101 1101 = −34.

Since the term *one's complement* can be used to describe both an operation (change all of the 1s to 0s, and all the 0s to 1s), and a positive and negative number system, some confusion can occur. To avoid this, remember that if you are asked to one's complement a number, it means to change all the 1s in the binary word to 0s and all 0s to 1s. If, on the other hand, you are asked to find the decimal equivalent of a one's complement number, you should decode the sign bit and binary value to determine the decimal equivalent value.

■ **INTEGRATED MATH APPLICATION:**

One's complement the following numbers.

    a. 0101 1111   **b.** 1110 0101

■   *Solution:*

To one's complement a number means to change all the 1s to 0s and all the 0s to 1s.

    a. 1010 0000   **b.** 0001 1010

■ **INTEGRATED MATH APPLICATION:**

Find the decimal equivalent of the following one's complement numbers.

    a. 0101 1111   **b.** 1110 0101

■   *Solution:*

    a. MSB = 0 (number is positive, so add together all the column weights that contain a 1). The number 0101 1111 = +95.
    b. MSB = 1 (number is negative, so add together all the column weights that contain a 0). The number 1110 0101 = −26.

Two's Complement Number System

A binary code used to represent positive and negative numbers, in which negative values are determined by inverting all of the binary digits of an equivalent positive value and then adding 1.

3. The Two's Complement (2's Complement) Number System

The **two's complement number system** is used almost exclusively in digital systems to represent positive and negative numbers.

Positive two's complement numbers are no different from positive sign–magnitude and one's complement numbers in that an MSB sign bit of 0 indicates a positive number, and the remaining bits of the word indicate the value of the number. Figure 8–34(a) shows two examples of positive two's complement numbers.

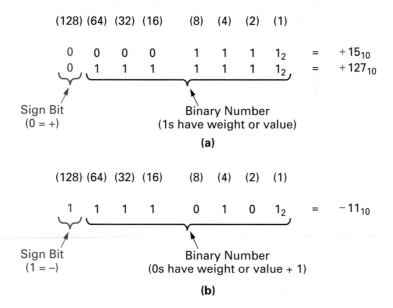

(a)

(b)

FIGURE 8–34   The Two's Complement Number System.

Negative two's complement numbers are determined by two's complementing a positive two's complement number code. To two's complement a number first one's complement it (change all the 1s to 0s and 0s to 1s), and then add 1. For example, if we were to two's complement the number 0000 0101 (+5), we would obtain the two's complement number code for −5. The procedure to follow to two's complement (+5) would be as follows:

0000	0101	Original number (+5)
1111	1010	One's complement (inverted)
+	1	Plus 1
1111	1011	Two's complement code for −5

The new two's complement code obtained after this two's complement operation is the negative equivalent of the original number. In our example, therefore:

Original number:	0000 0101 = +5
After two's complement operation:	1111 1011 = −5

Any negative equivalent two's complement code therefore can be obtained by simply two's complementing the same positive two's complement number. For example, to find the negative two's complement code for −125, we would perform the following operation.

Two's	0111 1101	Original number (+125)
complement →	1000 0010	One's complement (inverted)
operation	+ 1	Plus 1
	1000 0011	Two's complement code for −125

■   **INTEGRATED MATH APPLICATION:**

Find the negative two's complement codes for

a. −26   **b.** −67

■ *Solution:*

a. 0001 1010   Two's complement code for +26

| 1110 0101 | One's complement (inverted)
| +   1 | Plus 1

1110 0110   Two's complement code for −26

b. 0100 0011   Two's complement code for +67

| 1011 1100 | One's complement (inverted)
| +   1 | Plus 1

1011 1101   Two's complement code for −67

Performing the two's complement operation on a number twice will simply take the number back to its original value and polarity. For example, let's take the two's complement code for +6 (0000 0110) and two's complement it twice.

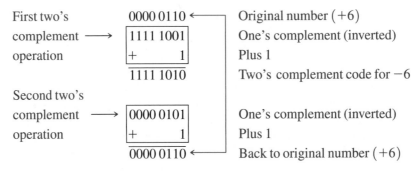

The decimal equivalent of a positive two's complement number (MSB = 0) is easy to determine, since the binary value following the sign bit applies the same as any binary number, as shown in Figure 8–34(a). On the other hand, the decimal equivalent of a negative two's complement number (MSB = 1) is not determined in the same way. To determine the value of a negative two's complement number, refer to the remaining bits following the sign bit, add the weight of all columns containing a 0, and then add 1 to the result. For example, the negative two's complement number in Figure 8–34(b) has 0s in the eights and twos columns, and therefore its value will equal the sum of 8 and 2, plus the additional 1 (due to two's complement). As a result:

$$1 \quad 1 \quad 1 \quad 1 \quad \overset{(8)}{0} \quad \overset{(4)}{1} \quad \overset{(2)}{0} \quad \overset{(1)}{1} = (8 + 2) + 1 = -11$$

■ INTEGRATED MATH APPLICATION:

Determine the decimal equivalent of the following negative two's complement numbers.

a. 1110 0001   **b.** 1011 1011   **c.** 1000 0000   **d.** 1111 1111

■ *Solution:*

a. $\overset{(\pm)}{1} \quad \overset{(64)}{1} \quad \overset{(32)}{1} \quad \overset{(16)}{0} \quad \overset{(8)}{0} \quad \overset{(4)}{0} \quad \overset{(2)}{0} \quad \overset{(1)}{1}_2 = (16 + 8 + 4 + 2) + 1 = -31_{10}$

b. $\overset{(\pm)}{1} \quad \overset{(64)}{0} \quad \overset{(32)}{1} \quad \overset{(16)}{1} \quad \overset{(8)}{1} \quad \overset{(4)}{0} \quad \overset{(2)}{1} \quad \overset{(1)}{1}_2 = (64 + 4) + 1 = -69_{10}$

c. $\overset{(\pm)}{1} \quad \overset{(64)}{0} \quad \overset{(32)}{0} \quad \overset{(16)}{0} \quad \overset{(8)}{0} \quad \overset{(4)}{0} \quad \overset{(2)}{0} \quad \overset{(1)}{0}_2 =$
$$(64 + 32 + 16 + 8 + 4 + 2 + 1) + 1 = -128_{10}$$

d. $\overset{(\pm)}{1} \quad \overset{(64)}{1} \quad \overset{(32)}{1} \quad \overset{(16)}{1} \quad \overset{(8)}{1} \quad \overset{(4)}{1} \quad \overset{(2)}{1} \quad \overset{(1)}{1}_2 = (0) + 1 = -1_{10}$

4-Bit Word	Standard Binary Value	2's Complement Value
0 0 0 0	0	0
0 0 0 1	1	+1
0 0 1 0	2	+2
0 0 1 1	3	+3 8 Positive
0 1 0 0	4	+4 Numbers
0 1 0 1	5	+5
0 1 1 0	6	+6
0 1 1 1	7	+7
1 0 0 0	8	−8
1 0 0 1	9	−7
1 0 1 0	10	−6
1 0 1 1	11	−5 8 Negative
1 1 0 0	12	−4 Numbers
1 1 0 1	13	−3
1 1 1 0	14	−2
1 1 1 1	15	−1

$2^4 =$ 16 Words

FIGURE 8–35 **Four-Bit Two's Complement Codes.**

As with the sign–magnitude and one's complement number systems, if you need to represent larger decimal numbers with a two's complement code, you will have to use a word containing more bits. For example, a total of 16 different combinations is available for a 4-bit word ($2^4 = 16$), as shown in Figure 8–35. With standard binary, these 16 different combinations give a count from 0 to 15, as shown in the center column in Figure 8–35. As for two's complement, we split these 16 different codes in half and end up with 8 positive two's complement codes and 8 negative two's complement codes, as shown in the right column in Figure 8–35. Since decimal zero uses one of the positive two's complement codes, the 16 different combinations give a range of -8_{10} to $+7_{10}$ (the maximum positive value, $+7$, is always one less than the maximum negative value, -8, since decimal zero uses one of the positive codes).

An 8-bit word will give 256 different codes ($2^8 = 256$), as shown in Figure 8–36. Dividing this total in half will give 128 positive two's complement codes and 128 negative two's complement codes. Since decimal zero uses one of the positive two's complement codes, the 256 different combinations give a range of -128_{10} to $+127_{10}$.

■ **INTEGRATED MATH APPLICATION:**

What would be the two's complement range for a 16-bit word?

■ *Solution:*

$2^{16} = 65{,}536$; half of $65{,}536 = 32{,}768$; therefore, the range will be $-32{,}768$ to $+32{,}767$.

8–7–3 *Two's Complement Arithmetic*

To perform its function, a computer's microprocessor contains several internal circuits, and one of these circuits is called the **arithmetic–logic unit,** or **ALU.** This unit is the number-crunching circuit of every digital system, and as its name implies, it can be controlled to perform either arithmetic operations (such as addition or subtraction) or logic operations (such as AND and OR). The arithmetic–logic unit has two parallel inputs, one parallel output, and a set of function select control lines, as shown in Figure 8–37.

The ALU has to be able to perform the four basic arithmetic operations, namely, addition, subtraction, multiplication, and division. One would expect, therefore, that an ALU

Arithmetic–Logic Unit (ALU)

The section of a computer that performs all arithmetic and logic operations.

	8-Bit Word	Standard Binary Value	2's Complement Value	
	0 0 0 0 0 0 0 0	0	0	
	0 0 0 0 0 0 0 1	1	+1	
	0 0 0 0 0 0 1 0	2	+2	
	0 0 0 0 0 0 1 1	3	+3	
	0 0 0 0 0 1 0 0	4	+4	
	0 0 0 0 0 1 0 1	5	+5	
	0 0 0 0 0 1 1 0	6	+6	
	0 0 0 0 0 1 1 1	7	+7	128 Positive Numbers
	0 1 1 1 1 0 1 1	123	+123	
	0 1 1 1 1 1 0 0	124	+124	
	0 1 1 1 1 1 0 1	125	+125	
	0 1 1 1 1 1 1 0	126	+126	
2^8 =	0 1 1 1 1 1 1 1	127	+127	
256 Words	1 0 0 0 0 0 0 0	128	−128	
	1 0 0 0 0 0 0 1	129	−127	
	1 0 0 0 0 0 1 0	130	−126	
	1 0 0 0 0 0 1 1	131	−125	
	1 0 0 0 0 1 0 0	132	−124	
	1 0 0 0 0 1 0 1	133	−123	128 Negative Numbers
	1 1 1 1 1 0 1 1	251	−5	
	1 1 1 1 1 1 0 0	252	−4	
	1 1 1 1 1 1 0 1	253	−3	
	1 1 1 1 1 1 1 0	254	−2	
	1 1 1 1 1 1 1 1	255	−1	

FIGURE 8–36   Eight-Bit Two's Complement Codes.

would contain a separate circuit for each of these operations. This is not so, since all four operations can be performed using a binary-adder circuit. To explain this point, because the binary-adder circuit's function is to perform addition, and since multiplication is simply repeated addition, the binary-adder circuit can also be used to perform any multiplication operations. For example, 3×4 means that the number 4 is added three times $(4 + 4 + 4 = 12$, therefore $3 \times 4 = 12)$. But what about subtraction and division? The answer to this problem is the two's complement number system. As you will see in this section, by representing all values as two's complement numbers, we can perform a subtraction operation by slightly modifying the second number and then adding the two numbers together. This means that the binary-adder circuit can also be used to perform subtraction operations, and since division is simply repeated subtraction, the subtraction operation of the binary-adder circuit can also be used to perform any division operation. For example, $12 \div 4$ means that you take the number 12 and see how many times you can subtract 4 from it $(12 - 4 - 4 - 4 = 0$; therefore, there are three 4s in 12, $12 \div 4 = 3)$.

By using two's complement numbers, therefore, we can use a binary-adder circuit within an ALU to add, subtract, multiply, and divide. Having one circuit able to perform all four mathematical operations means that the circuit will be small in size and fast to operate, will consume very little power, and will be cheap to manufacture. In this section we will see how we can perform all four mathematical operations by always adding two's complement numbers.

CHAPTER 8 / COMPUTER MATH

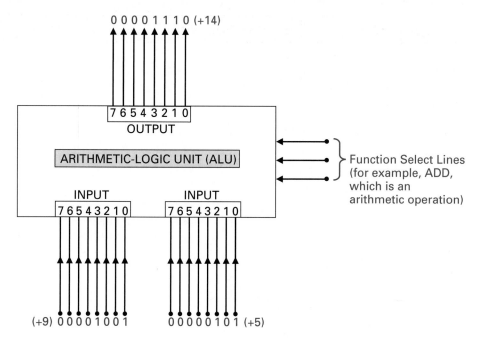

FIGURE 8–37 **Block Diagram of the Arithmetic-Logic Unit (ALU).**

1. Adding Positive Numbers

To begin, let us prove that the addition of two positive two's-complement numbers will render the correct result.

■ **INTEGRATED MATH APPLICATION:**

Add +9 (0000 1001) and +5 (0000 0101).

■ *Solution:*

$$
\begin{array}{r}
0000\ 1001 \\
+\ 0000\ 0101 \\
\hline
0000\ 1110
\end{array}
$$

MSB of 0 = positive number $\longrightarrow$

When both numbers to be added are positive two's complement numbers (both have an MSB of 0), it makes sense that the sum will have an MSB of 0, indicating a positive result. This is always true, unless the range of the two's complement word is exceeded. For example, the maximum positive number that can be represented by an 8-bit word is +127 (0111 1111). If the sum of the two positive two's complement numbers to be added exceeds the upper positive range limit, there will be a *two's complement overflow* into the sign bit. For instance:

	(±)	(64)	(32)	(16)	(8)	(4)	(2)	(1)	
	0	1	1	0	0	1	0	0	(+100)
+	0	0	0	1	1	1	1	0	+ (+30)
MSB of 1 = negative number $\longrightarrow$	1	0	0	0	0	0	1	0	(−126)

If two positive numbers are applied to an arithmetic circuit and it is told to add the two, the result should also be a positive number. Most arithmetic circuits are able to detect this problem if it occurs by simply monitoring the MSB of the two input words and the MSB of the output word, as shown in Figure 8–38. If the two input sign bits are the same, but the output sign bit is different, then a two's complement overflow has occurred.

FIGURE 8–38 Two's Complement Overflow Detection.

2. Adding Positive and Negative Numbers

By using the two's complement number system, we can also add numbers with unlike signs and obtain the correct result. As an example, let us add +9 (0000 1001) and −5 (1111 1011).

$$
\begin{array}{rcl}
 & 0000\ 1001 & (+9) \\
+ & 1111\ 1011 & +\ (-5) \\
\hline
\text{Ignore the final carry} \longrightarrow \quad 1 & 0000\ 0100 & (+4)
\end{array}
$$

A final carry will always be generated whenever the sum is a positive number; however, if this carry is ignored because it is beyond the 8-bit two's complement word, the answer will be correct.

■ **INTEGRATED MATH APPLICATION:**

Add +10 (0000 1010) and −12 (1111 0100).

■ *Solution:*

$$
\begin{array}{rcl}
 & 0000\ 1010 & (+10) \\
+ & 1111\ 0100 & +\ (-12) \\
\hline
 & 1111\ 1110 & (-2)
\end{array}
$$

---

3. Adding Negative Numbers

As a final addition test for the two's complement number system, let us add two negative numbers to see if we can obtain the correct negative sum.

■ INTEGRATED MATH APPLICATION:

Add −5 (1111 1011) and −4 (1111 1100).

■ *Solution:*

$$
\begin{array}{rr}
1111\,1011 & (-5) \\
+\ 1111\,1100 & +\ (-4) \\
\hline
1\quad 1111\,0111 & (-9)
\end{array}
$$

Ignore the final carry ⟶

The sum in this example is the correct two's complement code for −9.

Two's complement overflow can also occur in this condition if the sum of the two negative numbers exceeds the negative range of the word. For example, with an 8-bit two's complement word, the largest negative number is −128, and if we add −100 (1001 1100) and −30 (1110 0010), we will get an error, as follows:

$$
\begin{array}{rr}
1001\,1100 & (-100) \\
+\ 1110\,0010 & +\ (-30) \\
\hline
1\quad 0111\,1110 & (+126)
\end{array}
$$

Ignore the final carry ⟶

Once again, the two's complement overflow detect circuit, shown in Figure 8–38, will indicate such an error, since two negative numbers were applied to the input and a positive number appeared at the output.

In all the addition examples mentioned so far (adding positive numbers, adding positive and negative numbers, and adding negative numbers) you can see that an ALU added the bits of the input bytes without any regard for whether they were signed or unsigned numbers. This brings up a very important point, which is often misunderstood. The ALU adds the bits of the input words using the four basic rules of binary addition, and it does not know if these input and output bit patterns are coded two's complement words or standard unsigned binary words. It is therefore up to us to maintain a consistency, which means that if we are dealing with coded two's complement words at the input of an ALU, we must interpret the output of an ALU as a two-coded two's complement word. On the other hand, if we are dealing with standard unsigned binary words at the input of an ALU, we must interpret the output of an ALU as a standard unsigned binary word.

4. Subtraction through Addition

The key advantage of the two's complement number system is that it enables us to perform a subtraction operation using the adder circuit within the ALU. To explain how this is done, let us take an example and assume that we want to subtract 3 (0000 0011) from 8 (0000 1000), as follows:

Minuend:	0000 1000	(+8)
Subtrahend:	− 0000 0011	− (+3)
Difference:	0000 0101	+5

If we subtract the subtrahend of +3 from the minuend of +8, we will obtain the correct result of +5 (0000 0101); however, we must find the difference to this problem by using addition and not subtraction. The solution to this is achieved by *two's complementing the subtrahend, and then adding the result to the minuend.* The following display shows how this is done.

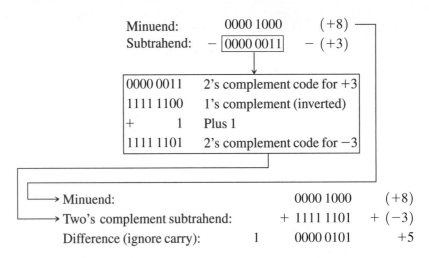

Minuend:	0000 1000	(+8)
Subtrahend:	− 0000 0011	− (+3)

0000 0011	2's complement code for +3
1111 1100	1's complement (inverted)
+ 1	Plus 1
1111 1101	2's complement code for −3

Minuend:	0000 1000	(+8)
Two's complement subtrahend:	+ 1111 1101	+ (−3)
Difference (ignore carry):	1 0000 0101	+5

As you can see from this operation, adding the −3 to +8 achieves exactly the same result as subtracting +3 from +8. We can therefore perform a subtraction by two's complementing the subtrahend of a subtraction problem and then adding the result to the minuend.

Digital electronic calculators and computer systems perform the previously described steps automatically. That is, they first instruct the ALU to one's complement the subtrahend input (function select control inputs = invert, which is a logic operation), then add 1 to the result (function select control inputs = +1, which is an arithmetic operation), and finally add the result of the two previous operations (which is the two's complemented subtrahend) to the minuend (function select = ADD, arithmetic operation). Although this step-by-step procedure or **algorithm** seems complicated, it is a lot simpler than having to include a separate digital electronic circuit to perform all subtraction operations. Having one circuit able to perform both addition and subtraction means that the circuit will be small in size and fast to operate, will consume very little power, and will be cheap to manufacture.

Flowcharts provide a graphic way of describing a step-by-step procedure (algorithm) or a program of instructions. Figure 8–39 shows the basic blocks used in flowcharts. Figure 8–40 shows a flowchart summarizing the subtraction through addition procedure.

5. Multiplication through Repeated Addition

Most digital electronic calculators and computers do not include a logic circuit that can multiply, since the adder circuit can also be used to perform this arithmetic operation. For example, to multiply 6 × 3, you simply add 6 three times.

Multiplicand:	6	6
Multiplier:	× 3 =	+ 6
Product:	18	+ 6
		18

As with subtraction through addition, an algorithm must be followed when a digital system needs to perform a multiplication operation. Figure 8–41 shows the flowchart for

Algorithm
A set of rules for the solution of a problem.

Flowchart
A graphic representation of a step-by-step procedure or program of instructions.

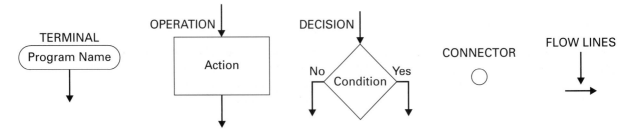

FIGURE 8–39 Flowchart Symbols.

FIGURE 8-40 Flowchart for Subtraction through Addition.

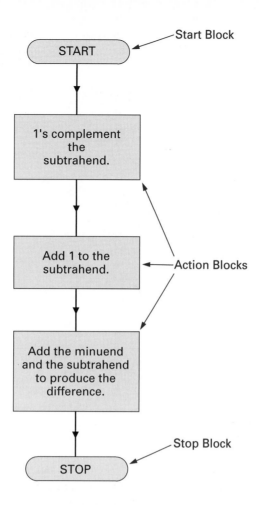

multiplication through repeated addition. Let us apply this procedure to our previous example of 6 × 3 and see how the multiplicand 6 is added three times to obtain a final product of 18. The steps are as follows:

Start:	multiplicand = 6, multiplier = 3, product = 0
After first pass:	multiplicand = 6, multiplier = 2, product = 6
After second pass:	multiplicand = 6, multiplier = 1, product = 12
After third pass:	multiplicand = 6, multiplier = 0, product = 18
Stop.	

In this example we cycled through the flowchart three times (until the multiplier was 0) and then finally obtained a product of 18.

Although this algorithm seems complicated, it is a lot simpler than having to include a separate digital circuit for multiplication. Having one circuit able to perform addition, subtraction, and multiplication means that the circuit will be small in size and fast to operate, will consume very little power, and will be cheap to manufacture.

6. Division through Repeated Subtraction

Up to this point we have seen how we can use the two's complement number system and the adder circuit within an ALU for addition, subtraction, and multiplication. Just as multiplication can be achieved through repeated addition, division can be achieved through repeated subtraction. For example, to divide 31 by 9 we simply see how many times 9 can be subtracted from 31, as follows:

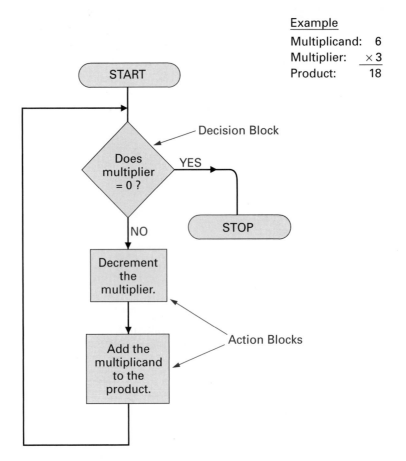

Example

Multiplicand:	6
Multiplier:	× 3
Product:	18

FIGURE 8–41   **Flowchart for Multiplication through Repeated Addition.**

$$
\begin{array}{cl}
& 3 \quad \text{Quotient} \\
\text{Divisor: } & 9\overline{)31} \quad \text{Dividend} \\
& 27 \\
& 4 \quad \text{Remainder}
\end{array}
$$

$$
\begin{array}{rl}
31 & \\
-9 & \text{First subtraction} \\
\hline
= \quad 22 & \\
-9 & \text{Second subtraction} \\
\hline
13 & \\
-9 & \text{Third subtraction} \\
\hline
4 &
\end{array}
$$

Since we were able to subtract 9 from 31 three times with a remainder of 4, then $31 \div 9 = 3$, remainder 4.

As with multiplication, an algorithm must be followed when a digital system needs to perform a division. Figure 8–42 shows the flowchart for division through repeated subtraction. Let us apply this procedure to our previous example of $31 \div 3$ and see how the quotient is incremented to 3, leaving a remainder of 4. The steps are as follows:

Start:	dividend = 31, divisor = 9, quotient = 0, remainder = 31
After first pass:	dividend = 22, divisor = 9, quotient = 1, remainder = 22
After second pass:	dividend = 13, divisor = 9, quotient = 2, remainder = 13
After third pass:	dividend = 4, divisor = 9, quotient = 3, remainder = 4
Stop.	

When this program of instructions cycles or loops back for the third pass, the divisor of 9 is subtracted from the remaining dividend of 4, resulting in -5. When the question, Is the dividend negative? is asked, the answer is yes, so the branch on the right is followed. This branch of the flowchart will add the divisor of 9 to the dividend of -5 to restore it to

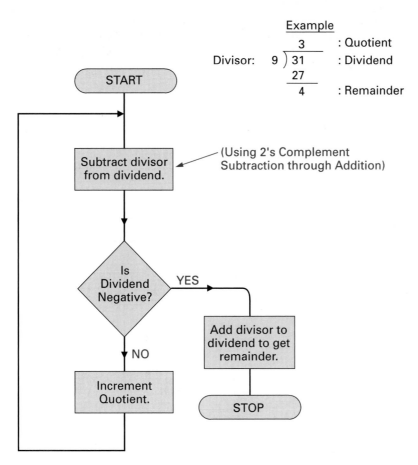

FIGURE 8–42   Flowchart for Division through Repeated Subtraction.

the correct remainder of 4 before the program stops. To summarize, these are the steps followed by this program:

$$
\begin{array}{ll}
(+31) & \\
-\ (+9) & \\
(+22) & \text{Quotient} = 1 \\
-\ (+9) & \\
(+13) & \text{Quotient} = 2 \\
-\ (+9) & \\
(+4) & \text{Quotient} = 3 \\
-\ (+9) & \\
(-5) & \\
-\ (+9) & \text{Add divisor to dividend} \\
(+4) & \text{Remainder} = 4
\end{array}
$$

When a digital system is instructed to perform a division, it will use the ALU for all arithmetic and logic operations. Since division is achieved through repeated subtraction, and subtraction is achieved through addition, the ALU's adder circuit can also be used for division.

Although this algorithm seems complicated, it is a lot simpler than having to include a separate digital circuit for division. As previously mentioned, having one circuit able to perform addition, subtraction, multiplication, and division means that the circuit will be small in size and fast to operate, will consume very little power, and will be cheap to manufacture.

Now that you have completed this section, you should be able to:

■ **Objective 13.** *Demonstrate how to add binary numbers.*

■ **Objective 14.** *Explain how the two's complement number system can be used to represent positive and negative numbers.*

■ **Objective 15.** *Demonstrate how two's complement numbers enable an adder circuit to add, subtract, multiply, and divide binary numbers.*

Use the following questions to test your understanding of Section 8–7.

1. Add the following: $1011 + 11101$.

2. Two's complement the following binary numbers.
   a. 0110 1011
   b. 1000 0111

3. Give the decimal equivalent of the following two's complement numbers.
   a. 1111 1011
   b. 0000 0110

REVIEW QUESTIONS

Multiple Choice Questions

1. Binary is a base _____ number system, decimal a base _____ number system, and hexadecimal a base _____ number system.
   a. 1, 2, 3   c. 10, 2, 3
   b. 2, 10, 16   d. 8, 4, 2

2. What is the binary equivalent of decimal 11?
   a. 1010   c. 1011
   b. 1100   d. 0111

3. Which of the following number systems is used as a shorthand for binary?
   a. Decimal   c. Binary
   b. Hexadecimal   d. Both a and b

4. What would be displayed on a hexadecimal counter if its count was advanced by one from $39FF_{16}$?
   a. 4000   c. 3900
   b. 3A00   d. 4A00

Communication Skill Questions

10. What are the three basic blocks of a computer? (8–1)

11. Give the full names of the following abbreviations and acronyms:
    a. Bit   c. ADC
    b. ASCII   d. DAC

12. What is the function of the ADC and the DAC? (8–2)

13. List the Boolean expression for the seven basic logic gates. (8–6)

14. Briefly describe how to: (8–3 and 8–4)
    a. Convert a binary number to a decimal number
    b. Convert a decimal number to a binary number
    c. Convert a hexadecimal number to a decimal number
    d. Convert a decimal number to a hexadecimal number
    e. Convert between binary and hexadecimal

15. Briefly describe how the following number systems are used to represent positive and negative binary numbers. (8–7)

5. What is the base of 101?
   a. 10   c. 16
   b. 2   d. Could be any base

6. Which of the following is an invalid BCD code?
   a. 1010   c. 0111
   b. 1000   d. 0000

7. What is the decimal equivalent of $1001\ 0001\ 1000\ 0111_{BCD}$?
   a. 9187   c. 8659
   b. 9A56   d. 254,345

8. DeMorgan's first theorem states that _____.
   a. $\overline{A \cdot B} = \overline{A} + \overline{B}$.   c. $\overline{A} + \overline{B} = \overline{A \cdot B}$.
   b. $\overline{A} + \overline{B} = \overline{A} + B$.   d. $\overline{A \cdot B} = A \cdot \overline{B}$.

9. DeMorgan's second theorem states that _____.
   a. $\overline{A \cdot B} = \overline{A} + \overline{B}$.   c. $\overline{A} + \overline{B} = \overline{A \cdot B}$.
   b. $\overline{A} + \overline{B} = \overline{A} + B$.   d. $\overline{A \cdot B} = A \cdot \overline{B}$.

   a. Sign–magnitude
   b. One's complement
   c. Two's complement

16. Why is the two's complement number system most frequently used to represent positive and negative numbers? (8–7)

17. How would you determine the maximum positive and maximum negative value of a 10-bit two's complement number? (8–7)

18. Describe how to two's complement a binary number. (8–7)

19. Using the two's complement number system, briefly describe how to perform each of the following operations. (8–7)
    a. Add two positive numbers
    b. Add two negative numbers
    c. Add a positive and a negative number
    d. Subtract through addition
    e. Multiply through repeated addition
    f. Divide through repeated subtraction

Practice Problems

20. Convert the following to their decimal equivalents.

   a. 110111_2
   b. $2F_{16}$
   c. 10110_{10}

21. Convert the following binary numbers to hexadecimal.

   a. 111101101001
   b. 1011
   c. 111000
   d. 1111111

22. Convert the following decimal numbers to BCD.

   a. 2365
   b. 24

23. Give the ASCII codes for the following:

   a. ?
   b. $
   c. 6

24. In the home security system shown in Figure 8–43(a), a two-input logic gate is needed to actuate an alarm (output *Y*) if either the window is opened (input *A*) or the door is opened (input *B*). Which decision-making logic gate should be used in this application?

25. In the office temperature control system shown in Figure 8–43(b), a thermostat (input *A*) is used to turn ON and OFF a fan (output *Y*). A thermostat-enable switch is connected to the other input of the logic gate (input *B*), and it will either enable thermostat control of the fan or disable thermostat control of the fan. Which decision-making logic gate should be used in this application?

26. Add the following 8-bit numbers.

   a. 0110 1000 and 0001 1111
   b. 0111 1111 and 0100 0011
   c. 0001 1010 and 0011 1010

27. Convert all the following two's complement numbers to their decimal equivalents.

   a. 0001 0000
   b. 1011 1110
   c. 1111 0101

28. Convert each of the following decimal numbers to an 8-bit two's complement number.

   a. +119
   b. −119
   c. −34

29. Two's complement the following numbers.

   a. 0101 1010
   b. 1000 0001
   c. 1111 1001

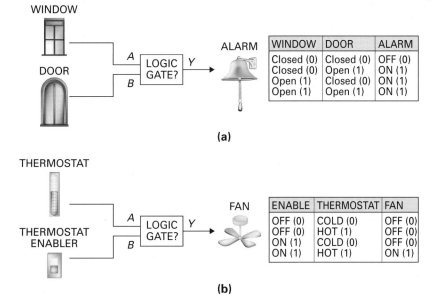

WINDOW	DOOR	ALARM
Closed (0)	Closed (0)	OFF (0)
Closed (0)	Open (1)	ON (1)
Open (1)	Closed (0)	ON (1)
Open (1)	Open (1)	ON (1)

(a)

ENABLE	THERMOSTAT	FAN
OFF (0)	COLD (0)	OFF (0)
OFF (0)	HOT (1)	OFF (0)
ON (1)	COLD (0)	OFF (0)
ON (1)	HOT (1)	ON (1)

(b)

FIGURE 8–43   **Decision-Making Logic Gates.**

Web Site Questions

Go to the Web site http://www.prenhall.com/cook, select the textbook *Introductory Mathematics*, select this chapter, and then follow the instructions when answering the multiple-choice practice problems.

Answers to Self-Test Evaluation Points

STEP 1-1

1. 10

2.
$$3 \times 10{,}000 = 30{,}000$$
$$5 \times 1{,}000 = 5{,}000$$
$$7 \times 100 = 700$$
$$2 \times 10 = 20$$
$$9 \times 1 = 9$$
$$35{,}729$$

3. See Section 1-1-3.

4.
499,997	
499,998	1
499,999	2
~500,000~	3
500,001	4
500,002	5

The units, tens, hundreds, thousands, and ten-thousands columns will reset and carry when the count is advanced by 3. The final count will be 500,002.

STEP 1-2

1. (a) $\dfrac{6}{8}$ (b) $\dfrac{4}{8}$ (c) $\dfrac{5}{16}$

2. (a) $\dfrac{2}{8}$ (b) $\dfrac{4}{8}$ (c) $\dfrac{11}{16}$

STEP 1-3

1. $\dfrac{4}{6} + \dfrac{1}{6} = \dfrac{4+1}{6} = \dfrac{5}{6}$

2. $\dfrac{3}{64} + \dfrac{2}{64} + \dfrac{25}{64} = \dfrac{3+2+25}{64} = \dfrac{30 \div 2}{64 \div 2} = \dfrac{15}{32}$

3. $\dfrac{3}{9} + \dfrac{6}{18} = \dfrac{6+6}{18} = \dfrac{12 \div 6}{18 \div 6} = \dfrac{2}{3}$

4. $\dfrac{1}{3} + \dfrac{1}{4} + \dfrac{3}{15} = ?$

$$
\begin{array}{c|ccc}
2 & 3 & 4 & 15 \\
2 & 3 & 2 & 15 \\
3 & 3 & 1 & 15 \\
5 & 1 & 1 & 5 \\
  & 1 & 1 & 1 \\
\end{array}
$$
$\longrightarrow 2 \times 2 \times 3 \times 5 = 60$

$\dfrac{1}{3} + \dfrac{1}{4} + \dfrac{3}{15} = \dfrac{20 + 15 + 12}{60} = \dfrac{47}{60}$

5. $\dfrac{3}{4} + \dfrac{4}{5} = \dfrac{15 + 16}{20} = \dfrac{31}{20} = 1\dfrac{11}{20}$

$$
\begin{array}{c|cc}
2 & 4 & 5 \\
2 & 2 & 5 \\
5 & 1 & 5 \\
  & 1 & 1 \\
\end{array}
$$
$\longrightarrow 2 \times 2 \times 5 = 20$

6. $\dfrac{9}{12} + \dfrac{4}{24} + \dfrac{3}{4} = \dfrac{18 + 4 + 18}{24} = \dfrac{40}{24} = 1\dfrac{16 \div 8}{24 \div 8} = 1\dfrac{2}{3}$

7. $\dfrac{7}{9} + \dfrac{5}{9} + \dfrac{4}{18} = \dfrac{14 + 10 + 4}{18} = \dfrac{28}{18} = 1\dfrac{10 \div 2}{18 \div 2} = 1\dfrac{5}{9}$

8. $\dfrac{15}{6} = 2\dfrac{3}{6} \dfrac{\div 3}{\div 3} = 2\dfrac{1}{2}$

9. $\dfrac{4}{24} = \dfrac{1}{6}$

10. $\dfrac{25 \div 25}{100 \div 25} = \dfrac{1}{4}$

STEP 1-4

1. $\dfrac{4}{32} - \dfrac{2}{32} = \dfrac{4-2}{32} = \dfrac{2 \div 2}{32 \div 2} = \dfrac{1}{16}$

2. $\dfrac{4}{5} - \dfrac{3}{10} = \dfrac{8-3}{10} = \dfrac{5 \div 5}{10 \div 5} = \dfrac{1}{2}$

3. $\dfrac{6}{12} - \dfrac{4}{16} = \dfrac{24-12}{48} = \dfrac{12}{48} = \dfrac{1}{4}$

$$
\begin{array}{c|cc}
2 & 12 & 16 \\
2 & 6 & 8 \\
2 & 3 & 4 \\
2 & 3 & 2 \\
3 & 3 & 1 \\
  & 1 & 1 \\
\end{array}
$$
$\longrightarrow 2 \times 2 \times 2 \times 2 \times 3 = 48$

4. $15\dfrac{1}{3} - 2\dfrac{4}{12} = 13\dfrac{4-4}{12} = 13\dfrac{0}{12} = 13$

5. $15\dfrac{15}{36} - 2\dfrac{31}{36} = 3\dfrac{15-31}{36} \leftarrow$ need to borrow

$= 4\left(\dfrac{36}{36}\right) + \dfrac{15}{36} - 2\dfrac{31}{36}$

$= 4\dfrac{51}{36} - 2\dfrac{31}{36} = 2\dfrac{51-31}{36} = 2\dfrac{20}{26} = 2\dfrac{5}{9}$

STEP 1-5

1. $\dfrac{1}{3} \times \dfrac{2}{3} = \dfrac{1 \times 2}{3 \times 3} = \dfrac{2}{9}$

2. $\dfrac{4}{8} \times \dfrac{2}{9} = \dfrac{8 \div 8}{72 \div 8} = \dfrac{1}{9}$

3. $\dfrac{4}{17} \times \dfrac{3}{34} = \dfrac{12}{578} = \dfrac{6}{289}$

4. $\dfrac{1}{16} \times 5\dfrac{3}{7} = \dfrac{1}{6} \times \dfrac{38}{7} = \dfrac{38}{112} = \dfrac{19}{56}$

5. $2\dfrac{1}{7} \times 7\dfrac{2}{3} = \dfrac{15}{7} \times \dfrac{23}{3} = \dfrac{345}{21} = 16\dfrac{9}{21} = 16\dfrac{3}{7}$

STEP 1-6

1. $\dfrac{2}{3} \div \dfrac{1}{12} = \dfrac{2}{3} \times \dfrac{12}{1} = \dfrac{24}{3} = 8$

2. $3\dfrac{1}{4} \div \dfrac{3}{4} = \dfrac{13}{4} \div \dfrac{3}{4} = \dfrac{13}{4} \times \dfrac{4}{3} = \dfrac{52}{12} = 4\dfrac{4}{12} = 4\dfrac{1}{3}$

3. $8\dfrac{16}{20} \div 4\dfrac{4}{20} = \dfrac{176}{20} \div \dfrac{84}{20} = \dfrac{176}{20} \times \dfrac{20}{84} = \dfrac{3520}{1680}$

$= 2\dfrac{160 \div 10}{1680 \div 10} = 2\dfrac{16 \div 4}{168 \div 4} = 2\dfrac{4 \div 2}{42 \div 2} = 2\dfrac{2}{21}$

4. $4\dfrac{3}{8} \div 2\dfrac{5}{16} = \dfrac{35}{8} \div \dfrac{37}{16} = \dfrac{35}{8} \times \dfrac{16}{37}$

$= \dfrac{560}{296} = 1\dfrac{264 \div 4}{296 \div 4} = 1\dfrac{66 \div 2}{74 \div 2} = 1\dfrac{33}{37}$

STEP 1-7

1. $\dfrac{16}{\cancel{27}} \times \dfrac{\cancel{27}}{240} = \dfrac{\cancel{16}}{1} \times \dfrac{1}{\cancel{240}} = \dfrac{1}{1} \times \dfrac{1}{15} = \dfrac{1}{15}$

2. $\dfrac{8}{\cancel{17}} \times \dfrac{3}{\cancel{4}} = \dfrac{2}{17} \times \dfrac{3}{1} = \dfrac{6}{17}$

3. $\dfrac{15}{32} \div \dfrac{30}{4} = \dfrac{\cancel{15}}{32} \times \dfrac{4}{\cancel{30}} = \dfrac{1}{\cancel{32}} \times \dfrac{\cancel{4}}{2}$

$= \dfrac{1}{8} \times \dfrac{1}{2} = \dfrac{1}{16}$

4. $6\dfrac{2}{3} \times 2\dfrac{7}{10} = \dfrac{\cancel{20}}{3} \times \dfrac{27}{\cancel{10}} = \dfrac{2}{\cancel{3}} \times \dfrac{\cancel{27}}{1}$

$= \dfrac{2}{1} \times \dfrac{9}{1} = \dfrac{18}{1} = 18$

5. $6\dfrac{1}{2} \div \dfrac{9}{32} = \dfrac{13}{\cancel{2}} \times \dfrac{\cancel{32}}{9} = \dfrac{13}{1} \times \dfrac{16}{9} = \dfrac{208}{9} = 23\dfrac{1}{9}$

STEP 2-1

1.
$$
\begin{array}{rcl}
1 \times & 100 & = 100 \\
7 \times & 10 & = 70 \\
8 \times & 1 & = 8 \\
6 \times & 0.1 & = 0.6 \\
4 \times & 0.01 & = 0.04 \\
9 \times & 0.001 & = \underline{0.009} \\
& & 178.649
\end{array}
$$

2. **(a)** $1\dfrac{5}{2} = 1 + (5 \div 2) = 1 + 2.5 = 3.5$

(b) $192\dfrac{3}{4} = 192 + (3 \div 4) = 192 + 0.75 = 192.75$

(c) $67\dfrac{6}{9} = 67 + (6 \div 9) = 67 + 0.667 = 67.667$

3. **(a)** $\dfrac{1}{2500} = 0.0004$

(b) $\dfrac{1}{0.25} = 4$

(c) $\dfrac{1}{0.000025} = 40,000$

4. **(a)** $7.25 = 7\dfrac{4}{16}\left(\dfrac{1}{16} = 0.0625\right)$

(b) $156.90625 = 156\dfrac{29}{32}\left(\dfrac{1}{32} = 0.03125\right)$

STEP 2-2

1. Addition, subtraction

2. **(a)** Subtraction **(c)** Division
  (b) Addition **(d)** Multiplication

3. **(a)**
$$
\begin{array}{r}
26.443 \\
197.1 \\
2.1103 \\
+0.004 \\
\hline
225.6573
\end{array}
$$

(c)
$$
\begin{array}{r}
894.357 \\
\times8.6 \\
\hline
5366142 \\
7154856 \\
\hline
7691.4702
\end{array}
$$

(b)
$$
\begin{array}{r}
19637.224 \\
-866.43 \\
\hline
18770.794
\end{array}
$$

(d) $0.015\overline{)1.3397}\,89.31333$

or $15\overline{)1339.7}$

4. $\dfrac{176 \div 8}{8 \div 8} = \dfrac{22}{1}$ or $22:1$ or 22.0 to 1 (ratio of twenty-two to one)

5. **(a)** 86.44 **(b)** 12,263,415.01 **(c)** 0.18

6. 86.43760 **(a)** 7
  (b) 6
  12,263,415.00510 **(a)** 13
  (b) 12
  0.176600 **(a)** 6
  (b) 4

7. **(a)** 0.075 **(b)** 220 **(c)** 0.235 **(d)** 19.2
8. **(a)** 35 **(b)** 4004

STEP 3-1

1. **(a)** $+3 - (-4) = +7$
  (b) $12 \times (+4) = +48$
  (c) $-5 \div (-7) = 0.7142$
  (d) $-0.63 \times (+6.4) = -4.032$
2. **(a)** $+6 + (-7)$
  (b) $-0.75 \times (18)$ or -0.75×18
3. **(a)** $\boxed{7}\ \boxed{.}\ \boxed{5}\ \boxed{+/-}\ \boxed{+}\ \boxed{4}\ \boxed{.}\ \boxed{6}\ \boxed{+/-}\ \boxed{=}$
  (b) $\boxed{5}\ \boxed{\times}\ \boxed{2}\ \boxed{+/-}\ \boxed{=}$
  (c) $\boxed{3}\ \boxed{1}\ \boxed{6}\ \boxed{.}\ \boxed{6}\ \boxed{2}\ \boxed{9}\ \boxed{\div}\ \boxed{1}\ \boxed{.}\ \boxed{4}\ \boxed{4}\ \boxed{+/-}\ \boxed{=}$

STEP 3-2

1. $+15 + (+4) = +19$
2. $-3 + (+45) = +42$
3. $+114 + (-111) = +3$
4. $-357 + (-74) = -431$
5. $17 + (15) = +32$
6. $-8 + (-177) = -185$
7. $+4600 + (-3400) = +1200$
8. $-6.25 + (+0.34) = -5.91$

STEP 3-3

1. $+18 - (+7) = +11$
2. $-3.4 - (-5.7) = +2.3$
3. $19,665 - (-5,031) = +24,696$
4. $-8 - (+5) = -13$
5. $467 - 223 = +244$
6. $-331 - (-2.6) = -328.4$
7. $8 - (+25) = -17$
8. $-0.64 - (-0.04) = -0.6$

STEP 3-4

1. $4 \times (+3) = +12$
2. $+17 \times (-2) = -34$
3. $-8 \times (+16) = -128$
4. $-8 \times (-5) = +40$
5. $+12.6 \times (+15) = +189$
6. $-3.3 \times (+1.4) = -4.62$
7. $+0.3 \times (-4) = -1.2$
8. $-4.6 \times (-3.3) = +15.18$

STEP 3-5

1. $+16.7 \div (+2.3) = 16.7 \div 2.3 = 7.26$
2. $+18 \div (+6) = 3$
3. $-6 \div (+2) = -3$
4. $+18 \div (-4) = -4.5$
5. $+2 \div (-8) = -0.25$
6. $-8 \div (+5) = -1.6$
7. $-15 \div (-5) = 3$
8. $0.664 \div (-0.2) = -3.32$

STEP 3-6

1. $+6 + (+3) + (-7) + (-5) = -3$
2. $+9 - (+2) - (-13) - (-4) = +24$
3. $-6 \times (-4) \times (-5) = -120$
4. $+8 \div (+2) \div (-5) = -0.8$

276 APPENDIX A / ANSWERS TO SELF-TEST EVALUATION POINTS

5. $-4 \div (-2) \times (+8) = +16$

6. $-9 + (+5) - (-7) = +3$

STEP 4-1

1. (a) $16^4 = 16 \times 16 \times 16 \times 16 = 65{,}536$

(b) $32^3 = 32 \times 32 \times 32 = 32{,}768$

(c) $112^2 = 112 \times 112 = 12{,}544$

(d) $15^6 = 15 \times 15 \times 15 \times 15 \times 15 \times 15 = 11{,}390{,}625$

(e) $2^3 = 2 \times 2 \times 2 = 8$

(f) $3^{12} = 3 \times 3 \times 3 \times 3 \times 3 \times 3 \times 3 \times 3 \times 3 \times 3 \times$

$3 \times 3 = 531{,}441$

2. (a) $\sqrt[2]{144} = 12$

(b) $\sqrt[3]{3375} = 15$

(c) $\sqrt[2]{20} = 4.47$

(d) $\sqrt[3]{9} = 2.08$

3. (a) $(9^2 + 14^2)^2 - \sqrt[3]{3 \times 7}$

$= (81 + 196)^2 - \sqrt[3]{21}$

$= 277^2 - 2.76$

$= 76{,}729 - 2.76 = 76{,}726.24$

(b) $\sqrt{3^2 \div 2^2} + \dfrac{151 - 9^2}{3.5^2}$

$= \sqrt{9 \div 4} + \dfrac{151 - 81}{12.25}$

$= \sqrt{2.25} + \dfrac{70}{12.25}$

$= 1.5 + 5.71 = 7.21$

STEP 4-2

1. (a) 10^2 **(d)** 10^6

(b) 10^0 **(e)** 10^{-3}

(c) 10^1 **(f)** 10^{-6}

2. (a) $6.3 \times 10^3 = 6.300. = 6300.0$ or 6300

(b) $114{,}000 \times 10^{-3} = 114.000. = 114.0$ or 114

(c) $7{,}114{,}632 \times 10^{-6} = 7.114632. = 7.114632$

(d) $6624 \times 10^6 = 6624.000000. = 6{,}624{,}000{,}000.0$

3. (a) $\sqrt{3 \times 10^6} = \sqrt{3{,}000{,}000} = 1732.05$

(b) $(2.6 \times 10^{-6}) - (9.7 \times 10^{-9}) = 0.0000025$ or 2.5×10^{-6}

(c) $\dfrac{(4.7 \times 10^3)^2}{3.6 \times 10^6} = (4.7 \times 10^3)^2 \div (3.6 \times 10^6) = 6.14$

4. (a) $47{,}000 = 47.000. = 47 \times 10^3$

(b) $0.00000025 = 0.000000250. = 250 \times 10^{-9}$

(c) $250{,}000{,}000 = 250.000000. = 250 \times 10^6$

(d) $0.0042 = 0.004.2 = 4.2 \times 10^{-3}$

STEP 4-3

1. (a) 10^3 **(d)** 10^6

(b) 10^{-2} **(e)** 10^{-6}

(c) 10^{-3}

2. (a) cm $\times$ 0.4 = inches, 15 cm $\times$ 0.4 = 6 inches

(b) kg $\times$ 2.2 = pounds, 23 kg $\times$ 2.2 = 50.6 pounds

(c) liters $\times$ 0.26 = gallons, 37 L $\times$ 0.26 = 9.62 gallons

(d) $\left(\dfrac{9}{5} \times {}^{\circ}\text{C}\right) + 32 = \left(\dfrac{9}{5} \times 23\right) + 32 = 73.4{}^{\circ}\text{F}$

3. (a) Miles/hour $\times$ 1.6 = kilometers/hour,

55 mph $\times$ 1.6 = 88 km/hour

(b) gallons $\times$ 3.8 = liters, 16 gallons $\times$ 3.8 = 60.8 liters

(c) square yards $\times$ 0.8 = square meters,

3 yd^2 $\times$ 0.8 = 2.4 m^2

(d) $\dfrac{5}{9} \times ({}^{\circ}\text{F} - 32) = \dfrac{5}{9} \times (92 - 32) = 33.33{}^{\circ}\text{C}$

4. (a) meter (m) **(f)** ampere (A)

(b) gram (g) **(g)** volt (V)

(c) degree Celsius ($^{\circ}$C) **(h)** liter (L)

(d) second (s) **(i)** joule (J)

(e) watt (W) **(j)** ohm (Ω)

5. (a) 25,000 volts = 25.000. $\times 10^3$ volts

= 25 kilovolts

(b) 0.014 watts = 0.014. = 14×10^{-3} watts

= 14 milliwatts

(c) 0.000016 microfarad = 0.000016×10^{-6} farad

= 0.000.016 $\times 10^{-9}$ farad

= 0.016×10^{-9} farad

= 0.016 nanofarad

STEP 5-1

1. Yes

2. No

3. $\dfrac{144}{12} \times \boxed{} = \dfrac{36}{6} \times 2 \times \boxed{} = 60$

$12 \times \boxed{} = 6 \times 2 \times \boxed{} = 60$

$12 \times 5 = 12 \times 5 = 60$

$\boxed{} = 5$

4. $\dfrac{(8 - 4) + 26}{5} = \dfrac{81 - 75}{2}$

$\dfrac{4 + 26}{5} = \dfrac{6}{2}$

$\dfrac{30}{5} = \dfrac{6}{2}$

$6 = 3$ (equation is not equal)

5. Yes

STEP 5-2

1. (a) $x + x = 2x$

(b) $x \times x = x^2$

(c) $7a + 4a = 11a$

(d) $2x - x = 1x$ or x

(e) $\dfrac{x}{x} = 1$

(f) $x - x = 0$

2. (a) $x + 14 = 30$

$x + 14 - 14 = 30 - 14$ (-14 from both sides)

$x = 30 - 14$

$x = 16$

(b) $8 \times x = \dfrac{80 - 40}{10} \times 12$

$8 \times x = 4 \times 12$

$8 \times x = 48$

$\dfrac{8 \times x}{8} = \dfrac{48}{8}$ ($\div 8$)

$x = \dfrac{48}{8}$

$x = 6$

(c)
$$y - 4 = 8$$
$$y - 4 + 4 = 8 + 4 \quad (+4)$$
$$y = 8 + 4$$
$$y = 12$$

(d)
$$(x \times 3) - 2 = \frac{26}{2}$$
$$(x \times 3) - 2 + 2 = \frac{26}{2} + 2 \quad (+2)$$
$$x \times 3 = \frac{26}{2} + 2$$
$$x \times 3 = 15$$
$$\frac{x \times 3}{3} = \frac{15}{3} \quad (\div 3)$$
$$x = \frac{15}{3}$$
$$x = 5$$

(e)
$$x^2 + 5 = 14$$
$$x^2 + 5 - 5 = 14 - 5 \quad (-5)$$
$$x^2 = 14 - 5$$
$$\sqrt{x^2} = \sqrt{14 - 5} \quad (\sqrt{\ })$$
$$x = \sqrt{14 - 5}$$
$$x = \sqrt{9}$$
$$x = 3$$

(f)
$$2(3 + 4x) = 2(x + 13)$$
$$6 + 8x = 2x + 26 \quad \text{(remove parentheses)}$$
$$6 + 8x - 2x = 2x + 26 - 2x \quad (-2x)$$
$$6 + (8x - 2x) = 26$$
$$6 + 6x = 26$$
$$6 + 6x - 6 = 26 - 6 \quad (-6)$$
$$6x = 26 - 6$$
$$6x = 20$$
$$\frac{6 \times x}{6} = \frac{20}{6} \quad (\div 6)$$
$$x = \frac{20}{6}$$
$$x = 3.3333$$

3. (a) $x + y = z, y = ?$
$$x + y - x = z - x \quad (-x)$$
$$y = z - x$$

(b) $Q = C \times V, C = ?$
$$\frac{Q}{V} = \frac{C \times V}{V} \quad (\div V)$$
$$\frac{Q}{V} = C$$
$$C = \frac{Q}{V}$$

(c) $X_L = 2 \times \pi \times f \times L, L = ?$
$$\frac{X_L}{2 \times \pi \times f} = \frac{2 \times \pi \times f \times L}{2 \times \pi \times f} \quad (\div 2 \times \pi \times f)$$
$$\frac{X_L}{2 \times \pi \times f} = L$$
$$L = \frac{X_L}{2 \times \pi \times f}$$

(d) $V = I \times R, R = ?$
$$\frac{V}{I} = \frac{I \times R}{I} \quad (\div I)$$
$$\frac{V}{I} = R$$
$$R = \frac{V}{I}$$

4. (a)
$$I^2 = 9$$
$$\sqrt{I^2} = \sqrt{9} \quad (\sqrt{\ })$$
$$I = \sqrt{9}$$
$$I = 3$$

(b)
$$\sqrt{Z} = 8$$
$$\sqrt{Z}^2 = 8^2$$
$$Z = 8^2$$
$$Z = 64$$

STEP 5-3

1. $x = y \times z$ and $a = x \times y, y = 14, z = 5, a = ?$
$$a = x \times y$$
$$a = y \times z \times y \quad \text{(substitute } y \times z \text{ for } x\text{)}$$
$$a = y^2 \times z \quad (y \times y = y^2)$$
$$a = 14^2 \times 5$$
$$a = 980$$

2. (a) For $I = \sqrt{\dfrac{P}{R}}$,
$$P = V \times I \text{ to } P = I^2 \times R$$
$$\frac{P}{R} = \frac{I^2 \times R}{R} \quad (\div R)$$
$$\frac{P}{R} = I^2$$
$$\frac{P}{R} = \sqrt{I^2} \quad (\sqrt{\ })$$
$$\sqrt{\frac{P}{R}} = I$$
$$I = \sqrt{\frac{P}{R}}$$

(b) For $I = \dfrac{P}{V}$,
$$P = V \times I$$
$$\frac{P}{V} = \frac{V \times I}{V} \quad (\div V)$$
$$\frac{P}{V} = I$$
$$I = \frac{P}{V}$$

(c) For $V = \sqrt{P \times R}$
$$P = V \times I \text{ to } P = \frac{V^2}{R}$$
$$P \times R = \frac{V^2 \times R}{R} \quad (\times R)$$
$$P \times R = V^2$$
$$\sqrt{P \times R} = \sqrt{V^2} \quad (\sqrt{\ })$$
$$\sqrt{P \times R} = V$$
$$V = \sqrt{P \times R}$$

(d) For $R = \dfrac{V^2}{P}$,
$$P = V \times I \text{ to } P = \frac{V^2}{R}$$
$$P \times R = \frac{V^2}{R} \times R \quad (\times R)$$
$$P \times R = V^2$$
$$\frac{R \times R}{R} = \frac{V^2}{P}$$
$$R = \frac{V^2}{P}$$

3. What percentage of 12 gives 2.5?

$$x \times 12 = 2.5$$
$$\frac{x \times 12}{12} = \frac{2.5}{12} \div 12$$
$$x = \frac{2.5}{12}$$
$$x = 0.2083 \quad (0.2083 \times 12 = 2.5)$$
$$x\% = 20.83\% \quad (20.83\% \times 12 = 2.5)$$

4. 60 is 22% of what number?

$$60 = 22\% \times x$$
$$60 = 0.22 \times x$$
$$\frac{60}{0.22} = \frac{0.22 \times x}{0.22} \quad (\div 0.22)$$
$$\frac{60}{0.22} = x$$
$$x = 272.73 \quad (60 = 0.22 \times 272.73$$
$$\text{or } 60 = 22\% \times 272.73)$$

STEP 5-4

1. $y = -2x + 3$
2. $(x - 5)(x - 3)$
3. $x = 0$ or $x = -3$
4. $16x^8y^4$

STEP 5-5

1. Parentheses, Exponents, Multiplication, Division, Addition, Subtraction.
2. Monomial: $3x$
Binomial: $3x + 4y$
Trinomial: $3x + 4y + 5z$
Polynomial: $2a + 2b + 2c + 2d + 2e$
3. **(a)** $a \div 1 = a$
(b) $a + a = 2a$
(c) $a - a = 0$
(d) $a \times a = a^2$
(e) $3x + b - 2x = x + b$
4. **(a)** $16 + a = 5 \times a$
$a = 4$
(b) $\frac{42}{b} = 3 \times 7$
$b = 2$

STEP 6-1

1. **(a)** 40° **(d)** 90° **(g)** 165° **(j)** 165°
(b) 60° **(e)** 115° **(h)** 180° **(k)** 65°
(c) 80° **(f)** 135° **(i)** 15° **(l)** 90°

2. **(a)** Acute **(d)** Right **(g)** Obtuse **(j)** Obtuse
(b) Acute **(e)** Obtuse **(h)** Obtuse **(k)** Acute
(c) Acute **(f)** Obtuse **(i)** Acute **(l)** Right

3. (See Figure 6-1)

STEP 6-2

1. (Section 6-2 Introduction)
2. **(a)** Figure 6-2 **(e)** Figure 6-6 **(i)** Figure 6-12
(b) Figure 6-3 **(f)** Figure 6-7 **(j)** Figure 6-13
(c) Figure 6-4 **(g)** Figure 6-10
(d) Figure 6-5 **(h)** Figure 6-11

3. **(a)** $A = \sqrt{C^2 - B^2}$
$= \sqrt{3^2 - 2^2}$
$= \sqrt{9 - 4}$
$= \sqrt{5}$
$= 2.24$ feet
(b) $B = \sqrt{C^2 - A^2}$
$= \sqrt{160^2 - 80^2}$

$= \sqrt{25,600 - 6400}$
$= \sqrt{19,200}$
$= 138.56$ km

(c) $C = \sqrt{A^2 + B^2}$
$= \sqrt{112^2 + 25^2}$
$= \sqrt{12,544 + 625}$
$= \sqrt{13,169}$
$= 114.76$ mm

4. **(a)** $x = \sqrt{A^2 + B^2}$
$= \sqrt{40^2 + 30^2}$
$= \sqrt{1600 + 900}$
$= \sqrt{2500}$
$= 50$ volts
(b) $x = \sqrt{A^2 + B^2}$
$= \sqrt{75^2 + 26^2}$
$= \sqrt{5625 + 676}$
$= \sqrt{6301}$
$= 79.38$ watts
(c) $x = \sqrt{A^2 + B^2}$
$= \sqrt{93^2 + 36^2}$
$= \sqrt{8649 + 1296}$
$= \sqrt{9945}$
$= 99.72$ mm
$y = \sqrt{A^2 + B^2}$
$= \sqrt{48^2 + 96^2}$
$= \sqrt{2304 + 9216}$
$= \sqrt{11,520}$
$= 107.33$ mm

STEP 6-3

1. **(a)** $O = 35$ mm $\qquad \theta = 36°$
$H = ?$

SOH or $\sin\theta = \dfrac{O}{H}$

$\sin 36° = \dfrac{35 \text{ mm}}{H}$

$0.59 = \dfrac{35}{H}$

$0.59 \times H = \dfrac{35}{H} \times H \quad (\times H)$

$\dfrac{0.59 \times H}{0.59} = \dfrac{35}{0.59} \quad (\div 0.59)$

$H = \dfrac{35}{0.59} = 59.32$ mm

(b) $H = 160$ km $\qquad \theta = 38°$
$A = ?$

CAH or $\cos\theta = \dfrac{A}{H}$

$\cos 38° = \dfrac{A}{160 \text{ km}}$

$0.79 = \dfrac{A}{160}$

$0.79 \times 160 = \dfrac{A}{160} \times 160 \quad (\times 160)$

$0.79 \times 160 = A$

$A = 126.4$ km

(c) $\textcircled{O}$= 163 cm $\qquad \theta = 72°$
$\textcircled{A}$= ?

$\downarrow$

$TOA \qquad$ or $\qquad \tan\theta = \dfrac{O}{A}$

$$\tan 72° = \dfrac{163\ \text{cm}}{A}$$

$$3.08 = \dfrac{163}{A}$$

$$3.08 \times A = \dfrac{163 \times \cancel{A}}{\cancel{A}} \qquad (\times A)$$

$$\dfrac{3.08 \times A}{3.08} = \dfrac{163}{3.08} \qquad (\div 3.08)$$

$$A = \dfrac{163}{3.08} = 52.92\ \text{cm}$$

2. (a) $\textcircled{H}$= 120 miles $\qquad \theta = ?$
$\textcircled{O}$= 38 miles

$\downarrow$

$SOH \qquad$ or $\qquad \sin\theta = \dfrac{O}{H}$

$$\sin\theta = \dfrac{38\ \text{miles}}{120\ \text{miles}}$$

$$\sin\theta = 0.32$$

$$\cancel{\text{invsin}} \times \cancel{\text{sin}}\ \theta = \text{invsin}\ 0.32 \quad (\times\text{invsin})$$

$$\theta = \text{invsin}\ 0.32$$

$$\theta = 18.66°$$

(b) $\textcircled{H}$= 25 feet $\qquad \theta = ?$
$\textcircled{A}$= 17 feet

$\downarrow$

$CAH \qquad$ or $\qquad \cos\theta = \dfrac{A}{H}$

$$\cos\theta = \dfrac{17\ \text{feet}}{25\ \text{feet}}$$

$$\cos\theta = 0.68$$

$$\cancel{\text{invcos}} \times \cancel{\text{cos}}\ \theta = \text{invcos}\ 0.68$$

$$\theta = \text{invcos}\ 0.68$$

$$\theta = 47.16°$$

(c) $\textcircled{A}$= 69 cm $\qquad \theta = ?$
$\textcircled{O}$= 51 cm

$\downarrow$

$TOA \qquad$ or $\qquad \tan\theta = \dfrac{O}{A}$

$$\tan\theta = \dfrac{51\ \text{cm}}{69\ \text{cm}}$$

$$\tan\theta = 0.74$$

$$\cancel{\text{invtan}} \times \cancel{\text{tan}}\ \theta = \text{invtan}\ 0.74$$

$$\theta = \text{invtan}\ 0.74$$

$$\theta = 36.5°$$

STEP 6-4

1. (Section 6-4 Introduction)
2. (Section 6-4-1)
3. (a) Figure 6-22 $\qquad$ **(c)** Figure 6-24 $\qquad$ **(e)** Figure 6-26
$\quad$ **(b)** Figure 6-23 $\qquad$ **(d)** Figure 6-25

4. A frustum is the base section of a solid pyramid or cone.

STEP 7-1

1. Exponent
2. (c) 4
3. (a) 1 $\qquad$ **(c)** 1.4 $\qquad$ **(e)** 4.02 $\qquad$ **(g)** -0.456
$\quad$ **(b)** 2 $\qquad$ **(d)** 2.176 $\qquad$ **(f)** 3.57 $\qquad$ **(h)** 0.243
4. (a) 56.23 $\qquad$ **(c)** 100,000 $\qquad$ **(e)** 316,227.76 **(g)** 1.26
$\quad$ **(b)** 223.87 $\qquad$ **(d)** 10 $\qquad$ **(f)** 1.78 $\qquad$ **(h)** 1^{10}

STEP 7-2

1.

FIGURE 7–2–1

2.

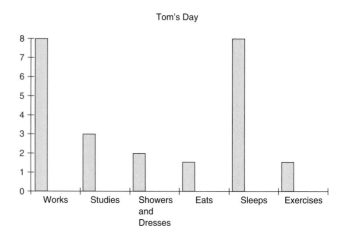

FIGURE 7–2–2

3. Graphing calculators have all the functions of scientific calculators, and in addition, they can graph functions.
4. (a) Y-Editor: Key is used to enter equation.
   (b) Table: Shows every possible result for all possible input values.
   (c) Graph: Displays the plotted data in graphical form.

STEP 8-1
1. Section 8-1-1
2. Section 8-1-1
3. Section 8-1-2
4. Circuit Simplicity and Accuracy
5. 2
6. Section 8-1-2

STEP 8-2
1. CPU, Memory, Input-Output
2. Analog to Digital Converter, Digital Analog Converter.
3. Analog, Digital
4. Digital, Analog

STEP 8-3
1. 26_{10}
2. 10111_2
3. LSB = least significant bit, MSB = most significant bit
4. 1101110_2

STEP 8-4
1. (a) Base 10   (b) Base 2   (c) Base 16
2. 21_{10} 15_{16}
3. BF7A
4. Binary equivalent = 100001
   Hexadecimal equivalent = 21

STEP 8-5
1. $0111\ 0110\ 0010\ 1001_{BCD}$
2. 11_{10}
3. Only one digit changes as you step to the next code group.
4. B a b b a g e

STEP 8-6
1. (a) AND   (c) OR   (e) NOR
   (b) XOR   (d) NAND   (f) NAND
2. $\overline{A \cdot B} = \overline{A} + \overline{B}$
3. (a) $\overline{AB}$ (b) $(\overline{A} \cdot \overline{B}) \cdot C$
4. (a)
   (b)
   (c)
   (d)

STEP 8-7
1. $1011 + 11101 = 101000$
2. 1001 0101
3. 0111 1001
4. −5
5. +6

Answers to Odd-Numbered Problems

Chapter 1
1. b
3. c
5. b
7. d
9. c
11. c

(The answers to Communication Skill Questions 13 through 23 can be found in the indicated sections that follow the questions.)

25. (a) 276   (b) 8070   (c) 20,049
27. (a) $3\frac{3}{4}$ (b) $2\frac{1}{3}$
29. (a) $\dfrac{5}{8} + \dfrac{2}{8} = \dfrac{5+2}{8} = \dfrac{7}{8}$

    (b) $\dfrac{47}{76} + \dfrac{15}{76} + \dfrac{1}{76} = \dfrac{63}{76}$

31. (a) $\dfrac{5}{2} = ?$, 2 into 5 = 2 with 1 remaining $\dfrac{5}{2} = 2\dfrac{1}{2}$

    (b) $\dfrac{17}{4} = 4\dfrac{1}{4}$

    (c) $\dfrac{25}{16} = 1\dfrac{9}{16}$

    (d) $\dfrac{37}{3} = 12\dfrac{1}{3}$

33. (a) $\dfrac{4 \div 4}{16 \div 4} = \dfrac{1}{4}$ (c) $\dfrac{74 \div 2}{128 \div 2} = \dfrac{37}{64}$

    (b) $\dfrac{16 \div 2}{18 \div 2} = \dfrac{8}{9}$ (d) $\dfrac{28}{45} = \dfrac{28}{45}$

35. (a) $\dfrac{1}{9} \times 4 = \dfrac{1 \times 4}{9 \times 1} = \dfrac{4}{9}$

    (b) $\dfrac{3}{6} \times \dfrac{4}{5} = \dfrac{3 \times 4}{6 \times 5} = \dfrac{12 \div 6}{30 \div 6} = \dfrac{2}{5}$

    (c) $\dfrac{1}{3} \times 4\dfrac{1}{2} = \dfrac{1}{3} \times \dfrac{9}{2} = \dfrac{9}{6} = 1\dfrac{3}{6} = 1\dfrac{1}{2}$

    (d) $2\dfrac{3}{4} \times 4\dfrac{4}{11} = \dfrac{\cancel{11}}{\cancel{4}} \times \dfrac{\cancel{48}}{\cancel{11}} = \dfrac{1}{1} \times \dfrac{12}{1} = 1 \times 12 = 12$

Chapter 2
1. a
3. b
5. a
7. c

(The answers to Communication Skill Questions 9 through 17 can be found in the indicated sections that follow the questions.)

19. (a) 2.3   (b) 0.507   (c) 9030.04
21. (a) $\dfrac{16}{32} \div \dfrac{2}{2} = \dfrac{1}{2} = 1 \div 2 = 0.5$

    (b) $3\dfrac{8}{9} = 3 + (8 \div 9) = 3 + 0.889 = 3.889$

    (c) $4\dfrac{9}{8} = 4 + \dfrac{8}{8} + \dfrac{1}{8} = 4 + 1 + \dfrac{1}{8} = 5\dfrac{1}{8} = 5 + (1 \div 8)$

    $= 5 + 0.125 = 5.125$

    (d) $195\dfrac{7}{3} = 195 + \dfrac{3}{3} + \dfrac{3}{3} + \dfrac{1}{3} = 197\dfrac{1}{3} = 197 + (1 \div 3)$

    $= 197 + 0.333 = 197.333$

23. (a) $0.777 = \dfrac{7}{9}$

    (b) $0.6149069 = \dfrac{99}{161}$

    (c) $43.125 = 43\dfrac{1}{8}$

25. (a) $\dfrac{20 \text{ ft}}{5 \text{ ft}} \leftarrow$ Reduce to lowest terms

    $\dfrac{20 \div 5}{5 \div 5} = \dfrac{4}{1}$

    The ratio of 20 ft to 5 ft is 4 to 1 (4:1).

    (b) Both quantities must be alike, so we must first convert minutes to seconds.

    $2\dfrac{1}{2} \text{ min} = 2\dfrac{1}{2} \times 60 \text{ s} = 150 \text{ s}$

    $\dfrac{150 \text{ s} \div}{30 \text{ s} \div} = \dfrac{5}{1}$

    The ratio of $2\dfrac{1}{2}$ min to 30 s is 5 to 1 (5:1).

27. 10.9
29. (a) 6

    (b) 17.25 or $17\dfrac{1}{4}$ s

    (c) 116 V

    (d) 505.75, or $505\dfrac{3}{4}$ Ω

    (e) 155 m

Chapter 3
1. c
3. a
5. a
7. b
9. a

(The answers to Communication Skill Questions 11 through 20 can be found in the indicated sections that follow the questions.)

21. (a) $+8 - (-5)$
    (b) -0.6×13
    (c) $-22.3 \div (-17)$
    (d) $(+4) \div (-9)$

23. (a) $+6 - (+8) = -2$
    (b) $+9 - (-6) = +15$
    (c) $-75 - (+62) = -137$
    (d) $-39 - (-112) = +73$

25. (a) $+19 \div (+3) = 6.33$
    (b) $+36 \div (-3) = -12$
    (c) $-80 \div (+5) = -16$
    (d) $-44 \div (-2) = 22$

27. $+15 \div (+5) \times (-3.5) = -10.5$
29. $-6 \div (-4) \div (-3) \times (-15) = 7.5$

Chapter 4
1. c
3. a
5. d
7. a
9. b

(The answers to Communication Skill Questions 11 through 20 can be found in the indicated sections that follow the questions.)

21. (a) $9^2 = 9 \times 9 = 81$
  (b) $6^2 = 6 \times 6 = 36$
  (c) $2^2 = 2 \times 2 = 4$
  (d) $0^2 = 0 \times 0 = 0$
  (e) $1^2 = 1 \times 1 = 1$
  (f) $12^2 = 12 \times 12 = 144$
23. (a) $9^3 = 9 \times 9 \times 9 = 729$
  (b) $10^4 = 10 \times 10 \times 10 \times 10 = 10,000$
  (c) $4^6 = 4 \times 4 \times 4 \times 4 \times 4 \times 4 = 4096$
  (d) $2.5^3 = 2.5 \times 2.5 \times 2.5 = 15.625$
25. (a) $\dfrac{1}{100} = 10^{-2}$
  (b) $1,000,000,000 = 10^9$
  (c) $\dfrac{1}{1000} = 10^{-3}$
  (d) $1000 = 10^3$

27.

	Scientific Notation Base of 1 to 10, plus power of ten	Engineering Notation Base greater than 1, plus $\times$ 3 power of ten
(a) 475	$= 4.75 \times 10^2$	475.0
(b) 8200	$= 8.2 \times 10^3$	8.2×10^3
(c) 0.07	$= 7 \times 10^{-2}$	70×10^{-3}
(d) 0.00045	$= 4.5 \times 10^{-4}$	450×10^{-6}

29. (a) km $\times$ 0.6 = mi, 100 km $\times$ 0.6 = 60 mi
  (b) $m^2 \times 1.2 = yd^2$, 29 $m^2 \times 1.2 = 34.8 \ yd^2$
  (c) kg $\times$ 2.2 = lb, 67 kg $\times$ 2.2 = 147.4 lb
  (d) L $\times$ 2.1 = pt, 2 L $\times$ 2.1 = 4.2 pt
31. (a) $\dfrac{5}{9} \times (°F - 32) = °C$

$= \dfrac{5}{9} \times (32°F - 32) = 0.5555 \times 0 = 0°C$

$32°F = 0°C$

  (b) $\dfrac{5}{9} \times (72°F - 32)$

$= \dfrac{5}{9} \times 40 = 22.22°C$

33. (a) meter
  (b) square meter
  (c) gram
  (d) cubic meter
  (e) liter
  (f) degrees Celsius
35. (a) 8000 ms = _____ μs
  $= 8000 \times 10^{-3} = $ _____ 10^{-6}
  $= 8000\underbrace{.000.} \times 10^{-6} = 8,000,000 \times 10^{-6}$ s
  $= 8000$ ms $= 8,000,000 \ \mu$s
  (b) 0.02 MV = _____ kV
  0.02×10^6 V = _____ $\times 10^3$ V
  $0.\underbrace{020.} \times 10^3$ V $= 20 \times 10^3$ V
  0.02 MV = 20 kV
  (c) 10 km = _____
  10×10^3 m = _____ $\times$ 100 m
  $10.\underbrace{000.} \times 100$ m = 10,000 m
  10 km = 10,000 m
  (d) 250 mm = _____ cm
  250×10^{-3} m = _____ $\times 10^{-2}$ m
  $25.\underbrace{0.} \times 10^{-3}$ m $= 25 \times 10^{-2}$ m
  250 mm = 25 cm

Chapter 5
1. a
3. d
5. c
7. c
9. b

(The answers to Communication Skill Questions 11 through 25 can be found in the indicated sections that follow the questions.)

27. (a) $4x = 11$

$4 \times x = 11$

$\dfrac{\cancel{4} \times x}{\cancel{4}} = \dfrac{11}{4} \qquad (\div 4)$

$x = \dfrac{11}{4}$

$x = 2.75$

  (b) $6a + 4a = 70$

$10a = 70$

$10 \times a = 70$

$\dfrac{\cancel{10} \times a}{\cancel{10}} = \dfrac{70}{10} \qquad (\div 10)$

$a = \dfrac{70}{10}$

$a = 7$

  (c) $5b - 4b = \dfrac{7.5}{1.25}$

$1b = \dfrac{7.5}{1.25} \qquad (1b = 1 \times b = b)$

$b = \dfrac{7.5}{1.25}$

$b = 6$

  (d) $\dfrac{2z \times 3z}{4.5} = 2z$
  $\left[\begin{array}{l} 2z \times 3z = 2 \times z \times 3 \times z \\ \qquad = (2 \times 3) \times (z \times z) \\ \qquad = 6 \times z^2 \\ \qquad = 6z^2 \end{array}\right]$

$\dfrac{6z^2}{4.5} = 2z$

$\dfrac{6z^2}{\cancel{4.5}} \times \cancel{4.5} = 2z \times 4.5 \qquad (\times 4.5)$

$6z^2 = 2z \times 4.5$

$\dfrac{{}^3\cancel{6} \times \cancel{z} \times z}{\cancel{2} \times \cancel{z}} = \dfrac{(2 \times z) \times 4.5}{2 \times z} \qquad (\div 2z)$

$3z = 4.5$

$\dfrac{\cancel{3} \times z}{\cancel{3}} = \dfrac{4.5}{3} \qquad (\div 3)$

$z = \dfrac{4.5}{3}$

$z = 1.5$

29. (a) Power (P) = voltage $(V) \times$ current (I)
  1500 watts = 120 volts $\times$?
  Transpose formula to solve for I.
  $P = V \times I$
  $\dfrac{P}{V} = \dfrac{\cancel{V} \times I}{\cancel{V}} \qquad (\div V)$
  $\dfrac{P}{V} = I$
  $I = \dfrac{P}{V}$

Current (I) in amperes $= \dfrac{\text{power } (P) \text{ in watts}}{\text{voltage } (V) \text{ in volts}}$

$I = \dfrac{1500 \text{ W}}{120 \text{ V}}$

Current $(I) = 12.5$ amperes

(b) Voltage (V) in volts $=$ current (I) in amperes $\times$ resistance (R) in ohms

$$120 \text{ volts} = 12.5 \text{ amperes} \times \text{?}$$

Transpose formula to solve for R.

$$V = I \times R$$

$$\frac{V}{I} = \frac{\cancel{I} \times R}{\cancel{I}}$$

$$\frac{V}{I} = R$$

$$R = \frac{V}{I}$$

$$\text{Resistance } (R) \text{ in ohms} = \frac{\text{voltage } (V) \text{ in volts}}{\text{current } (I) \text{ in amperes}}$$

$$R = \frac{120 \text{ V}}{12.5 \text{ A}}$$

$$\text{Resistance } (R) = 9.6 \text{ ohms}$$

Chapter 6

1. b	**9.** c
3. c	**11.** d
5. d	**13.** a
7. b	**15.** d

(The answers to Communication Skill Questions 17 through 25 can be found in the indicated sections that follow the questions.)

27. (a) $C = \sqrt{A^2 + B^2}$
$= \sqrt{20^2 + 53^2}$
$= \sqrt{400 + 2809}$
$= \sqrt{3209}$
$= 56.65 \text{ mi}$

(b) $C = \sqrt{2^2 + 3^2}$
$= 3.6 \text{ km}$

(c) $C = \sqrt{4^2 + 3^2}$
$= 5 \text{ inches}$

(d) $C = \sqrt{12^2 + 12^2}$
$= 16.97 \text{ mm}$

29. (a) $\sin 0° = 0$
(b) $\sin 30° = 0.5$
(c) $\sin 45° = 0.707$
(d) $\sin 60° = 0.866$
(e) $\sin 90° = 1.0$
(f) $\cos 0° = 1.0$
(g) $\cos 30° = 0.866$
(h) $\cos 45° = 0.707$
(i) $\cos 60° = 0.5$
(j) $\cos 90° = 0$
(k) $\tan 0° = 0$
(l) $\tan 30° = 0.577$
(m) $\tan 45° = 1.0$
(n) $\tan 60° = 1.73$
(o) $\tan 90° = \infty$ (infinity)

31. (a)

Known values	Unknown values
$H = 27$ miles	$O = \text{?}$
$\theta = 37°$	$A = \text{?}$

We must calculate the length of the opposite and adjacent sides. To achieve this we can use either H and θ to calculate O (SOH), or H and θ to calculate A (CAH).

$$\sin \theta = \frac{O}{H}$$

$$\sin 37° = \frac{O}{27}$$

$$0.6 = \frac{O}{27}$$

$$0.6 \times 27 = \frac{O}{\cancel{27}} \times \cancel{27}$$

$$O = 0.6 \times 27$$

$$\text{Opposite} = 16.2 \text{ mi}$$

The next step is to use the Pythagorean theorem to calculate the length of the unknown side now that the length of two sides are known.

$$B = \sqrt{C^2 - A^2}$$
$$= \sqrt{27^2 - 16.2^2}$$
$$= \sqrt{729 - 262.44}$$
$$= \sqrt{466.56}$$
$$\text{Adjacent or } B = 21.6 \text{ mi}$$

(b)

Known values	Unknown values
$O = 29$ cm	$H = \text{?}$
$\theta = 21°$	$A = \text{?}$

We can use either O and θ to calculate H (SOH), or O and θ to calculate A (TOA).

$$\tan \theta = \frac{O}{A}$$

$$\tan 21° = \frac{29 \text{ cm}}{A}$$

$$0.384 = \frac{29}{A}$$

$$0.384 \times A = \frac{29}{\cancel{A}} \times \cancel{A} \, (\times A)$$

$$\frac{0.384 \times A}{0.384} = \frac{29}{0.384} \, (\div 0.384)$$

$$A = \frac{29}{0.384}$$

$$A = 75.5 \text{ cm}$$

Now that O and A are known, we can calculate H.

$$C \text{ (or } H) = \sqrt{A^2 + B^2}$$
$$= \sqrt{75.5^2 - 29^2}$$
$$= \sqrt{5700.25 + 841}$$
$$= \sqrt{6541.25}$$
$$= 80.88 \text{ cm}$$

(c)

Known values	Unknown values
$H = 34$ volts	$\theta = ?$
$A = 18.5$ volts	$O = ?$

Because both H and A are known, we can use cosine to calculate θ (*CAH*)

$$\cos \theta = \frac{A}{H}$$
$$\cos \theta = \frac{18.5 \text{ V}}{34 \text{ V}}$$
$$\cos \theta = 0.544$$
$$\cancel{\text{invcos}} \times \cancel{\cos} \theta = \text{invcos } 0.544 \qquad (\times \text{invcos})$$
$$\theta = \text{invcos } 0.544$$
$$\theta = 57°$$

To calculate the length of the unknown side, we can use the Pythagorean theorem.

$$B = \sqrt{C^2 - A^2}$$
$$= \sqrt{34^2 - 18.5^2}$$
$$= \sqrt{813.75}$$
$$= 28.5 \text{ V}$$

33. $S = \pi d^2 = \pi \times 2^2 = 12.57 \text{ in}^2, V = \frac{1}{6}\pi \times d^3 = 4.189 \text{ in}^3$

35. $L = \frac{1}{2}ps$

$$= \frac{1}{2} \times 24 \times 8 = 96 \text{ in}^2$$

Chapter 7

1. b	**7.** b
3. c	**9.** c
5. a	

(The answers to Communication Skill Questions 10 through 17 can be found in the indicated sections that follow the questions.)

21.
(a) 1	**(g)** −0.3
(b) 4	**(h)** −0.125
(c) 6.09	**(i)** −0.9
(d) 1.7	**(j)** −0.02
(e) 0.176	**(k)** 1.088
(f) −0.6	**(l)** 1.88

23. *Step 1:* Find the log of each number.
Step 2: Change each number to base 10.
Step 3: Add the base 10 exponents.
Step 4: Convert the log answer back to a number using antilog.

25. *Step 1:* Find the log of the dividend and divisor.
Step 2: Change each number to base 10.
Step 3: Subtract the log of the divisor from the log of the dividend.
Step 4: Convert the log answer back to a number using antilog.

27. (a) Line graph **(b)** Bar graph

29. (a) Children are able to access educational material.
　(b) 1. Children are able to access educational material.
　　2. Video conferencing
　　3. Accessing information on hobbies or interests
　　4. Accessing information for work
　　5. Sending notes to friends and family

CHAPTER 8

1. b	**7.** a
3. b	**9.** c
5. d	

(The answers to Communication Skill Questions 10 through 19 can be found in the indicated sections that follow the questions.)

21. (a) F69
　(b) B
　(c) 38
　(d) 7F
23. ASCII codes are:
　(a) ? = 0111111
　(b) $ = 0100100
　(c) b = 0110110
25. AND gate
27. (a) +16
　(b) −66
　(c) −11
29. (a) 1010 0110
　(b) 0111 1111
　(c) 0000 0011

Index

Calculator Keys
Add key, 43
Change-sign key, 66
Common antilogarithm, 211
Common logarithm, 209
Conversion functions, 41
Convert key, 109
Converting number bases, 242
Cosine key, 191
Decimal-point key, 33
Divide key, 43
Equals key, 43
Expand function, 134
Exponent key, 95
Factor function, 133
Inverse cosine, 192
Inverse sine, 190
Inverse tangent, 194
Multiply key, 43
Normal, scientific, engineering modes, 99
Order of operations, 80
Percent key, 58
Pi key, 39
Reciprocal key, 38
Round function, 56
Sine key, 189
Solve function key, 144
Square key, 88
Square root key, 89
Subtract key, 43
Tangent key, 193
Xth root of the Y key, 90
Y to the X power key, 88

A
Acute angle, 174
Addend, 45
Adding fractions, 9
Adding mixed numbers, 14
Adding positive and negative numbers, 67
Addition, 43
Adjacent, 188
Algebra rules, 163
Algebra, 118
Alphanumeric code, 245
Ammeter, 113
Analog data, 226
Analog readout, 229
Analog, 227
Analog-to-digital converter, 231, 237
AND gate, 248
Antilogarithm, 211
Apex, 201

Arabic number system, 32
Arcsine, 190
Arithmetic-logic unit, 263
Armstrong, Neil, 224
ASCII, 245
Associative law of addition, 44
Associative law of multiplication, 49
Augend, 45
Averages, 59

B
Babbage, Charles, 118
Bar graph, 215
Base, 4
Basics of algebra, 120
Binary addition, 256
Binary coded decimal, 244
Binary codes, 243
Binary number system, 232
Binary point, 232
Binary to decimal, 234
Binary to hexidecimal, 241
Binary, 229
Bit, 229
Boolean, 247
Borrowing, 19

C
Calculating in decimal, 42
Calculator's table of common logarithms, 209
Canceling fractions, 26
Carry, 44
Cartesian coordinate system, 154
Circle formula example, 146
Circle, 185
Circumference, 146
Coefficient, 154
Common denominator, 10
Common logarithms, 208
Commutative law of addition, 44
Commutative law of multiplication, 49
Computer math, 224
Cone, 201
Converting decimal fractions to written fractions, 39
Converting to powers of ten, 95
Converting written fractions to decimal fractions, 35
Cosine of theta, 191
Customary system, 100
Cylinder, 199

D
Decimal calculation, 30
Decimal fraction, 32
Decimal point, 32
Decimal to binary, 235
Decimal to hexadecimal, 240
Decimus, 32
Degrees Celsius, 107
Degrees Fahrenheit, 107
Delta, 156
DeMorgan's first theorem, 251
DeMorgan's second theorem, 253
Denominator, 7
Descartes, René, 84
Diameter, 146
Difference of perfect cubes, 161
Difference of perfect squares, 161
Difference, 46
Digital data, 227
Digital readout, 229
Digital-to-analog converter, 231, 237
Digitize, 229
Directly proportional, 122
Distributing, 132
Distributive law of multiplication, 133
Dividend, 52
Dividing fractions, 23
Dividing mixed fractions, 24
Dividing positive and negative numbers, 76
Division of exponents, 159
Division, 52
Divisor, 52

E
Electrical conversions, 108
Electrical prefixes, 108
Electrical units, 108
Engineering notation, 97
Equality on both sides of the equal sign, 120
Equation, 120
Equations, 153
Equilateral, 184
Exponents raised to a power, 159
Exponents, 84, 158
Expressing positive and negative numbers, 65
Extra zeroes, 35

F
Factor, 12
Factoring patterns, 160

Factoring, 132
Fermat, Pierre de, 170
Floating point number system, 98
Formula, 120
Fraction bar, 7
Fractional power, 159
Fractions, 6
Frequency, 165
Frustum, 201

G

Gauss, Carl, 2
Geometric terms, 172
Geometry, 170
Graphing calculator, 218
Graphing linear equations, 155
Graphing, 153
Graphs, 213

H

Hardware, 247
Hexadecimal number system, 238
Hexadecimal to decimal, 238
Hypotenuse, 188

I

Improper fractions, 13
Increment, 5
Input port, 231
Integer, 154
Inverse sine, 190
Inversely proportional, 123
Isosceles triangle, 184

L

Lateral surface area, 198
Line graph, 215
Linear circuit, 227
Linear equations, 153
Linear response, 218
Literal number, 120
Literal number, 153
Logarithmic response, 218
Logarithms, 206
Logic gates, 247
Lowest common denominator, 11

M

Measuring devices, 110
Measuring power, 115
Metric prefixes, 101
Metric system, 101
Metric unit of area, 105
Metric unit of length, 103
Metric unit of temperature, 107

Metric unit of volume, 106
Metric unit of weight, 106
Metric units, 101
Micrometer, 110
Minuend, 46
Mixed number, 13
Multimeter, 113
Multiplicand, 49
Multiplication of exponents, 158
Multiplication, 48
Multiplier, 49
Multiplying fractions, 20
Multiplying mixed fractions, 21
Multiplying positive and negative
     numbers, 74

N

NAND gate, 252
Napier, John, 206
Natural logarithms, 208
Negative exponents, 159
Newton, Isaac, 30
NOR gate, 250
Numerator, 7

O

Obtuse angle, 174
Ohmmeter, 114
Ohms law example, 139
One's complement number system, 259
Opposite, 188
OR gate, 248
Order of operations, 79
Output port, 231

P

Parallelogram, 175
Pascal, Blaise, 62
Percentages, 57
Period, 165
Pi, 38
Pie chart, 216
Plane figures, 174
Plotting a graph, 216
Point-slope form, 157
Polygon, 175
Polygons, 186
Positive and negative numbers, 62
Power formula example, 143, 163
Powers and roots in combination, 90
Powers of ten, 94
Powers of ten in combination, 96
Prime number, 12
Prism, 198
Product, 49
Proportion, 122
Protractor, 172

Pure binary, 243
Pyramid, 201
Pythagorean theorem, 178

Q

Quadratic equations, 162
Quadratic formula, 162
Quadrilaterals, 175
Quotient, 52

R

Radius, 146
Radix, 4
Raising a base number to a higher power,
     86
Ratio, 122
Ratios, 53
Reading graphs, 213
Reciprocals, 37
Rectangle, 176
Reducing fractions, 16
Remainder, 52
Representing fractions, 6
Reset and carry, 5
Resultant vector, 181
Rhombus, 177
Right triangle, 180
Right-angle triangle, 176
Rise, 156
Root of a number, 89
Root power, 159
Rounding off, 55
Run, 156

S

Scalene triangle, 183
Scientific notation, 97
Second-degree equations, 162
Sides, 172
Significant places, 57
Sine and magnitude system, 257
Sine of theta, 188
Sine wave, 195
Slope intercept form, 156
Slope of align, 155
Software, 247
Solid figure, 198
Sphere, 200
Square and square root, 136
Square of a number, 87
Square, 175
Standard form, 153
Statistics, 59
Story problems, 130
Subscript, 4
Substitution, 146
Subtracting fractions, 18

Subtracting mixed fractions, 18
Subtracting positive and negative numbers,
  70
Subtraction, 46
Subtrahend, 46
Sum of perfect cubes, 161
Sum, 45

T
Tangent of theta, 192
Theta, 188
Total surface area, 198
Transposing formulas, 136

Transposition, 126
Trapezoid, 178
Treating both sides equally, 121
Trigonometry, 187
Two's complement number system, 260
Types of graphs, 215

V
Vector addition, 181
Vector diagrams, 181
Vernier caliper, 111
Vertex, 172
Vinculum, 7

Voltmeter, 114
Volume, 198

W
Weights and measure, 100
Whole numbers, 4
Word, 241

X
XNOR Gate, 254
XOR Gate, 253